Surface Treatment in Bonding Technology

Surface Treatment in Bonding Technology

Anna Rudawska
Faculty of Mechanical Engineering
Lublin University of Technology
Lublin, Poland

ELSEVIER

ACADEMIC PRESS
An imprint of Elsevier

Academic Press is an imprint of Elsevier
125 London Wall, London EC2Y 5AS, United Kingdom
525 B Street, Suite 1650, San Diego, CA 92101, United States
50 Hampshire Street, 5th Floor, Cambridge, MA 02139, United States
The Boulevard, Langford Lane, Kidlington, Oxford OX5 1GB, United Kingdom

The project/research was financed in the framework of the project Lublin University of Technology - Regional Excellence Initiative, funded by the Polish Ministry of Science and Higher Education (contract no.030/RID/2018/19)

Notices

Knowledge and best practice in this field are constantly changing. As new research and experience broaden our understanding, changes in research methods, professional practices, or medical treatment may become necessary.

Practitioners and researchers must always rely on their own experience and knowledge in evaluating and using any information, methods, compounds, or experiments described herein. In using such information or methods they should be mindful of their own safety and the safety of others, including parties for whom they have a professional responsibility.

To the fullest extent of the law, neither the Publisher nor the authors, contributors, or editors, assume any liability for any injury and/or damage to persons or property as a matter of products liability, negligence or otherwise, or from any use or operation of any methods, products, instructions, or ideas contained in the material herein.

Library of Congress Cataloging-in-Publication Data
A catalog record for this book is available from the Library of Congress

British Library Cataloguing-in-Publication Data
A catalogue record for this book is available from the British Library

ISBN 978-0-12-817010-6

Publisher: Matthew Deans
Acquisition Editor: Brian Guerin
Editorial Project Manager: Leticia M. Lima
Production Project Manager: Nirmala Arumugam
Cover Designer: Greg Harris

Typeset by SPi Global, India

Contents

Introduction

1

Bonding technology plays a vital role in the modern industry and is becoming an indispensible feature in a growing number of industrial sectors. Its presence is noticeable in such fields as the household and personal products industry, machine engineering, or the aerospace industry, where high-precision technology is a key component of aircraft and spacecraft machines [1–4].

Since its introduction as a new form of machine part assembly in industrial manufacturing, adhesive joining, along traditional processes, such as welding, riveting, or soldering, has become one of the fastest-developing methods for joining finished and semifinished products. The rapid development of the adhesive joining technology is attributed to the progress in the fields of adhesive joints, or properties of adhesives, which have led to the emergence of novel adhesives of superior properties and consequently to proliferation of studies into the adhesion-related phenomena.

The surge of interest in adhesive bonding as a joining method could not have taken place without the advances in the field of structural adhesives, which have opened new possibilities in forming cutting-edge compositions, providing high performance in structures exposed to various unfavourable conditions. Another factor contributing to the increasing popularity of the method in question is the significant progress in the field of materials engineering, which has thus enabled introducing significant modifications in adhesive joining. These improvements have been encouraged by findings concerning processes accompanying adhesive bonding and by developments in quality control by means of modern diagnostic tools.

The use of adhesive joining technology in large-scale applications brings unparalleled technical and economical advantages, such as a substantial reduction of joint production cost compared to other fasteners (e.g. threaded joints).

Adhesively bonded joints offer high-strength fastening of virtually any pair of substrates (similar and dissimilar joints) [5–10]:

- metal-to-metal,
- metal-to-polymer,
- metal-to-composite,
- metal-to-glass,
- polymer-to-polymer,
- composite-to-composite,
- glass-to-polymer,
- rubber-to-rubber,
- rubber-to-metal, etc.

The scope of applications of adhesives goes beyond the aspect of forming adhesively bonded joints and includes:

- coating,
- sealing,

Surface Treatment in Bonding Technology. https://doi.org/10.1016/B978-0-12-817010-6.00001-1

- impregnation,
- securing and sealing threaded fittings,
- fixing coaxial metal components, and
- regeneration and filling of local cracks and defects.

The numerous advantages of adhesive bonding technology determine its suitability for the machinery industry applications; nevertheless, the role of adhesives and adhesive joints in machine assembly is the subject of constant development and change. There are numerous advantages of adhesively bonded joints described in the literature [1, 2, 6]:

- weight reduction of constructions,
- joining homogeneous and heterogeneous substrate materials of similar or different geometry,
- joining substrates of different thickness,
- joining adherends of small dimensions,
- forming visually aesthetic joint,
- reducing the number of joint elements compared to other methods; adhesive bonding does not require mechanical fasteners, such as screws or rivets, etc.,
- relatively low joint assembly cost, as little or no specialist equipment is required,
- good sealing properties,
- short joint assembly time compared to other joining methods,
- resistance to galvanic corrosion, water, oil, grease, etc.,
- can be included in hybrid joint assemblies to improve their properties,
- vibration dampening,
- uniformity of adhesive material structure, unattainable for welding or pressure welding, and
- uniform stress distribution due to lack of holes on the surface, which are common in other joint types, e.g. screw joints.

Along with the numerous advantages, there are certain drawbacks involved with the use of adhesive bonding technology. The flaws of adhesive bonding may necessitate further analysis of the joint design, or even introducing changes to the joint. The reported adverse effects that adhesive bonding is associated with are listed below [1, 2]:

- susceptibility to joint-strength-inhibiting ageing processes, caused by, e.g. fluctuations of atmospheric conditions,
- once adhesive cures, no mechanical treatment is possible,
- some adhesives cure at elevated temperature and at considerable pressure,
- creep may occur at high fixed loads,
- limited range of joint service temperatures,
- the number and labour consumption of quality control operations, and
- the need for different surface treatments in the case of certain substrates.

Adhesive bonding embraces a number of adhesive-joining-related issues, including various joint designs and assembly methods, however, what is of fundamental importance to producing good-quality joints is the surface pretreatment prior to adhesive bonding.

Proper preparation of substrate surface layer is critical to forming an adhesive bond, which is particularly important in joints expected to develop specified properties [7, 11–14]. Surface pretreatment is expected to stimulate bonding mechanisms of the physical, the mechanical, and the chemical nature. This initial stage of adhesive joining technology is intended to impart forming strong bonds in adhesive joints [15–19].

There are a number of factors that have a bearing on the selection of particular surface pretreatment method. Fundamentally, technological operations performed as surface preparation must ensure that the produced surface of adherends is of the right texture and energy properties. Therefore, it is the type of substrate which is frequently the central factor deciding on the quality of surface treatment. Certain adherends will not only require removal of various debris from their surface but also modification of their surface layer. Processes modifying the surface layer are carried out to modify the physical or chemical structure of the surface layer [20–22], increase the surface roughness [23], improve wettability and adhesion, or to increase the surface free energy [16, 24–30]. The modification usually consists in reinforcing the adherend surface layer as a result of cross-linking processes and emergence of function groups, which form strong physical and chemical bonds with other materials (particularly polar groups constituting active polar centres).

Surface pretreatment involves such processes as [1, 10, 15]

- removal of surface debris, i.e. contaminants remaining on the surface after processing (e.g. press forming, rolling, forging, and machine cutting), serving as corrosion protection during storage and transport, substances providing resistance against biological influences, or in polymer materials, plasticisers, or stabilisers migrating to the interface,
- modification of surface geometry by increasing its roughness, and thus the effective contact surface area for wetting, and
- modification of the range of saturation of intermolecular forces, which results in the increase of the surface free energy (also affecting the wetting process).

Surface pretreatment may involve a variety of operations [7, 15, 17, 31–34], such as: degreasing [35], mechanical [4, 6, 23, 32, 36], chemical or electrochemical treatment (e.g. anodising and chromatising) [3, 9, 37–39], rinsing, drying, special operations involving application of adhesion promoters [8, 18, 40, 41], special operations modifying the surface layer of polymers, e.g. corona discharge treatment [27, 29], flame treatment [29, 42], plasma treatment [20, 21, 25, 26, 30, 43–45], chemical treatment [1, 24], ozone oxidation [46], acting with different particles on the surface of substrates [5, 46, 47] (by means of radiation), including laser treatment [28, 48]. It ought to be mentioned that the methods employed in treatment of polymers are frequently employed in pretreatment of other materials such as: aluminium and aluminium alloys [13, 19, 24, 33], titanium [22], or steel [4].

The critical importance of surface pretreatment in adhesive joining (including bonded joints) is highlighted and descried in numerous works [12–14, 24, 34, 37, 38, 49]. Its significance is primarily ascribed to its enhancing bonding performance, which increases joint strength. There are several reasons why surface treatment has been the subject of much systematic investigation, such as rapid advances in material engineering, developments in surface treatment methods and equipment, and measurement instrumentation. Nevertheless, when selecting suitable pretreatment, it is important to consider all the factors affecting the surface treatment stage to ensure that the joint is assembled correctly and maintains its properties for the expected service lifetime.

This work is intended to provide a reference regarding the basics of adhesive joining, surface treatments of material for bonding and other adhesive processes. The theoretical presentation is furthermore supplemented with the test results data, providing the basis for comparison of different pretreatments with regard to designing the adhesive joining technology. The preparation of surfaces for bonding would not be possible without the knowledge of adhesion and wetting [10, 16, 23, 29, 35, 39].

References

[1] R.D. Adams, Adhesive Bonding. Science, Technology and Applications, Woodhead Publishing Limited, Oxford, 2010.

[2] W. Brockmann, P.L. Geiß, J. Klingen, B. Schröder, Adhesive bonding. Materials, Applications and Technology, Wiley-Vch Press, Weinheim, 2009.

[3] L. Domingues, J.C.S. Fernandes, M.D.C. Belo, M.G.S. Ferreira, L. Guerra-Rosa, Anodising of Al 2024-T3 in a modified sulphuric acid/boric acid bath for aeronautical applications, Corros. Sci. 45 (2003) 149–160.

[4] G. Rotella, M. Alfano, T. Schiefer, I. Jansen, Evaluation of mechanical and laser surface pre-treatments on the strength of adhesive bonded steel joints for automotive industry, J. Adhes. Sci. Technol. 30 (2016) 747–758.

[5] S. Bhowmik, P.K. Ghosh, S. Ray, K. Barthwal, Surface modification of high density polyethylene and polypropylene by DC glow discharge and adhesive bonding to steel, J. Adhes. Sci. Technol. 12 (1998) 1181–1204.

[6] N. Brack, A.N. Rider, The influence of mechanical and chemical treatments on the environmental resistance of epoxy adhesive bonds to titanium, Int. J. Adhes. Adhes. 44 (2014) 20–27.

[7] G.W. Critchlow, D.M. Brewis, Review of surface pretreatments for alluminium alloys, Int. J. Adhes. Adhes. 16 (1996) 255–275.

[8] J. Aakkula, O. Saarela, Silane based field level surface treatment methods for aluminium, titanium and steel bonding, Int. J. Adhes. Adhes. 48 (2014) 268–279.

[9] P. Hass, O. Kläusler, S. Schlegel, et al., Effects of mechanical and chemical surface preparation on adhesively bonded wooden joints, Int. J. Adhes. Adhes. 51 (2014) 95–102.

[10] A.J. Kinloch, Adhesion and Adhesives, Chapman and Hall, New York, 1987.

[11] R.D. Bhatt, V. Radhakrishnan, Evaluation of some machining process as adherend surface treatments in bonding aluminium, J. Adhes. Sci. Technol. 3 (1989) 383–396.

[12] A.M. Guedes Pinto, A.G. Magalhães, F. Gomes da Silva, A.P. Monteiro Baptista, Shear strength of adhesively bonded polyolefins with minimal surface preparation, Int. J. Adhes. Adhes. 28 (2008) 452–456.

[13] R. Rechner, I. Jansen, E. Beyer, Influence on the strength and aging resistance of aluminium joints by laser pre-treatment and surface modification, Int. J. Adhes. Adhes. 30 (2010) 595–601.

[14] P.R. Underhill, A.N. Rider, D.L. DuQuesnay, The effect of warm surface treatments on the fatigue life in shear of aluminium joints, Int. J. Adhes. Adhes. 26 (2006) 199–205.

[15] R.M. Podhajny, Comparing surface treatments, Converting 8 (1990) 46–52.

[16] G. Rozovskis, J. Vinkevičius, Plasma surface modification of polyimide for improving adhesion to elekctroless copper coatings, J. Adhes. Sci. Technol. 10 (1996) 399–406.

[17] O. Lunder, F. Lapique, B. Johnsen, K. Nisancioglu, Effect of pre-treatment on the durability of epoxy-bonded AA6060 aluminium joints, Int. J. Adhes. Adhes. 24 (2004) 107–117.

[18] P. He, M. Huang, S. Fisher, et al., Effects of primer and annealing treatment on the shear strength between anodized Ti6Al4V and epoxy, Int. J. Adhes. Adhes. 57 (2015) 49–56.
[19] C. Sperandio, J. Bardon, A. Laachachi, H. Aubiert, D. Ruch, Influence of plasma surface treatment on bond strength behaviour of an adhesively bonded aluminium-epoxy system, Int. J. Adhes. Adhes. 30 (2010) 720–728.
[20] G.Z. Xiao, Effects of solvent on the surface properties of oxygen plasma-treatment polyethylene and polypropylene films, J. Adhes. Sci. Technol. 11 (1997) 655–663.
[21] Z.Z. You, J.Y. Dong, Effect of oxygen plasma treatment on the surface properties often-doped indium oxide substrates for polymer LEDs, J. Colloid Interface Sci. 300 (2006) 697–703.
[22] A. Acar, Ö. İnan, S. Halkaci, Effects of airborne-particle abrasion, sodium hydroxide anodization, and electrical discharge machining on porcelain adherence to cast commercially pure titanium, J. Biomed. Mater. Res. B Appl. Biomater. 82B (2007) 267–274.
[23] A.F. Harris, A. Beevers, The effect of grit-blasting on the surface properties for adhesion, Int. J. Adhes. Adhes. 19 (1999) 445–452.
[24] S. Ebnesajjad, C. Ebnesajjad, Surface Treatment of Materials for Adhesive Bonding, second ed., William Andrew Inc., Norwich, NY, 2013.
[25] L. Sorrentino, L. Carrino, Influence of process parameters of oxygen cold plasma treatment on wettability ageing time of 2024 aluminium alloy, Int. J. Adhes. Adhes. 29 (2009) 136–143.
[26] G.S. Sheu, S.S. Shyu, Surface modification of Kevlar 149 fibers by gas plasma treatment. Part II. Improved interfacial adhesion to epoxy resin, J. Adhes. Sci. Technol. 8 (1994) 1027–1042.
[27] J. Gołębiewski, M. Żenkiewicz, Influence of corona treatment energy and storage time on the changes of the contact angle of activated polysteryrene film, Polimery 48 (2003) 134–137.
[28] Q. Bénard, M. Fois, M. Grisel, P. Laurens, Surface treatment of carbon/epoxy composites with an excimer laser beam, Int. J. Adhes. Adhes. 26 (2006) 534–549.
[29] J. Custódio, J. Broughton, H. Cruz, P. Winfield, Activation of timber surface by flame and corona treatments to improve adhesion, Int. J. Adhes. Adhes. 29 (2009) 167–172.
[30] Y. Kusano, H. Mortensen, B. Stenum, S. Goutianos, S. Mitra, Atmospheric pressure plasma treatment of glassy carbon for adhesion improvement, Int. J. Adhes. Adhes. 27 (2007) 402–408.
[31] B.B. Johnsen, F. Lapique, A. Bjørgum, The durability of bonded aluminium joints: a comparison of AC and DC anodizing pretreatments, Int. J. Adhes. Adhes. 24 (2004) 153–161.
[32] L. Kumosa, D. Armentrout, D. Kumosa, The effect of sandblasting on the initiation of stress corrosion cracking in unidirectional E-glass/polymer composites used in high voltage composite (non-ceramic) insulators, Compos. Sci. Technol. 62 (2002) 1999–2015.
[33] W. Polini, L. Sorrentino, Adhesion of a protective coating on a surface of aluminium alloy treated by air cold plasma, Int. J. Adhes. Adhes. 27 (2007) 1–8.
[34] S.G. Prolongo, A. Ureña, Effect of surface pre-treatment on the adhesive strength of epoxy-aluminium joints, Int. J. Adhes. Adhes. 29 (2009) 23–31.
[35] A. Rudawska, Surface free energy and 7075 aluminium bonded joint strength following degreasing only and without any prior treatment, J. Adhes. Sci. Technol. 26 (2012) 1233–1247.
[36] A. Rudawska, I. Danczak, M. Müller, P. Valasek, The effect of sandblasting on surface properties for adhesion, Int. J. Adhes. Adhes. 70 (2016) 176–190.

[37] G.W. Critchlow, K.A. Yendall, D. Bahrani, A. Quinn, F. Andrews, Strategies for the replacement of chromic acid anodizing for the structural bonding of aluminium alloys, Int. J. Adhes. Adhes. 26 (2006) 419–453.
[38] A.C. Kennedy, R. Kohler, P. Poole, A sodium hydroxide anodize surface pretreatment for the adhesive bonding of titanium alloys, Int. J. Adhes. Adhes. 3 (1983) 133–139.
[39] A. Rudawska, I. Danczak, Adhesive properties and adhesive joint strength of aluminium alloys following surface pre-treatment, Recent Pat. Mech. Eng. 6 (2013) 152–165.
[40] J.A. Bishopp, M.J. Parker, T.A. O'Reilly, The use of statistical design procedures in the development and "robust" manufacture of a water-based, corrosion-inhibiting adhesive primer, Int. J. Adhes. Adhes. 21 (2001) 473–480.
[41] G.A. Wade, W.J. Cantwell, Adhesive bonding and wettability of plasma treated, glass fiber-reinforced nylon-6,6 composites, J. Mater. Sci. Lett. 19 (2000) 1829–1832.
[42] C.D. Volpe, A. Deimichei, T. Riccò, A multiliquid approach to the surface free energy determination of flame-treated surfaces of rubber-toughened polypropylene, J. Adhes. Sci. Technol. 12 (1998) 1141–1180.
[43] N. Encinas, J. Abenojar, M.A. Martínez, Development of improved polypropylene adhesive bonding by abrasion and atmospheric plasma surface modifications, Int. J. Adhes. Adhes. 33 (2012) 1–6.
[44] M.-H. Jung, H.-S. Choi, Surface treatment and characterization of ITO thin films using atmospheric pressure plasma for organic light emitting diodes, J. Colloid Interface Sci. 310 (2007) 550–558.
[45] Y. Nakamura, Y. Suzuki, Y. Watanabe, Effect of oxygen plasma etching on adhesion between polyimide films and metal, Thin Solid Films 290–291 (1996) 367–369.
[46] M. Żenkiewicz, Effect of electron radiation dose on oxidation of surface layers of polyethylene, polypropylene and poly(ethylene terephthalate) films, J. Adhes. Sci. Technol. 18 (2004) 607–616.
[47] M. Żenkiewicz, P. Rytlewski, J. Czupryńska, J. Polański, T. Karasiewicz, W. Engelgard, Contact angle and surface free energy of electron-beam irradiated polymer composites, Polimery 53 (2008) 446–451.
[48] E.G. Babraj, D. Starikov, J. Evans, G.A. Shafeev, A. Bensaoula, Enhancement of adhesive joint strength by laser surface modification, Int. J. Adhes. Adhes. 27 (2007) 268–276.
[49] H. Moghadamzadeh, H. Rahimi, M. Asadollahzadeh, A.R. Hemmati, Surface treatment of wood polymer composites for adhesive bonding, Int. J. Adhes. Adhes. 31 (2011) 816–821.

Further reading

[50] L.B. Cohen, Zircoaluminate adhesion promoters, J. Adhes. Sci. Technol. 5 (1991) 439–448.

Bonding technology

2

2.1 Bonding process

Bonding is a technique of joining materials by means of a nonmetallic substance—adhesive, which bonds the materials at their surface through forced adhesion, while retaining adequate internal strength through cohesion [1–3]. By implementing adhesive bonding, we ensure that the stress is distributed uniformly along the bonded area, which in turn increases the service lifetime of the assembly. Adhesively bonded joints are capable of withstanding higher dynamic loads than other types of fastening methods and offer higher fatigue resistance. However, in order for the joint to develop its full performance, the contact between the surfaces of both adherends must be remarkably close and consistent throughout the bondline, particularly at the edge of the overlap. Otherwise, the joint is bound to develop a fraction of the strength of the correctly assembled joint.

There are three basic steps to producing an adhesive bond [4, 5]:

- application of the adhesive on the adherend surfaces,
- joint assembly, and
- curing the adhesive until the adherends are fully bonded.

If assembled correctly, adhesive joint will carry substantial to enormous loads without the need for additional reinforcement with rivets or bolts. Adhesive joint may be also applied as an auxiliary fastener in, e.g. screwing or during other production processes [1]. In adhesive joining it is crucial that the surface treatment conditions are repeatable, otherwise the strength characteristics of thus produced joints may show substantial discrepancy. Even the slightest scratch on the surface of adherends could lead to significant disparity in joints performance as a result of local weakening of the material. Furthermore, a cured adhesive should exhibit a slightly lower modulus of elasticity than the adherend, i.e. should be to a certain degree stiffer. If not, the unfavourable stress distribution could lead to emergence of stresses along the overlap that the joint might not be capable of withstanding [1, 5].

The mechanism of bonding involves shrinkage of adhesive, which, if excessive, may lead to developing internal stresses in the adhesive. The strength of the bond should match that of the adherend, in order for the cohesive and adhesive bond not to fail before the adherend in the case of joint failure [1, 2].

Two types of forces are responsible for the bonding process, i.e. the forces of adhesion and cohesion (Fig. 2.1).

It is the ratio of adhesion and cohesion forces that is a key determinant of the adhesive bond quality and hence its applicability. Furthermore, it is indicative of the type of joint failure that will occur under external loading, i.e. debonding at the adhesive interface (due to low adhesion forces), cohesive failure in the adhesive (as a result of low cohesion forces in the adhesive or in the adherend). Adhesive failure

Surface Treatment in Bonding Technology. https://doi.org/10.1016/B978-0-12-817010-6.00002-3

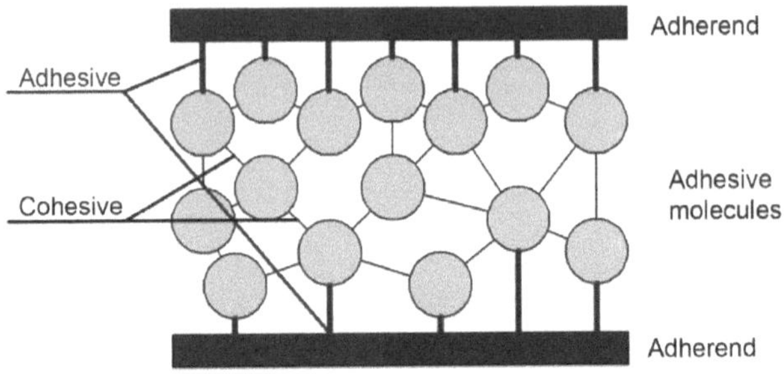

Fig. 2.1 Cohesion and adhesion forces in the adhesive joint.

takes place when the forces of adhesion are lower than the forces of cohesion and the force of the external load [1].

Adhesion is a force of attraction between materials, which essentially involves intermolecular forces at the surface layer of solids [6–9]. The presence of these forces causes the materials which come in contact with to join. Adhesion is a force which is either of the mechanical nature (mechanical adhesion), i.e. when the adhesive becomes interlocked to the rough projections of the adherend surface or proper adhesion; a sum of intermolecular and interatomic interactions. The force field between the surface layers of two adherends decreases with distance. The measure of adhesion is the work per unit area, governed by the forces of physical attraction and adsorption, which constitute Van der Waals forces. These forces are effective in the interatomic distance, and decrease proportionally to the distance [3, 6]. Their strength will be also affected if the adhesive fails to properly penetrate the rough substrate surface, which highlights the importance of good wetting of the surface by the adhesive [10–14].

The wettability of surfaces solely depends on the surface energy of adherends [3, 15–17]. The material will be properly wetted on the condition that its surface tension is equal to or even lower than its surface energy, which is the reason why adhesive joining of polymers proves considerably more problematic than that of metals [16, 18, 19]. The lower the value of the contact angle θ, the wettability of a given surface is better (Fig. 2.2).

The contact angle value depends on the surface tension of a solid substrate in the air γ_S and in the liquid γ_L, as well as the surface tension at the liquid–solid interface γ_{SL} (Fig. 2.3) [20–23]. The function takes the following form [18]:

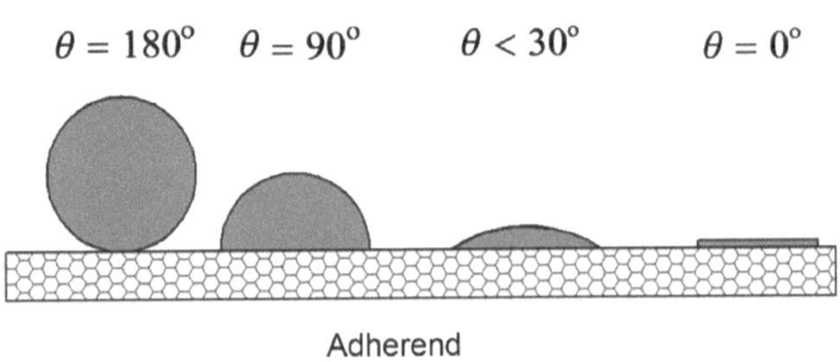

Fig. 2.2 Various cases of wetting of adherend surface.

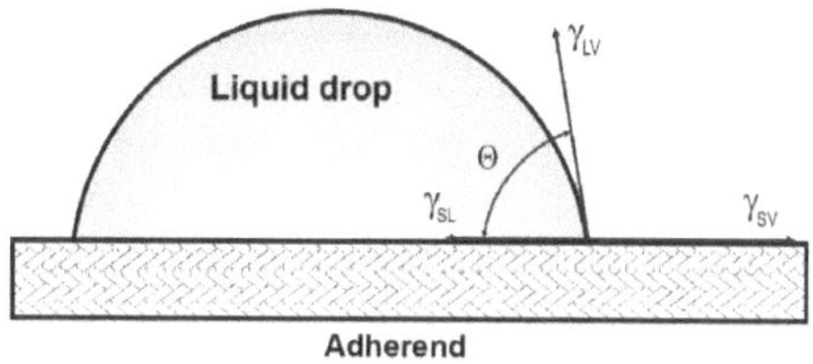

Fig. 2.3 Schematic representation of the dependence between contact angle components.

$$\gamma_{SV} = \gamma_{SL} + \gamma_{LV} \cos\theta \tag{2.1}$$

where

γ_{SV}—interfacial surface free energy for solid–vapour,
γ_{SL}—interfacial surface free energy for solid–liquid,
γ_{LV}—interfacial surface free energy for liquid–vapour.

The presence of contamination in the adhesive or at the adherend surface is another factor which may adversely affect wettability [24–27].

The contact angle was also used to determine the surface free energy [28–32].

Cohesion is the measure of integrity of an adhesive bond, which describes its mechanical strength [2, 3]. This strength results from the mutual attraction between molecules of the adhesive and this is the force that is responsible for the cohesion of the adhesive. Cohesion is interpreted as the force of resistance of the material subjected to destructive forces. Therefore, the measure of cohesion is the work required to break the material, divided by the surface area of the failed adhesive bond. The force of cohesion is also dependent on the state of matter: the more organised the particles building the material are, the higher the forces of cohesion. These forces grow with the increase in intermolecular forces per unit volume. Furthermore, cohesion may be to a certain extent determined by the microstructure of a given material, as it is influenced by such phenomena as microcracking or cavitation [1, 3].

Adhesive joint failure is a consequence of one of the three occurrences: adhesive failure (the adhesive detaches from the adherend—rupture at the adhesive-adherend interface), cohesive failure—intermolecular forces in the adhesive yield to cohesive forces (rupture and delamination occurs in the adhesive) and external load forces, or failure in the adherend [1, 3, 22, 23].

The international standard *EN ISO 10365: Adhesives—Designation of main failure patterns*, defines adhesive failure as "rupture of an adhesive bond in which the separation appears visually to be at the adhesive/adherend interface", and cohesive failure as: "rupture of a bonded assembly in which the separation appears, visually, to be in the adhesive or the adherend".

2.2 Strength of adhesive joints

There are two types of strength considered in adhesive joint strength testing: the static and the ultimate strength. The former is relatively easy to measure; however, it does not provide a great amount of relevant information. Nevertheless, it is the one most frequently analysed in strength tests and is measured under loading time from several to several dozen seconds [1, 5, 33].

The static strength of an assembled adhesive joint depends, to a considerable extent, on the applied surface treatment method, the type of adhesive and its cure, the thickness of adherends and dimensions of the bond, the type and value of loading, and on the design of the joint [34–40]. If the given joint application permits, the bond should be arranged along the stresses, so that they are carried at shear, simultaneously preventing unfavourable peel stresses [41]. Fig. 2.4 shows most commonly used lap and strap adhesive joints designs [2, 35, 42–45].

The joint type which is most commonly used is the single-lap joint as it is capable of carrying considerable shear stresses at light weight. In service, both lap and strap joints are susceptible to nonuniform stress distribution along the direction of applied loads. The reason is the disparity in physical properties of adherends and the difference in their dimensions and the size of the adhesive joint [35, 44].

The long-term joint strength is determined from the time in which the bond is capable of handling expected loads. Its value depends on the design of the joint and certain external factors; nevertheless, it tends to develop generally lower values than the static strength [46–51].

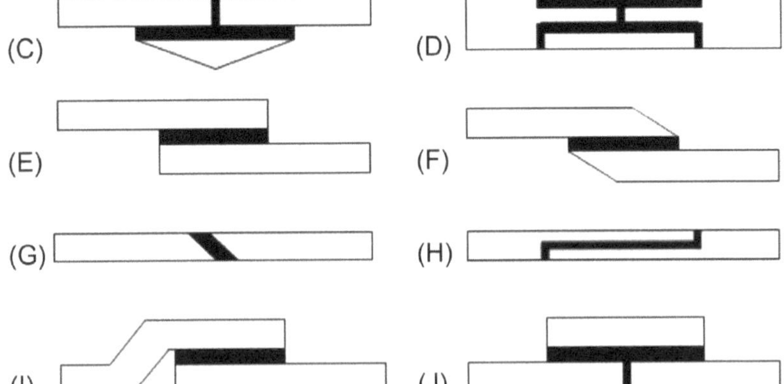

Fig. 2.4 Common lap and strap joints: (A) symmetrical double-lap joint, (B) double-strap joint, (C) tapered-double strap joint, (D) recessed double-strap joint, (E) single-lap joint, (F) tapered single-lap joint, (G) scarf joint, (H) symmetrical single-lap joint, (I) joggle-lap joint, and (J) single-strap joint.

The third type of strength considered in strength tests is fatigue strength. Fatigue strength is measured at dynamic stress, and the tests lead to damaging or even disintegration of the adhesive bond. The failure occurs suddenly at loads lower than in the case of the static strength. The value of fatigue strength is determined by a number of factors, including: mechanical properties of adhesive, type of substrate, surface pretreatment, and adhesive layer thickness [51–57].

Factors relevant to strength of bonded assemblies may be classified into four groups: materials [16, 18, 35, 58–60], technological [35, 61–64], structural [5, 33, 35–37, 39, 65], and environmental [35, 38, 47, 48, 52] which are presented in Fig. 2.5.

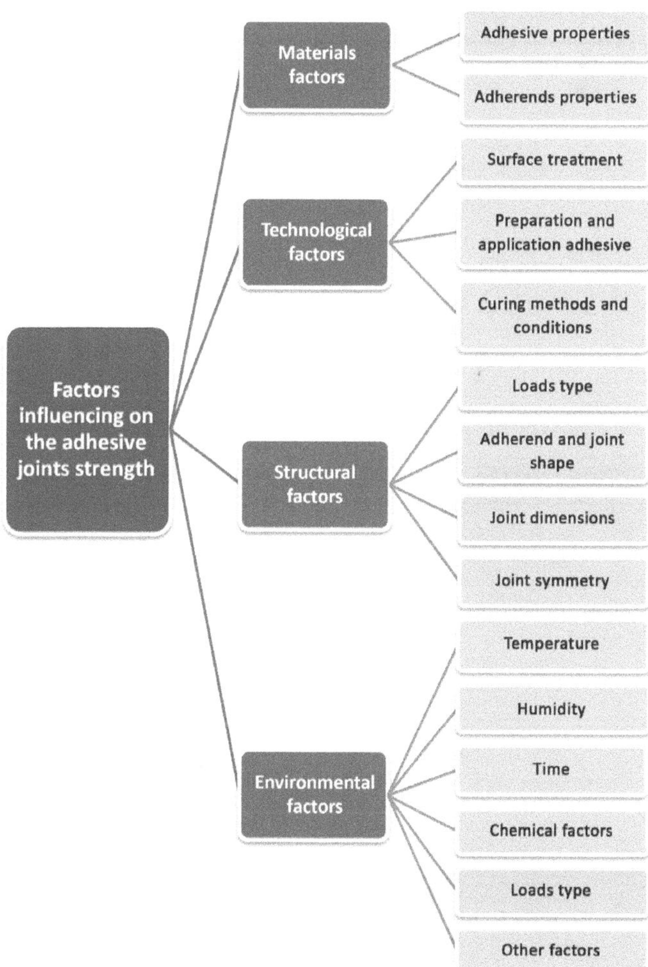

Fig. 2.5 Factors impacting on strength of adhesive joints.

2.3 Bonding technology operations

The technological process of bonding includes a number of operations leading to assembling the adhesive joint as per design, such as: pretreatment of adherend surface for bonding, preparation and mixing the adhesive, adjusting the position of and assembling the adherends, curing adhesive bond, joint quality control, and finishing operations [1, 2, 4] (Fig. 2.6).

In order to perform the listed operations we employ different means and measures, as well as various equipment, devices, and special or specialised tools.

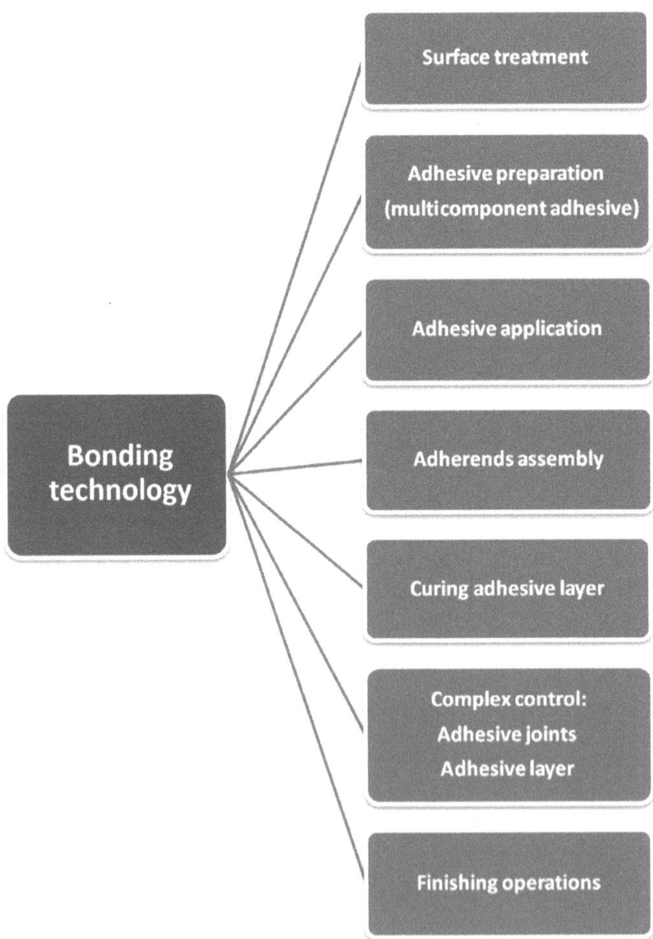

Fig. 2.6 Bonding process operations.

2.3.1 Surface treatment

The first and by far the most relevant stage of adhesive joining technology is the preparation of adherend surfaces [5, 18, 40, 66–68]. The process should exhibit high effectiveness in order to prevent partial bonding, which could have an adverse effect on joint strength. Proper surface treatment is crucial to fostering good adhesive bonding and producing a joint of desired properties [35, 69–73]. A well-prepared surface should promote all bonding mechanisms, be it physical, mechanical, or chemical by nature. It will, furthermore, ensure that cohesion is lower than adhesion, which is critical to adhesive joint strength and resistance to environmental conditions.

This initial stage of adhesive bonding technology should produce maximum strength adhesive bonds in the joint, and will therefore require, e.g. Refs. [4, 18, 66]:

- removing contamination from the surface of adherends (such as dust, oil, lubricants, microorganisms, gas bubbles, and loosely attached corrosion products),
- developing desired surface texture of adherends, and
- producing a well-activated surface of adherends.

The surface preparation process involves different operations (Fig. 2.7). Key surface pretreatment operations include: degreasing, surface cleaning in acidic or alkaline bath, rinsing, drying, special treatment (mechanical, chemical, electrochemical, and others), and additional operations like application of primers [1, 40, 63, 67, 69–76].

There are several factors that have bearing on the selection of particular surface treatments, such as:

- the type, the properties, and the stereometric structure of substrate surface,
- properties and the dimensions of adherends,

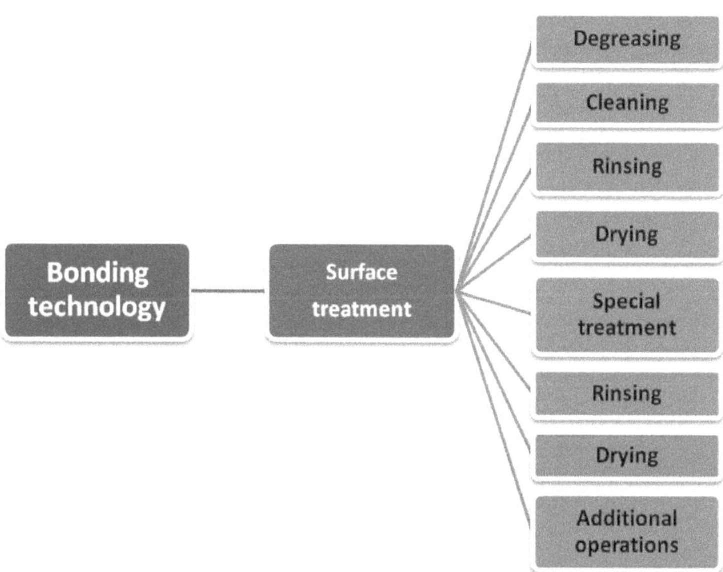

Fig. 2.7 Example of surface preparation for bonding.

- the technical/technological conditions of bonding, and
- the type of production.

In addition, these factors are of certain relevance to technological parameters of pretreatment operations as well.

2.3.2 Preparation of adhesives

In addition to the surface, this is also the adhesive that requires preparation to bonding. Its preparation depends on numerous factors, including the type and form, the number of its components, or the method of dispensing [4].

Multicomponent adhesives are predominantly found in the form of a liquid or semifluid, and may consist of various components, e.g. a resin (the adhesive substance), a hardener, an activator or a cross-linking agent, modifiers, and other additives. Prior to application on the surface, all the components of the adhesive must be added and mixed thoroughly to produce an adhesive composition [77, 78]. In addition, in the case of hot melt adhesives, available as films, granules, or wires, the adhesive must be heated prior to application in the form of molten rolls or by extrusion in the molten state.

Fig. 2.8 shows typical operations performed in preparation of adhesives.

The preparation of the adhesive will predominantly depend on its market form. Adhesives are available in three states: solid, semifluid, and liquid. Solid adhesives are generally sold as single-component, readymade solutions, however, the majority

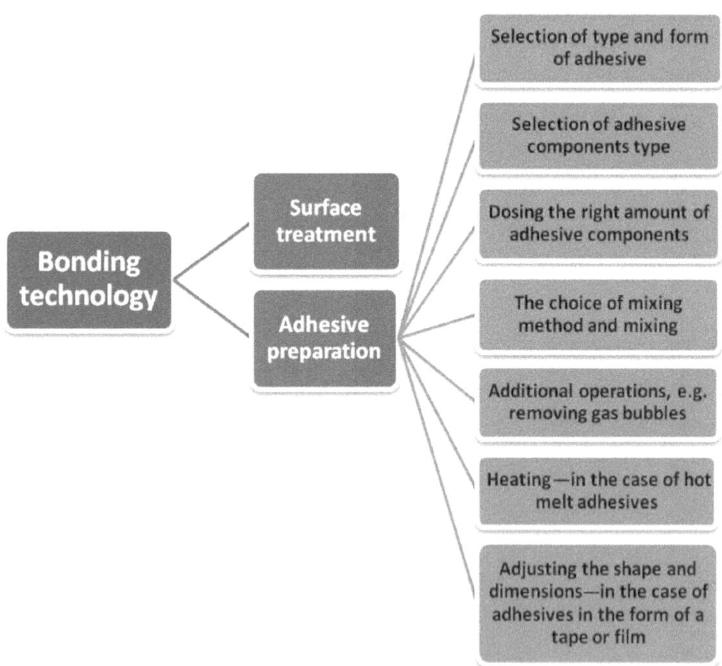

Fig. 2.8 Typical adhesive preparation operations.

of available multicomponent adhesive compositions require mixing with additional substances, such as solvents, fillers, hardeners, or other additives.

Preparation of liquid adhesives mainly concerns the multicomponent ones, in semi-fluid and liquid forms, which are made up of several components, e.g. a resin (the adhesive substance), a hardener, an activator or a cross-linking agent, modifiers and other substances [4, 79–86]. The ingredients of the adhesive composition should be mixed by the specified proportion. Any divergence, whether shortage or excess, from the specified amount of curing agent could lead to improper cross-linking of polymer in which case the produced adhesive joint will exhibit diminished strength.

The process of producing the adhesive mix frequently involves adding various modifiers to the mix, alongside resin and hardener (Table 2.1). Modifiers play an active role in the adhesive curing stage, and are classified into different categories, such as: plasticisers, coalescing agents, surfactants, thixotropic agents, antioxidants, special-purpose fillers, etc. [58, 77, 87–95].

The adhesive composition is commonly supplemented with fillers, whose role consists in increasing chemical and thermal resistance of the polymer, hence of the adhesive bond as well, and in decreasing the raw material input cost. Apart from typically chemical fillers, we can also distinguish fillers such as fibre tapes, fabrics, mats, and meshes made of metal, polymer, glass fibre, or carbon fibre, which serve as structural reinforcement of the adhesive joint [80–83, 92, 95–97].

Adhesive compositions are also enriched with adhesion promoters, whose structure enables them to diffuse to the adhesive-adherend interphase, where they physically or chemically act on the adherend surface [4, 58, 77, 95, 98–100].

One of the several key aspects of preparing the adhesive composition is to avoid trapping air bubbles inside the mix. Their presence during curing may contribute to the emergence of defects in the bond, and may consequently lead to compromising adhesive joint strength.

Work life of the adhesive is another significant factor to consider in bonding. This property of the adhesive describes the time span during which a mixed adhesive retains its properties.

Properties of the majority of adhesives are subject to change immediately upon opening the packaging (e.g. solvent-based adhesives or humidity-curing adhesives

Table 2.1 Examples of adhesive modifiers and their basic properties

Modifier	Properties
Plasticisers (softeners)	Increase workability and flexibility, decrease hardness, reduce tensile strength
Surfactants	Improve wettability of solids
Thixotropic agents	Impart thixotropic properties
Stabilisers	Improve stability—resistance to ageing over the service lifetime, processing, and storing
Fillers	Affect a variety of properties, e.g.: reduce the exothermic reaction, extend the work life of adhesive, reduce cost
Antioxidants	Increase resistance to oxidative ageing

setting by absorbing moisture from the air), or after adding curing agents, which is the case with the majority of two-component adhesives.

There exists a range of adhesive mixers and agitators available on the market, including mechanical or continuous-type mixers and mixer heads [4]. We distinguish static and dynamic mixers. The selection of a suitable mixer and agitator predominantly depends on the viscosity of the composition and the specified efficiency of the process.

Multicomponent adhesives are mainly available in the form of a liquid or semifluid, and combine various components, e.g. a resin (the adhesive substance), a hardener, an activator or a cross-linking agent, modifiers and other additives. Prior to application on the surface, all the components of the adhesive must be added and mixed thoroughly to produce a homogenous adhesive mix.

Mixing of adhesives is carried out by means of different kinds of mixers supplied with mechanical agitators. The agitator, which comes in a number of different shapes, such as paddles, propellers, anchors, or screws, performs a circular movement at a desired speed, thus producing a chemically homogeneous mix of adhesive components [4, 101]. Fig. 2.9 shows the classification.

In preparation of an adhesive mix, particularly in the case of multicomponent compositions, it is important to maintain the required mixing ratio. To this end the market offers a number of available solutions, such as: dispensers, dual cartridge guns with static mixers, scales, and other mixing and dosing devices, which frequently perform both dosing and dispensing functions [4, 101].

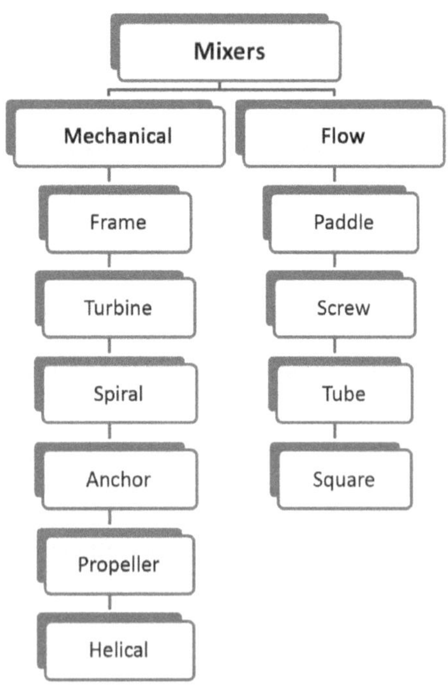

Fig. 2.9 Classification of mixers used in adhesive preparation operations.

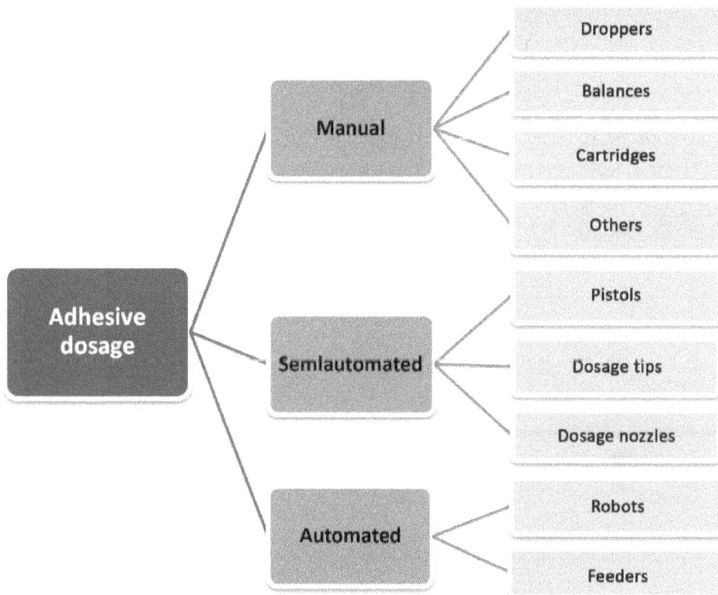

Fig. 2.10 Classification of dosing methods and example equipment.

Three basic dosing methods are distinguished (Fig. 2.10) as follows:

- manual,
- semiautomated, and
- automated.

Manual dosing consists of preparing the formulation by hand, using such equipment as technical scales, mixing vessels, droppers, and mixers. The manual process, however, tends to be relatively time-consuming and problematic in terms of observing the technical conditions for adhesive mix preparation (e.g. mixing time, temperature of the adhesive, humidity, etc.), and is therefore implemented in piece production, repair, or regeneration of machine parts in workshop conditions.

2.3.3 Application of adhesives

A number of factors have a bearing on the adhesive application method, such as [4, 90, 96, 101]:

- the form/state of the adhesive (liquid, film, granules, etc.),
- viscosity and other properties,
- the access to the joint,
- the assembly method, and
- the type of structure.

Typical operations involved in adhesive application are shown in Fig. 2.11.

There is a wide range of adhesive application solutions available; however, their design depends on the abovementioned factors.

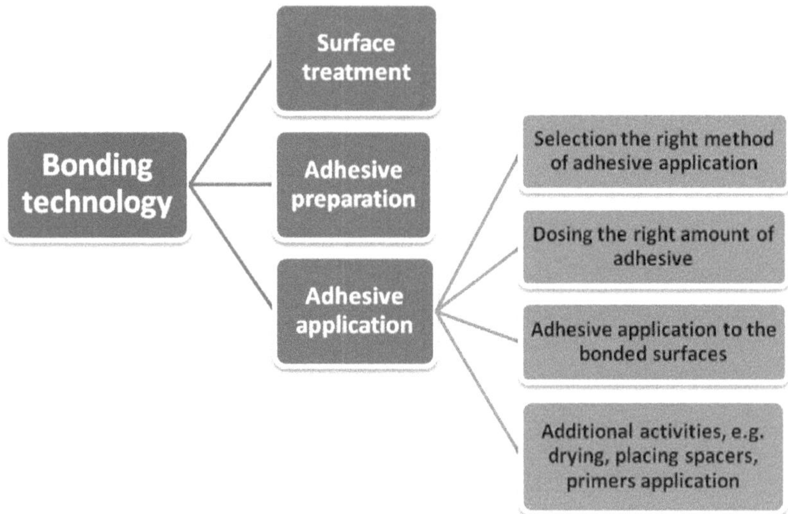

Fig. 2.11 Typical adhesive application operations.

The process of application, the selection of the application method, and implemented equipment is determined by the following elements [90, 96]:

- the number of components,
- the form/state of the adhesive,
- viscosity and other properties,
- the dimensions, size, and surface structure of adherends,
- the access to the joint,
- the type of production (piece, batch, or mass production), and
- the workshop conditions.

Two main methods of dispensing adhesives are distinguished [96, 101] as follows:

(1) depending on application:
- spot application,
- continuous line,
- continuous bead, and
- surface mount;

(2) depending on viscosity:
- spraying,
- immersion,
- droplet dispensing,
- pouring,
- with a roller,
- with a brush,
- with a trowel,
- with a doctor knife, and
- with a stamp.

One of the factors considered when choosing the adhesive application method is the form/state of the adhesive. Table 2.2 presents typical dispensing methods for

Table 2.2 Liquid adhesive application methods

Method	**Schematic representation**
Spraying	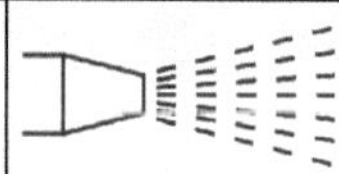
Immersion	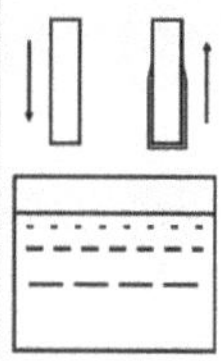
Droplet dispensing	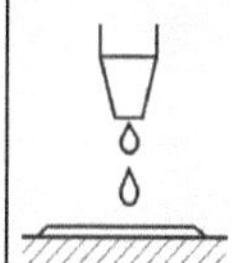
Roller	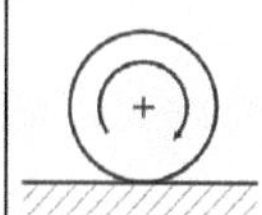
Ribbon	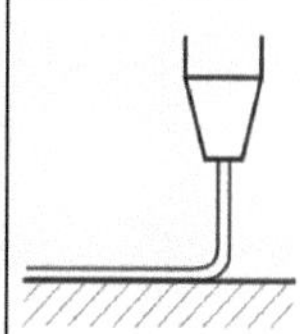
Brush	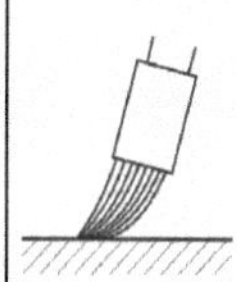
Trowel: Regular and notched	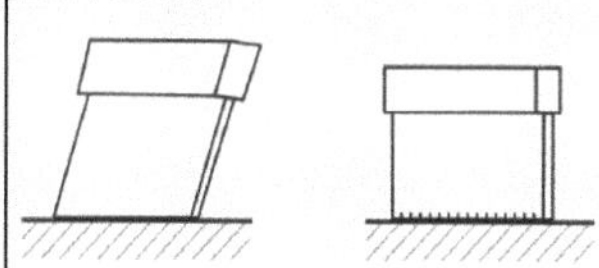
Sponge	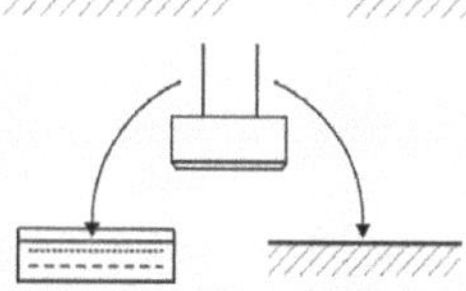

Source: own work based on G. Habenicht, Applied Adhesive Bonding, Wiley-VCH Press, Weinheim, 2006.

liquid adhesives. In Figs 2.12–2.17 selected methods for the application of liquid two-component adhesives are shown. There are several procedures for producing of such joints: components of such adhesives may be either applied mixed on both or one of the adherends, alternatively, components are dispensed individually on separate adherends and become mixed upon assembly of the joint to initiate the curing reaction.

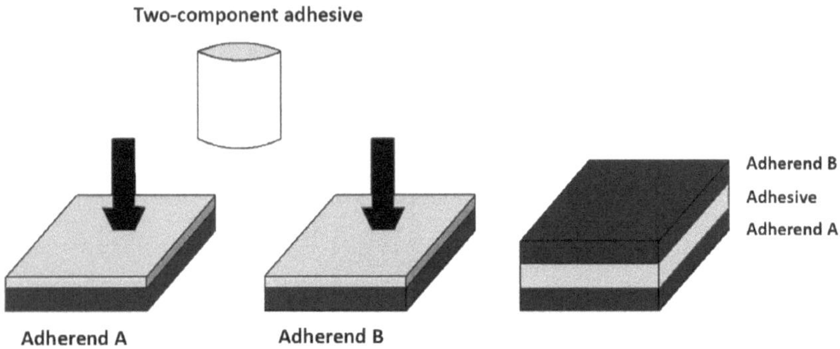

Fig. 2.12 Application of a two-component adhesive on two adherends surfaces—mixing the components before application.

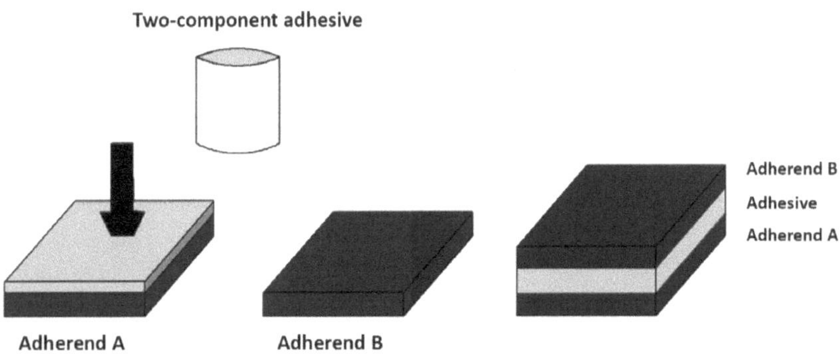

Fig. 2.13 Application of a two-component adhesive on one adherend surface—mixing the components before application.

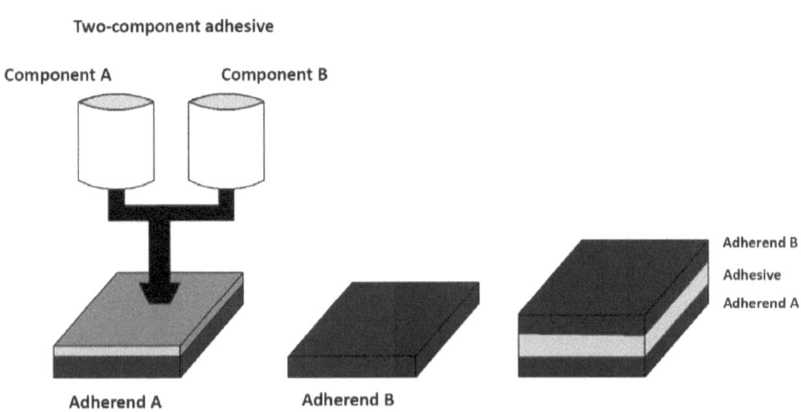

Fig. 2.14 Application of a two-component adhesive by using of a dosing device on one adherend surface.

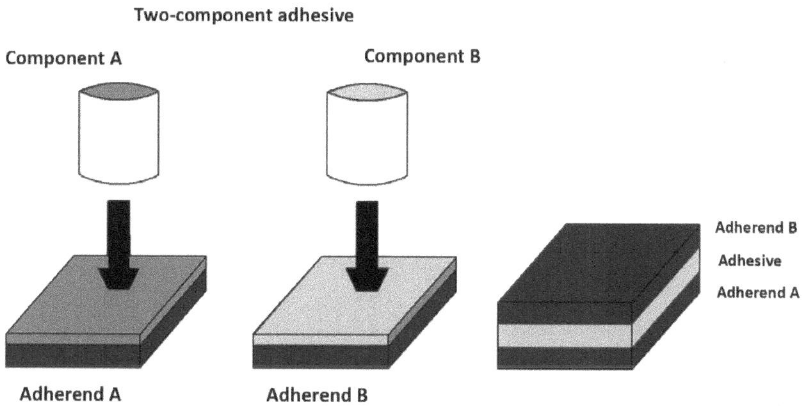

Fig. 2.15 Application of a two-component adhesive with separate dosing components of adhesives on separate adherends surfaces.

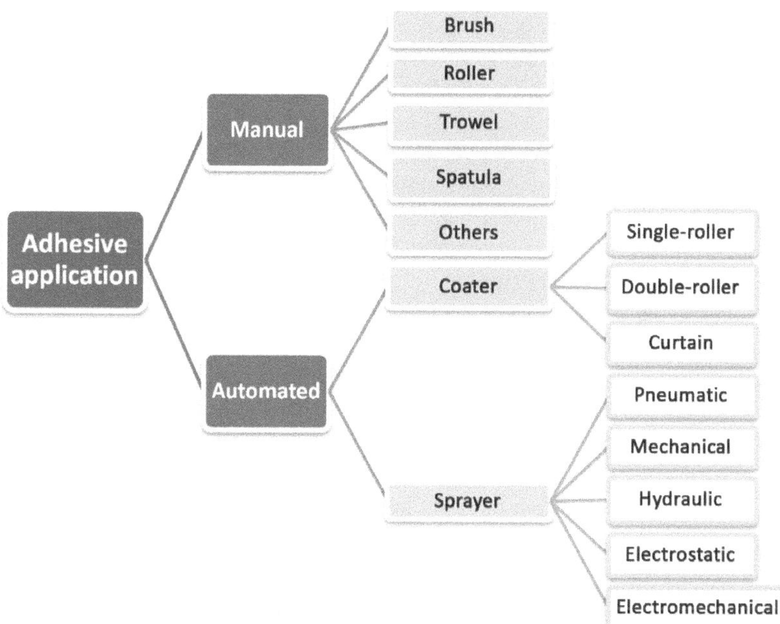

Fig. 2.16 Adhesive application method and example equipment.

Adhesives in the solid state are available in a range of forms, e.g. rods, wires, films or powders, and their common characteristic is that they turn into liquid when heated or on reaching the melting point, which is when they become wettable. The process is reversible on the condition that the adhesive is heated for a short period of time. However, maintaining temperature higher than the melting point of a given adhesive for an extended period of time will initiate the solidification process, hence the adhesive will begin to cure.

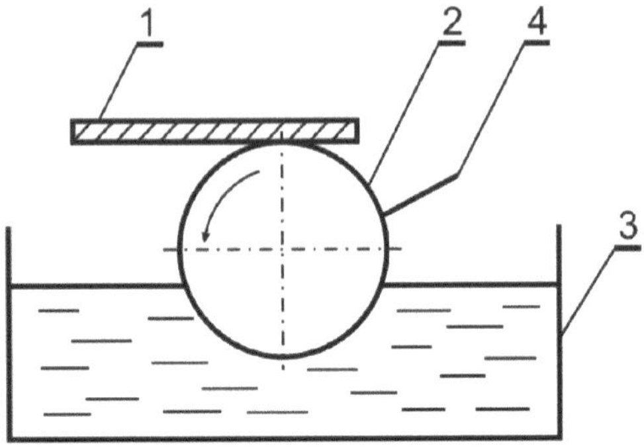

Fig. 2.17 Single-roller adhesive coating machine: (1) substrate, (2) roller, (3) adhesive container, (4) scraper [102].

When applied in the form of a stick, the adhesive is rubbed against the heated adherend to form a liquid layer on its surface. This method, however, does not guarantee that the adhesive is uniformly dispensed; therefore, it is more common to apply ground adhesive material in the form of powder. The adhesive is sprinkled over the surface of adherends. The surface of adherends may be hot or cold, and in the case of the latter, the adhesive is dispensed in a heat chamber, such as a drier, and subjected to heat treatment until it melts [4, 90, 96, 101].

Adhesives in the form of wires are applied onto the surface in the process of flame spraying, by means of special spray applicators. The end of the rod is melted in the flame, and propelled with the stream of compressed air it is released under the form of fine droplets onto the bonding surface.

Adhesives in the solid state are most efficiently handled and dispensed in the form a film or adhesive transfer tape.

Adhesive films are sold as tapes in a wide range of width options, which are suitable for different purposes and may be instantly applied onto the surface for bonding without prior preparation. In addition, the film is cut out to the shape of bonded elements, which further increases its ease of application.

Adhesive films are available in the following forms:

- a clear adhesive film,
- a reinforced adhesive film (with, e.g. polyester mesh or cloth),
- a film laminated to blotting paper, plastic-based, and other materials, and
- a film with a paper or plastic-based release liner.

The type and the amount of adhesive (basis weight), the reinforcement, and the release liner material depend on the specific purpose.

Adhesive transfer tapes are applied to bonded surface with a special transfer tape applicator, which removes the protective layer and cuts the tape to a specified dimension.

A number of reinforced adhesive films are available on the market. The reinforcement material, e.g. glass fibre mesh, is incorporated into the film to impart the cohesive strength of the bond, furthermore enabling to regulate its thickness. In the curing stage, when adherends are loaded with sufficient pressure, the adhesive film melts at the curing temperature which brings the adherend surfaces to an intimate contact, limited by the threads of the fabric material; hence the thickness of the mesh, rather than the amount of loading, is the decisive factor of the bond thickness. In addition, the reinforcement fabric counteracts excessive flash of adhesive from the joint area.

The application of adhesive on the surface for bonding may be carried out manually or in an automated process (Fig. 2.16). Manual application is popular in low-surface-area bonding, piece production, and in workshop conditions. Liquid adhesives are applied manually with rollers or flat paint brushes with medium-stiff natural bristle. The adhesive mass should be applied with one stroke to prevent introducing air bubbles into the composition, whereas the speed of brush stroke should be adjusted to the viscosity of a given adhesive.

In the case of adhesives characterised by higher viscosity, such as pastes, the best spreading solution is a trowel, which is available in a range of edge shapes. The most uniform dispensing of adhesive is obtained with trowels with a notched edge.

The implementation of different nozzles in mixing guns enables dosing and dispensing adhesives on surfaces varying in dimensions or geometry, which would otherwise be difficult for the adhesive to coat.

In the lot or mass production the most popular dispensing solutions are rollers, slot spray nozzles, single- or double-roller coating machines, curtain coating machines, and hydraulic, pneumatic, or electrostatic adhesive spray applicators.

Application of adhesives by means of rollers or roller adhesive spreaders is possible for flat objects of constant unidirectional curvature. In automated assembly and in the case of joining large surfaces, adhesives are applied by various kinds of coating machines, such as roller or curtain coating machines. Fig. 2.17 shows the single-roller machine and its principles of operation.

Application of the adhesive on the surface of a substrate (1) is conducted by means of a turning roller (2), which is partly submerged in the adhesive container (3). The scraper (4) is responsible for even distribution of the adhesive mass on the roller, and for removing the remaining set adhesive off the roller surface. The major limitation of the method consists in its incapability of simultaneous coating both sides of the substrate surface with the adhesive.

In applications requiring simultaneous application of the adhesive on both sides of substrate material, this is a double-roller coating machine that offers higher efficiency of application. Fig. 2.18 shows a coating machine with two horizontal applicator rollers (1) turning in the opposite directions. The substrate material (3) is transmitted between the rollers, which take the adhesive (2) from the containers (4).

Despite their efficiency, the use of roller coating machines is limited to flat substrate applications. In addition, the adhesive should be spread evenly on the entire surface of the roller, whose surface must exhibit high wettability. In order to provide good wettability of the rollers, they are often coated with a layer of rubber, which additionally compensates for potential cylindricity errors of roller applicators.

The abovementioned faults of roller coaters are not observed in the case of curtain coating machines (Fig. 2.19). The design of these applicators enables coating the substrate (2) with the adhesive (3) dispensed by the regulated head (4). The substrate is

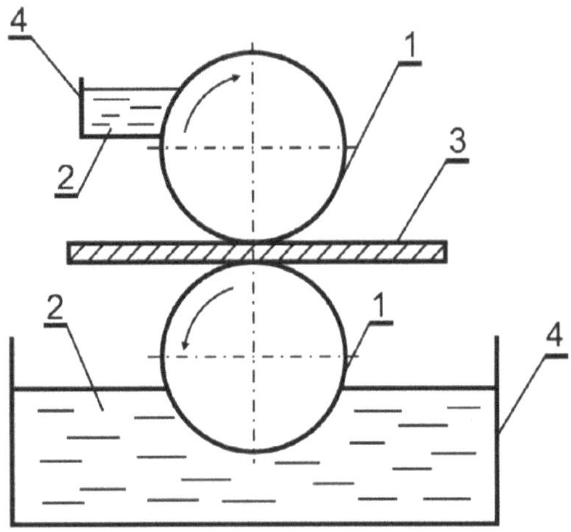

Fig. 2.18 Double-roller adhesive coating machine: (1) applicator rollers, (2) adhesive, (3) substrate, and (4) adhesive containers [102].

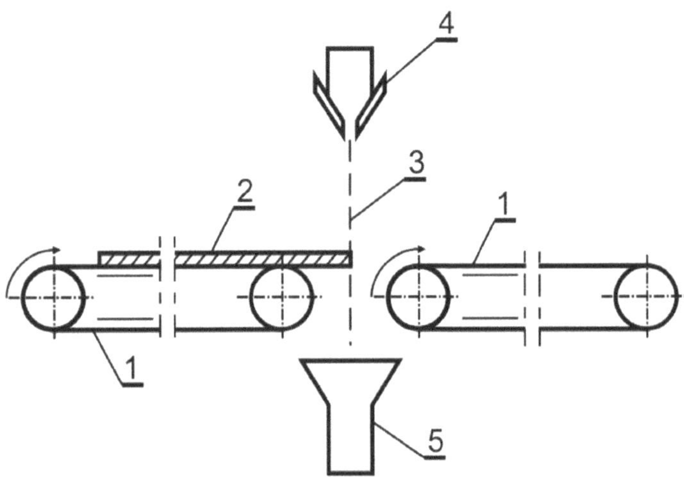

Fig. 2.19 Curtain coating machine: (1) conveyor belt, (2) substrate material, (3) adhesive, (4) regulated head, and (5) adhesive excess container [102].

fed into the applicators by the conveyor belt (1), whereas the excess of the adhesive is removed to the adhesive excess container (5).

Curtain coating machines provide higher cost and time efficiency than roller coaters, simultaneously ensuring higher adhesive layer dispensing uniformity. Curtain coaters dispensing apparatus is the coating head with a regulated slot width, through which the adhesive is dispensed onto the substrate material being transported at a desired speed on the conveyor belt.

The final category of adhesive dispensing systems is spray applicators, which utilise a range of adhesive propelling technologies such as: pneumatic, mechanical, hydraulic, or electrostatic. Spray applicators are employed in dispensing liquid or powder adhesives. Spraying ensures equal coverage of the substrate surface with the adhesive. The dispensing procedure usually involves moving the spraying nozzle along the substrate surface. Although they are of the most robust design, pneumatic systems are rarely applied owing to the numerous disadvantages of the method, including: significant adhesive loss, high energy consumption, surface contamination, and considerable solvent use.

In mechanical spray applicators (Fig. 2.20) the adhesive is transferred by the flow nozzle (1) to the rotating cylindrical element (2) where it moves under the centrifugal force. Simultaneously, the pressure in the tube pushes the adhesive towards the nozzle mouth (3). This method reduces the use of adhesive by 20%, compared to the pneumatic spraying.

The principle of operation of electrostatic spray systems is that the high-voltage charge is generated to propel the adhesive onto the substrate surface. In order to apply these systems, specific requirements must be fulfilled regarding the dielectric properties of the adhesive and its viscosity. Two out of three major spray application methods have been presented: electrostatic and pneumatic. The third established method of spray dispensing technologies is the electromechanical method.

The electromechanical method employs rotating spray nozzles, which simultaneously generate ionising field between the spraying heads and the substrate. The adhesive is dispensed under inertia forces, which are induced by the rotating movement of the spray nozzle heads. The centrifugal force propels the particles of the adhesive out of the spray nozzle onto the ring body of the spray nozzle heads, where the adhesive acquires the electrical charge, leaves the nozzle in the form of an ionised powder and is loosely deposited on the surface of adherend.

Electrostatic applicators deposit the adhesive material solely by the forces of the electrostatic field, which is induced in the space between the spraying nozzle and the substrate. Electrostatic spraying is performed with the use of static slot applicators located over the moving substrate material.

An important advantage of this method is that the adhesive loss is virtually eliminated. Slot applicators are used to coat flat surface of simple-geometry adherends.

The electrostatic method is not only suitable for the liquid adhesives but also it may be employed in powder adhesive applications. Fig. 2.21 shows a schematic representation of the powder adhesive application method.

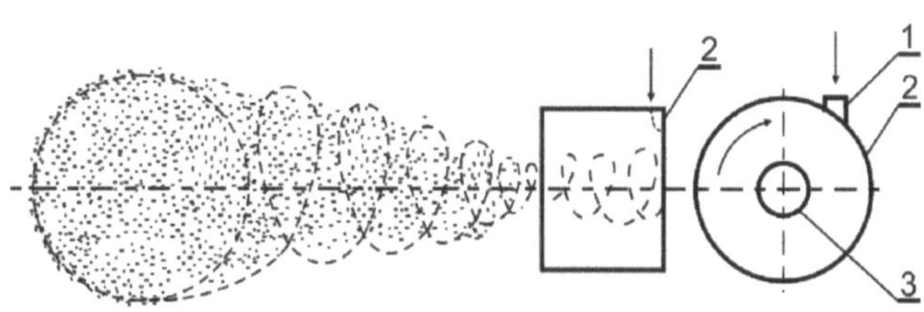

Fig. 2.20 Schematic representation of mechanical adhesive spray applicator operating principles: (1) feed nozzle, (2) rotating cylinder, and (3) adhesive outlet [102].

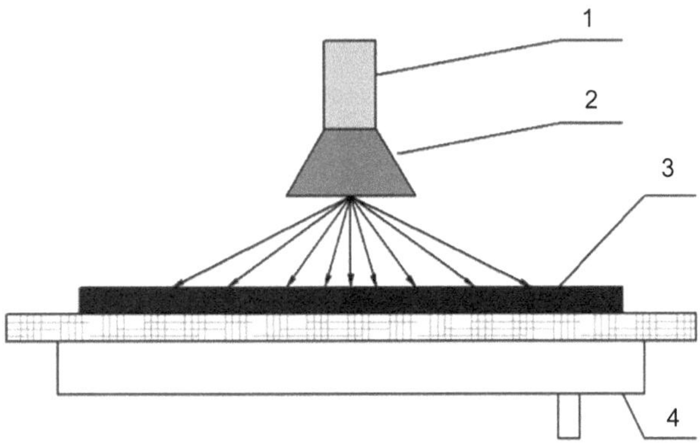

Fig. 2.21 A schematic representation of electrostatic application of powder adhesive: (1) adhesive dispensed under high voltage, (2) to-and-fro motion of the spray nozzle, (3) adherend covered with powder adhesive, and (4) adhesive excess collector.

In pneumatic spray systems a mixture of highly compressed mixture of gas and adhesive is forced out of the nozzle. Particular components of the adhesive frequently become diluted in the stream of air, which leads to thickening of the adhesive, which becomes more viscous and having lost its properties becomes unsuitable from the viewpoint of the technological process. Although adhesive loss is comparably insignificant, the method is not frequently applied. The literature in the field of adhesive bonding technology describes a number of other structural solutions for dosing and applying the adhesives mass, which is found in [4, 96, 101].

2.3.4 Assembly of adherends

Proper assembly of adhesive joints requires high precision and accuracy when aligning and fixing the adherend surfaces. One of the key technological requirements in joint assembly is that the surfaces of substrates to be bonded must be in an intimate contact and at the same time high pressures should be avoided. Examples of adherend fixing and assembly operations are shown in Fig. 2.22.

Fixing and bonding of adherends are of crucial significance in adhesive joint assembly. Only by proper aligning and bonding adherends do we ensure that the produced joint exhibits high quality and strength. What is more, the shape and dimensional accuracy of adhesive joints is an equally important factor contributing to their strength properties, and as a result, the service lifetime and the reliability of the assembly. Applying a proper fixture solution is necessary for the assembly to hold in place, particularly due to the presence of a liquid adhesive which allows the adherends to move freely until it cures. The correct assembly of joints is further encumbered by the fact that in certain cases the adherends are of considerable dimensions, hence requiring much effort in aligning and fixing. It is important to remark that proper alignment of substrates to be bonded should also ensure that the thickness of the bondline is uniform along the whole length of the joint, unless specified otherwise.

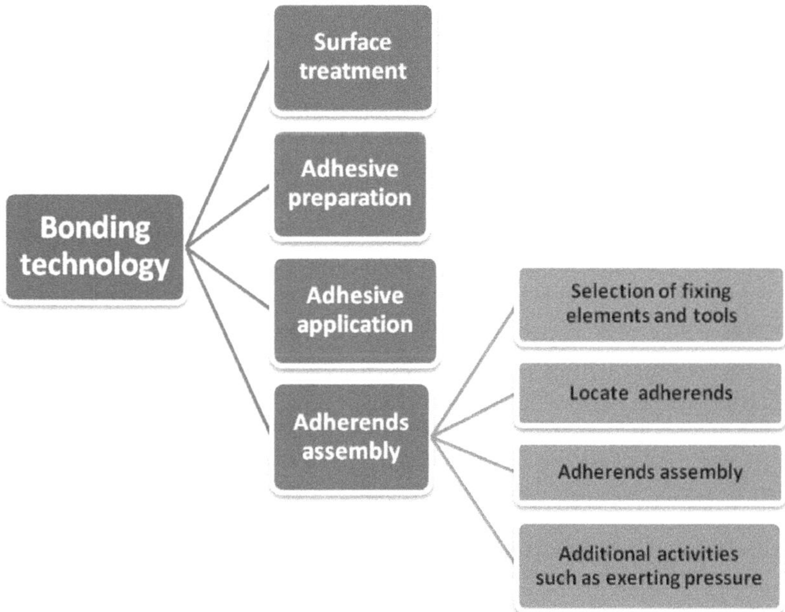

Fig. 2.22 Examples of adherend fixing and assembly operations.

The considered stage of joint assembly is specified by a number of criteria, e.g. [1, 4, 101, 103, 104]:

- the surface preparation,
- the dimension and shape of adherends,
- the type of substrate material,
- the location of the joint in the structure,
- the access to the joint,
- the type of assembler joint,
- the type and properties of the adhesive,
- the type of production, and
- the technical requirements and technology.

One technological criterion fundamental to the quality of assembled joints is to align the adherends so that their surfaces maintain intimate contact without great pressure. All cracks, waviness, warping, pores, and other defects are handled by applying additional pressure, which may however become the source of adverse stresses causing decrease in the strength of adhesive joints and the strength of the entire structure. With regard to surface roughness, it ought to be remarked that in extreme cases excessive values of the height parameters lead to misalignment of adherends.

When fixing and assembling adhesive joints, the shape of adherends (flat/cylindrical) and the type of joint (lap joint, butt joint, etc.) play an important role, as they determine, e.g. the selection of fixing jigs and their layout. Furthermore, it is necessary to analyse the specified location of the joint in the structure, as it may require developing a special purpose-built tooling.

The technical requirements and the technology are equally important, for the reason that certain adherend alignment operations employ tooling previously applied in the assembly of other connectors. Most commonly, adherends are arranged by means of special jigging/fixture tools equipped with fastening or clamping elements, which will ensure that the adhered pieces remain in close contact securing adequate load distribution, and will not move while the adhesive cures. These fixtures are furthermore responsible for producing joints of repeatable dimensions [1, 90, 101]. There is a wide range of fixture elements used in adhesive joining, such as: locating pins, rivets, screws, pins, bolts, vee locations or vee clamps, etc.

Adherends require aligning and fixing also when bonded in adhesive coating machines and systems. For that purpose, standard locating elements are employed, e.g.:

- flat surface fixture elements and systems (e.g. locating pins or plates),
- fixture elements and systems for external cylindrical surfaces (e.g. vee locations or diagonal plates),
- fixture elements and systems for internal cylindrical surfaces (e.g. locating pins and pivots),
- external cylindrical surface fixture elements and systems (e.g. vee locations, diagonal plates, and jaw chucks), and
- internal cylindrical surface fixture elements and systems (e.g. pins and dowels).

2.3.5 Curing adhesives

Adhesive curing is a process of transforming adhesive material into a strong and insoluble adhesive bond. Adhesive hardens as a result of cross-linking of the resin (or the monomer) at the presence of a curing agent, a catalyst, or at elevated temperature [95, 96, 105]. Anaerobic adhesives, which are predominantly one-component adhesives, cure at the absence of oxygen, whereas solvent-based adhesives form the impervious bond when the solvent (i.e. water, alcohol, or an organic solvent) evaporates from the mass.

The following technological parameters are considered in adhesive curing (Fig. 2.23) [78, 106]:

- loading (curing pressure),
- curing temperature, and
- curing time.

The curing process of solvent-based adhesives is characterised by the following parameters [4, 95, 101]:

- open time, i.e. the time of partial solvent evaporation,
- set time,
- ambient temperature,
- air circulation intensity, and
- permeability of the substrate material (solvent adsorption capacity).

Pressure applied during curing constitutes one of the technological parameters of adhesive hardening. The selection of the pressure value and type to be applied during curing depends on such factors as [4, 84, 90, 101, 107–110]:

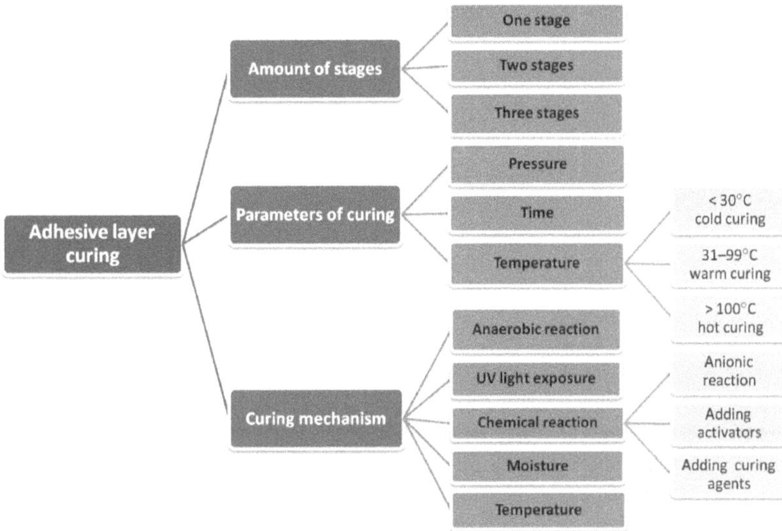

Fig. 2.23 Type of curing process and the factors affecting the curing process.

- the type and properties of the adhesive, including its density, viscosity, thermal shrinkage, or chemical cure shrinkage,
- the type of substrate,
- the adherend surface preparation for bonding,
- the shape of adherends,
- the location of the joint in the structure, and
- the type and the shape of additives.

Pressure in adhesive bonding fosters wetting, particularly in filling micropores and surface micro-irregularities. It creates favourable conditions for spreading the adhesive and fixing the adherends. The applied pressure should ensure that the adherend surfaces are in an intimate contact, and the bondline thickness is optimal (0.05–0.20 mm).

An excessive amount of pressure applied to the assembly, or improper layout of clamping tools may result in the deformation of the joint. Alternatively, the disproportionately high pressure could produce the bond of inadequate thickness, where the adherends are at risk of potentially coming into direct contact. On the other hand, the insufficient load applied to the joint during cure may result in obtaining a bond of excessive thickness and porous texture. In both scenarios, the assembled joint exhibits drop in strength.

The amount of pressure during cure will predominantly depend on the cure shrinkage characteristics of a given adhesive. The loading applied to the joint should be smaller, usually less than 0.1 MPa (contact pressure) when the shrinkage does not exceed a fraction of a percent. In the case of higher shrinkage cure rates (of several percent), the force acting on the joint will be in the range of 0.5–2.0 MPa.

Curing time is closely connected with the curing temperature. In cold-setting adhesives, the strength of adhesive joints rises with the curing time. Intermediate-temperature-setting and hot-setting adhesives have a specific setting time, which ensures the highest joint strength [77, 78, 90, 96].

The curing temperature adhesives determines whether they are considered as cold-setting (ambient temperature <30°C), intermediate-temperature-setting (t=31–99°C), or hot-setting (t>100°C), (Fig. 2.25). Maximum strength of adhesive joints may be obtained under the condition that the optimal curing temperature for a given adhesive is retained throughout the process; curing temperatures are provided by manufacturers.

Depending on the curing method, adhesives are classified into the following categories [77, 78, 95, 96, 111–118]:

- cured by anaerobic reaction,
- cured by UV light exposure,
- cured by anionic reaction (e.g. cyanoacrilates),
- cured by adding activators (e.g. modified acrylates),
- cured by moisture (e.g. silicones, urethanes), and
- cured by heat (e.g. epoxy adhesives).

Anaerobics are single-component adhesives that cure at room temperature in the absence of oxygen. The curing catalyst activates when the atmospheric oxygen is removed from the bond, particularly when remaining in a simultaneous contact with metal in assembly of parts and components.

The cross-linking of the resin in UV-cured adhesives is regulated by the UV light intensity and wavelength. The radiation penetrating the joint causes photoinitiators, i.e. substances initiating polymerisation of the adhesive, to split, which in turn leads to the emergence of bonds between the adhesive molecules.

Adhesives curing in an anionic reaction begin to link together in a polymerisation reaction when they come into contact with alkaline surfaces. The stabilising agent, composing the adhesive mass, acts to neutralise the humidity level on the surface of adherends, maintaining it within a certain range. Ensuring adhesive mass humidity levels within 40%–60%, guarantees optimal strength of adhered surfaces. Lower humidity leads to slower curing and the higher humidity accelerates this process.

The process of adhesive cure by means of activators is initiated once the activating agent is supplied to the adhesive mass, which enables conducting the assembly of machine parts in a wide range of temperatures.

Hot-setting adhesives, mainly epoxy resins, cure at temperatures exceeding 100°C, however, their exact thermal properties are specific to the particular type of adhesive material. The curing time of adhesives depends on the temperature of cure: the higher the latter the quicker the adhesive will cure.

Adhesive bonds may be cured in a two-stage process. After initial cold curing, the joint is subjected to short-time exposure to elevated temperature (intermediate-temperature-setting and hot-setting). The second curing stage is referred to as post-curing of the adhesive. In the first stage the joint is brought to solidify at ambient temperature (15–20°C) under curing pressures specific to the given type of adhesive. The solid adhesive reaches at this time approximately 60%–70% of its final strength. At the second step, the adhesive joint becomes fully set without the use of curing pressure, which most commonly consists in adhesive joint detention for several hours at an elevated temperature (approximately 50–100°C), to produce the joint exhibiting full strength and full cure. The two-stage curing process allows reducing the total joint setting time by several times, as compared to curing at room temperature.

Cold-setting adhesives may be subjected to a two-stage curing, or alternatively to a single-stage curing, carried out at elevated temperatures. The process of thermal curing is conducted at a temperature of 60–120°C, under suitable curing pressure. Compared to the two-stage curing, it provides several or even a dozen-fold reduction of the curing time.

Adhesive heat-curing takes place in three stages. The first step is designed to bring the adhesive to the curing temperature. The time span of this stage and the heating intensity depend on the mass of adhesive assembly and the type of the adhesive joint. The second step consists in exposing the adhesive to the curing temperature, and is carried out in autoclaves with electronic systems of programming and control of gluing cycle parameters. The third step of adhesive heat curing is slow cooling the assembly to room temperature, where the important factor is the cooling time, most frequently determined experimentally. Most importantly, the time should be long enough so as not to induce shrinkage stress and short enough not to prolong the production cycle; as a result, the curing time is generally within a range of 10–20 min.

There is a wide range of appliances used in the curing process, which are responsible for generating and applying pressure or thermal treatment of adhesives.

The pressure applied to the joint assembly during cure may be produced by different methods. There exists a number of solutions for applying pressure during setting, including the following [1, 4, 101]:

- levers,
- springs,
- hydraulic mechanisms,
- vacuum equipment,
- pressure chambers, and
- magnetic chucks.

The types of equipment above employ various mechanisms and technologies and are designed for different applications, however, each of them must ensure that the basic requirements are fulfilled, i.e.

- to generate precise amount of loading force,
- to guarantee stability of assembled adherends throughout the cure,
- to ensure repeatability of joint assembly dimensions, and
- to make sure that the applied pressure acts on the entire adhesive layer rather than locally.

All the listed functional criteria are important indicators of applicability of particular pressure-loading systems; nevertheless, it appears that high precision in adjusting the value of such pressure is key to producing high strength joints (Fig. 2.24). The pressure must ensure that the adherends remain in an intimate contact, simultaneously not causing any deformation of the entire assembly. Excessive pressure may cause the adherends to come in a direct contact as a result of removing the adhesive layer from the joint. Insufficient pressure, on the other hand, causes the produced adhesive bond of disproportionate and irregular thickness, and of porous structure. All these are indicative of a weak adhesive joint.

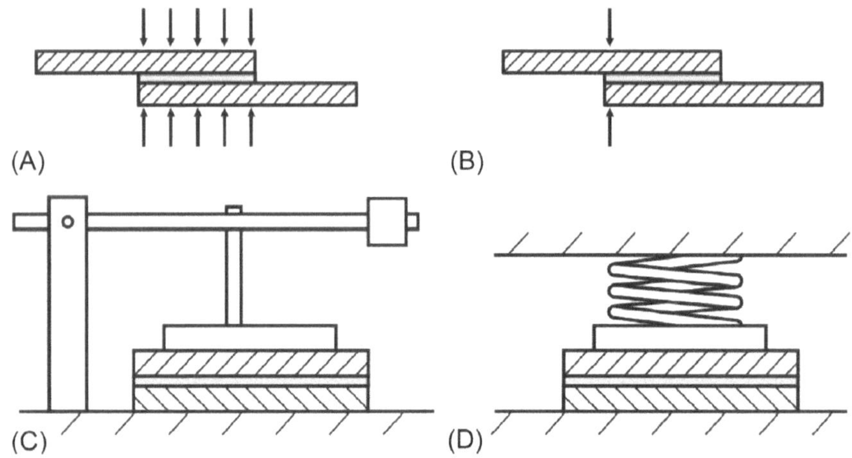

Fig. 2.24 Methods of applying pressure during curing: (A) right, (B) wrong, (C) lever system, and (D) spring system.

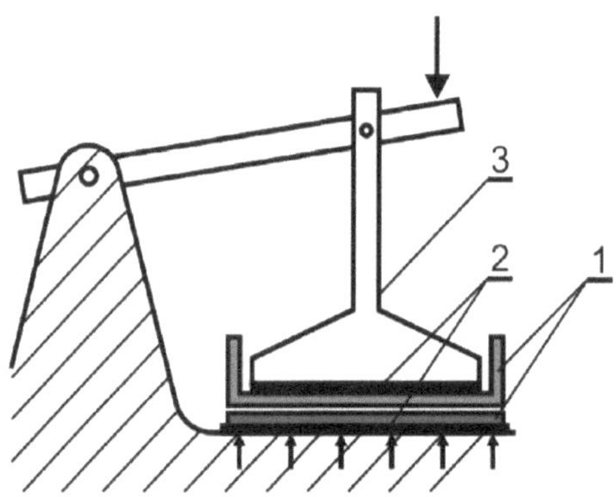

Fig. 2.25 Schematic representation of lever mechanism for applying pressure during cure: (1) adherends, (2) rubber pads, and (3) weight-loading element [102].

Lever mechanisms are generally preferred in applications where the adherends require dissimilar pressure over different lengths of joint. A schematic representation of a lever mechanism is shown in Fig. 2.25.

Metal and rubber elements play a crucial role in lever systems. Metal elements provide protection against crushing, whereas rubber pads are responsible for uniform distribution of the load over the entire adherend. The rubber pads simultaneously compensate for the changes in loading on the adherends, which may result from design or production errors of the weight-loading element and of other parts of the system.

Vacuum bag systems employ pressure on the adhered surfaces by means of an elastic membrane. The standard material for the membrane is rubber, characterised by high thermal resistance. In order to increase the air-tightness of the system, the membranes are additionally covered on one or two sides with special balloon films (Fig. 2.26).

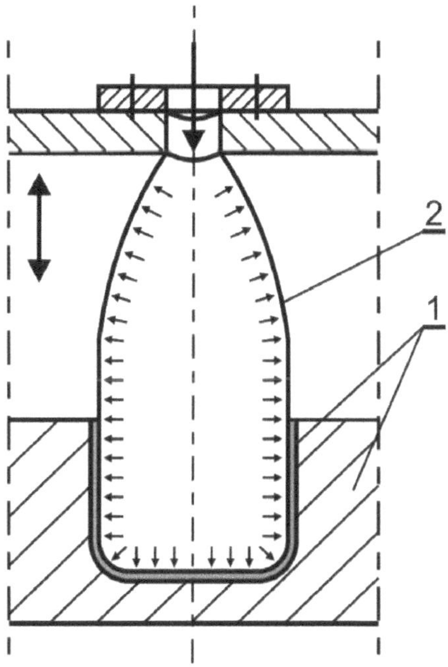

Fig. 2.26 Schematic representation of vacuum bagging method: (1) adherends, (2) rubber membrane [102].

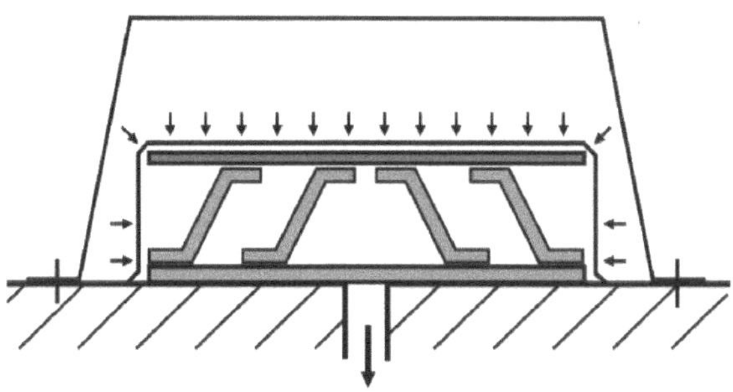

Fig. 2.27 Schematic representation of autoclave for generating pressure on the adhesive joint [102].

The pressure of the membrane to the adherends is generated by compressed air, injected by the vacuum pumping system. The vacuum method offers an excellent range of applications, as it is suitable for curing joint assemblies of complex geometries (e.g. round/wavy), which are impossible to process using standard mechanical weight-loading systems. Autoclaves constitute a certain variety of instruments implementing the vacuum bagging system for adhesive joint setting.

These pressure vessels exert loading on the elastic membrane by means of pressure difference generated in the chamber. A schematic representation of the method is shown in Fig. 2.27.

During assembly, particular elements of the adhesive joint are placed in hermetic elastic bags. The vacuum system of the autoclave generates the subatmospheric pressure in the zone confined by the elastic bags, whereas the pressure system fills the vacuum chamber with gas of specified pressure and maintains the steady level of the pressure throughout the adhesive joining process.

The equipment presented above plays an important role in assembly of adhesive joints. Another group includes various types of presses, such as: hydraulic, screw, or pneumatic tools. Presses have a large base plate, which enables simultaneous assembly of a considerable number of joints, or adhesive joining of large surfaces. In addition, the process may be streamlined by placing stacks of adherends in different sections of the press.

Ensuring proper temperature is vital for curing hot-setting adhesives. There exist many systems and heating units controlling the temperature of the joint during assembly, and their applicability is dependent on the requirements, the production capabilities, the type of production, the joint shape, or the location of the joint in the structure. Typical examples of such systems are presented in Fig. 2.28.

Not only are technological factors considered in the selection of a particular heating method and equipment but also the choice is often dictated by the required properties of the cured adhesive layer. Therefore, the curing method is pivotal to the end properties of the assembled joint, e.g. its strength, reliability, service lifetime, etc.

2.3.6 Adhesive joints control

The process of adhesive joint control involves a number of operations focused on assessing a variety of factors involved in the technology of adhesive joining and the end effects—the assembled joints. Adhesive joining is regarded as a special process requiring nonstandard approach in control, which may necessitate applying a variety of testing methods in order to control the assembled joint or the process of joining [101, 119, 120]. In the majority of cases complex adhesive joint control is required, which may involve (Fig. 2.29) [1, 4, 105, 121–124]:

- the visual inspection of the joint shape and dimensional accuracy with a naked eye or with the aid of appropriate equipment,
- the adhesive bond control (defect detection in the adhesive layer itself),

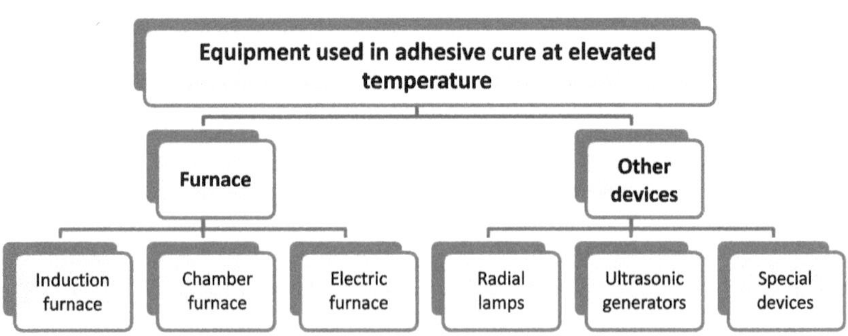

Fig. 2.28 Examples of equipment used in adhesive cure at elevated temperature.

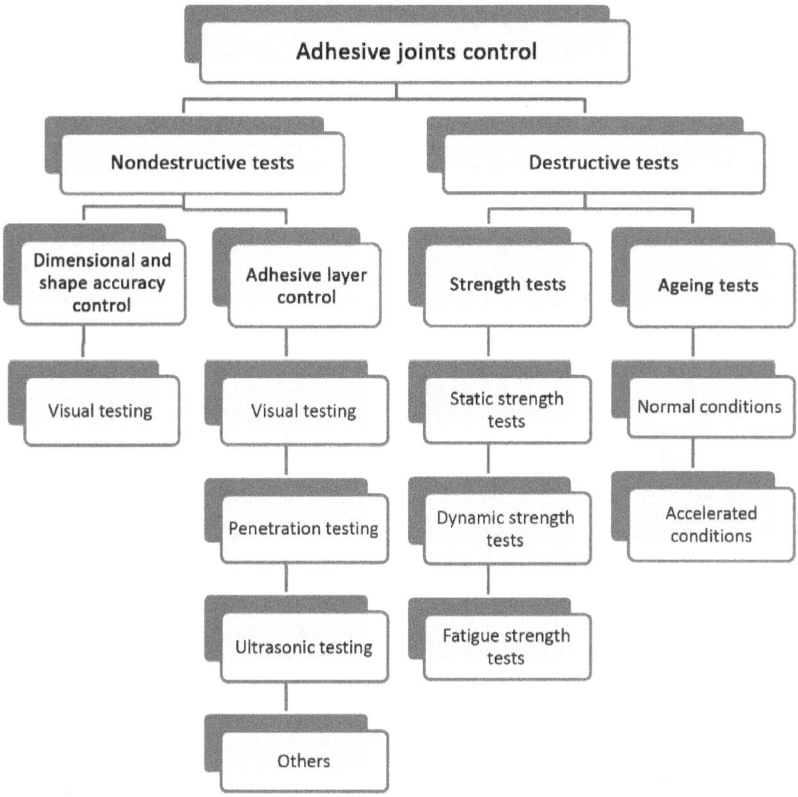

Fig. 2.29 Examples of joint control methods.

- strength tests, and
- in-service tests.

Comprehensive joint control should involve the visual inspection, the testing of adhesive technology quality by means of nondestructive methods (detection of defects in the bond) and strength tests on joint specimens (destructive testing).

Evaluation of joint quality with the use of nondestructive methods does not lead to joint failure or damage observed in the bond or in the adherends. Nevertheless, not all methods are applicable in every instance of adhesive joint control. There exists an extensive body of literature regarding the nondestructive methods [119, 124, 125].

Tests carried out within the scope of joint quality control are to determine the location and the extent of adhesive layer problems and to detect technological defects, such as: presence of gas bubbles, contamination in the adhesive layer or the adherend, surface porosity, nonuniform bondline thickness, adhesive layer discontinuity, delamination, and tightness of joints. Major methods and techniques implemented in nondestructive testing are as follows [90, 126–142]:

- visual methods,
- ultrasonic methods,

- acoustic methods,
- radiography methods,
- thermographic methods,
- holographic methods, etc.

Each method requires specialised measuring equipment and tools, which is a factor of certain relevance to the selection of a particular joint quality control method, whether in the destructive or the nondestructive testing application.

However, the key considerations when choosing the joint testing method are [1, 90, 101, 119, 120, 125, 134–137, 143]:

- type and thickness of adherends,
- heterogeneity of substrates,
- joint geometry,
- location of the joint in the structure, particularly from the viewpoint of assembly access,
- structural characteristics of the tested product,
- costs of operation,
- type of production,
- available testing equipment,
- testing conditions.
- exclusion of destructive testing (joint must be intact after tests),
- safety factor requirements,
- the degree of importance of the joint, and
- other factors.

The factors mentioned above are considered when selecting specific destructive/nondestructive tests. Adhesive joints subjected to nondestructive testing do not succumb to failure or damage in the adhesive layer nor the adherends, and are therefore ready for service upon the end of testing. In the case of destructive testing methods, the joint assembly becomes damaged or destroyed, which is why the tests are conducted on a test batch of joints. Among destructive testing methods, we may distinguish various strength tests (static and dynamic, short or long term, fatigue, under various loading conditions), and ageing tests.

Specific methods require specific tools and instrumentation, which take into consideration the factors determining the choice between the destructive/nondestructive methods.

Visual method

Visual method is one of the nondestructive testing methods (NDT). Visual testing is defined as a process of observation conducted with a naked eye (direct method) or aided (indirect method) [1, 120, 138, 139, 144]. The process aims at verification of conformity of the test object with specifications described in, e.g. norms, acceptance requirements, and other regulations.

Commonly assisted by other NDT, it constitutes the first basic control stage, consisting in visual inspection for discrepancies observable with a naked eye (direct visual inspection) or with the aid of appropriate tools (indirect visual inspection), usually optic, frequently preceded by the measurement of their characteristic dimensions.

Visual inspection requires compliance with basic conditions regarding the correctness of control, such as: having required qualifications, good eyesight, appropriate luminous intensity (500 lux), as well as the ability to distinguish and interpret inaccuracies.

Penetrant testing

Liquid penetration method employs the capillary action. Any surface cracks act as capillaries, which are filled with a penetrant. The popularity of the method may be attributed to its quick assessment of large surfaces for the presence of surface discontinuities [130]. Penetration method employs ready-made liquids, which according to norms and standards should be attested and represent either colour contrast or fluorescent types.

Depending on the type of penetration liquid, three inspection techniques are distinguished:

- colour contrast,
- fluorescent, and
- dual purpose (fluorescent and colour contrast).

Ultrasonic method

Ultrasonic testing is recognised as one of the most important methods and is found in a range of applications in nondestructive testing [1, 125, 134–138, 140, 145]. Ultrasonic testing utilises propagation of ultrasonic energy in solids. The ultrasonic wave is generated by piezoelectric transducer connected to high-frequency voltage (usually between 0.5 and 15 MHz). The generation of waves is done by means of a set-up composed of an ultrasonic flaw detector (defectoscope) and a head.

Ultrasonic waves are excited by the ultrasonic head into the inspected object through a film of contact liquid (e.g. oil, water, and a solution of wallpaper adhesive) in the process of acoustic coupling. It is then reflected by the surface or internal discontinuities of the material and displayed on the device screen in the form of bottom echo or echo from defect. Given the geometry of the object and the characteristics of the ultrasonic head we may pinpoint the actual location of the defect from the echo from defect.

Acoustic emission

Acoustic emission (AE) is categorised as one of the ultrasonic testing methods [1, 119, 131, 132, 142]. It consists in generating stress waves (acoustic or vibroacoustic), which are excited when intermolecular bond energy is released as a result of, e.g. deformation, cracking, or phase changes.

AE apparatus does not emit signals or affect the physical condition of a test object; instead it registers the physical effects that are self-generated in the monitored object. When analysing adhesive joints, the signals that are picked up are generated by the defects found in the adhesive bond, which are the source of elastic waves propagating in the object. The elastic waves, originating in the source of AE, propagate in all directions from the source, and having reached the AE sensor, they are transmitted to the AE analyser in the form of the electric current changes.

Radiography

Radiography enables detection of internal, surface, and subsurface discontinuities in the analysed object. The method consists in exposing the test objects to ionising radiation (by means of, e.g. X-ray radiation from the X-ray tube, or γ-ray radiation from artificial isotope sources) and registering the shadowgraph of the test object discontinuities on radiograms. This method is capable of detecting both volumetric and flat discontinuities, provided that the specific testing procedures are followed [118, 134, 141].

Radiodefectoscopes employ the principle of exposure to gamma-ray radiation in testing adhesive joint assemblies.

Laser holography

Holography is a branch of optics that deals with producing 3D images with coherent light and is distinctly different from conventional lens imaging [118, 133].

Laser holography as a representative of NDT methods harnesses an impulse or continuous light source in the visible region, which allows us to record and reproduce holographic images on photosensitive films of required resolution. The laser beam is projected onto the film and superimposed in a form close to the original image. When reproduced, the image gives an impression of being the real object, on account of the visual effect of giving the impression of seeing different sides of the object from different perspectives.

Thermography

Infrared thermography is a branch of science concerned with detection, registration, processing, and visualisation of the invisible infrared radiation emitted by an object. The image formed with the aid of this technique is a thermogram, representing the temperature distribution on the surface of the test object [118, 134, 136, 139, 140].

Nondestructive testing by means of infrared thermography is carried out in passive or active configurations. In passive IR the test object is characterised by its naturally emitted temperature field. That is why passive testing is widely popular in in situ nondestructive testing of machines and their elements during work or immediately after, when the disproportionate temperature distribution on their surface may be indicative of defects. Test objects are mechanically or thermally loaded during operation, defects emit or absorb that thermal energy, and thus can be diagnosed by means of the passive methods.

In active infrared thermography, another source of thermal stimulation is employed to heat or cool the test object. Prior to testing, material defects are of uniform temperature, which is equal to the ambient temperature, and therefore do not emit signals that could be registered by the system. Hence the test object requires additional thermal stimulation for its surface to show dynamic temperature field. Since there exist certain dependencies between the results of examination and the time of observation, special data processing procedures must frequently be employed to account for this and other factors affecting test results.

The main instrument implemented in the active infrared thermography method is a thermographic camera equipped in a microbolometer operating in the long range of infrared wavelengths.

Active thermographic testing methods are classified according to various criteria, such as

- the mutual position of the source of thermal stimulation and the detector,
- the type of thermal stimulation source,
- the shape and dimensions of the thermal stimulation and temperature reading zones.

The first of the listed types of IR methods classification (i.e. according to the mutual position of the heat source and the detector, is also the most widely applied. In this category there the following IR methods are distinguished:

- infrared reflection method,
- infrared transmission method, and
- internal heat source-based analysis.

The mutual position of the heat source and the thermal detector is crucial to the detection of defects and constitutes an important determinant of the practical realisation of IR testing. Thermal stimulation of the test object is realised by heating or cooling, which from the viewpoint of thermodynamics carry equivalent significance, given that the amount of heat stream is identical. However, considering the density of the heat stream and bearing in mind the technological considerations and potential hindrances, heating appears more practical.

References

[1] R.D. Adams (Ed.), Adhesive Bonding. Science, Technology and Applications, Woodhead Publishing Limited, Cambridge, 2010.
[2] W. Brockmann, P.L. Geiß, J. Klingen, B. Schröder, Adhesive Bonding. Materials, Applications and Technology, Wiley-VCH Press, Weinheim, 2009.
[3] D.E. Packham (Ed.), Handbook of Adhesion, Longman, New York, 1992.
[4] G. Habenicht, Applied Adhesive Bonding, Wiley-VCH Press, Weinheim, 2006.
[5] R.D. Adams, J. Comyn, W.C. Wake, Structural adhesive joints in engineering, second ed., Chapman & Hall, London, 1997.
[6] A. Baldan, Adhesion phenomena in bonded joints, Int. J. Adhes. Adhes. 38 (2012) 95–116.
[7] R.D. Hazlett, On surface roughness effects in wetting phenomena, in: K.L. Mittal (Ed.), Contact Angle, Wettability and Adhesion, VSP, Zeist, 1993, pp. 173–181.
[8] Y.L. Chen, C.A. Helm, J.N. Israelachvili, Molecular mechanisms associated with adhesion and contact angle hysteresis of monolayer surfaces, J. Phys. Chem. 95 (1991) 10736–10747.
[9] J. Comyn, Diffusion and adhesion, in: J. Comyn (Ed.), Polymer Permeability, Chapman & Hall, London, 1985, pp. 177–215.
[10] D.E. Packham, Surface energy, surface topography and adhesion, Int. J. Adhes. Adhes. 23 (2003) 437–448.
[11] K.M. Hay, M.I. Dragila, J. Liburdy, Theoretical model for the wetting of a rough surface, J. Colloid Interface Sci. 325 (2008) 472–477.

[12] R.D. Hazlett, On surface roughness effects in wetting phenomena, J. Adhes. Sci. Technol. 6 (1992) 625–633.
[13] L.H. Lee, Roles of molecular interactions in adhesion, adsorption, contact angle and wettability, J. Adhes. Sci. Technol. 7 (1993) 538–634.
[14] A.J. Kinloch, Adhesion and Adhesives, Chapman & Hall, New York, 1987.
[15] K.L. Mittal, The role of the interface in adhesion phenomena, Polym. Eng. Sci. 17 (1997) 467–473.
[16] P. Dalet, E. Papon, J.-J. Villenave, Surface free energy of polymeric materials: relevancy of conventional contact angle data analyses, J. Adhes. Sci. Technol. 13 (1999) 857–870.
[17] F.M. Fowkes, Determination of interfacial tensions, contact angles, and dispersion forces in surfaces by assuming additivity of intermolecular interactions in surfaces, J. Phys. Chem. 66 (1962) 382.
[18] F. Awaja, M. Gilbert, G. Kelly, B. Fox, P.J. Pigram, Adhesion of polymers, Prog. Polym. Sci. 34 (2009) 948–968.
[19] C.J. Van Oss, R.J. Good, Surface tension and the solubility of polymers and biopolymers, J. Macromol. Sci. Chem. A26 (1989) 1183–1203.
[20] D.S. Rimal, L.P. Demejo, R.C. Bowen, Mechanics of particle adhesion, J. Adhes. Sci. Technol. 8 (1994) 1333–1355.
[21] F. Barberis, M. Capurro, Wetting in the nanoscale: a continuum mechanics approach, J. Colloid Interface Sci. 326 (2008) 201–210.
[22] K.L. Mittal, Interfacial chemistry and adhesion: recent developments and prospects, Pure Appl. Chem. 52 (1980) 1295–1305.
[23] W.C. Wake, Adhesion and the Formation of Adhesives, second ed., Applied Publishers, London, 1982.
[24] C.J. van Oss, R.J. Good, M.K. Chaudhury, Additive and nonadditive surface tension components and the interpretation of contact angles, Langmuir 4 (1998) 884–891.
[25] A. Burdzik, M. Stähler, M. Carmo, D. Stolten, Impact of reference values used for surface free energy determination: an uncertainty analysis, Int. J. Adhes. Adhes. 82 (2018) 1–7.
[26] C.W. Extrand, Y. Kumagai, An experimental study of contact angle hysteresis, J. Colloid Interface Sci. 191 (1997) 378–383.
[27] C.J. Van Oss, Interfacial forces in aqueous, media, Dekker, New York, 1994.
[28] A. Rudawska, E. Jacniacka, Evaluating uncertainty of surface free energy measurement by the van Oss-Chaudhury-Good method, Int. J. Adhes. Adhes. 82 (2018) 139–145.
[29] A.J. Kinloch, G.K.A. Kdokian, On the calculation of dispersion and polar force components of the surface free energy, J. Adhes. 34 (1991) 41–44.
[30] A.W. Neumann, R.J. Good, Thermodynamics of contact angles. I. Heterogeneous solid surfaces, J. Colloid Interface Sci. 38 (1972) 341–358.
[31] C. Della Volpe, S. Siboni, Some reflections on acid-base solid surface free energy theories, J. Colloid Interface Sci. 195 (1997) 121–136.
[32] J. Shang, M. Flury, J.B. Harsh, R.L. Zollars, Comparison of different methods to measure contact angles of soil colloids, J. Colloid Interface Sci. 328 (2008) 229–307.
[33] G. Bresson, J. Jumel, M.E.R. Shanahan, P. Serin, Strength of adhesively bonded joints under mixed axial and shear loading, Int. J. Adhes. Adhes. 35 (2012) 27–35.
[34] A.D. Crocomble, Y.X. Hua, W.K. Loh, M.A. Wahab, I.A. Ashcroft, Predicting the residual strength for environmentally degraded adhesive lap joints, Int. J. Adhes. Adhes. 26 (2006) 325–336.
[35] L.F.M. da Silva, R.J.C. Carbas, G.W. Critchlow, M.A.V. Figueiredo, K. Brown, Effect of material, geometry, surface treatment and environment on the shear strength of single lap joints, Int. J. Adhes. Adhes. 29 (2009) 621–632.

[36] J.M. Arenas, J.J. Narbón, C. Alía, Optimum adhesive thickness in the structural adhesives joint using statistical techniques based on Weibull distribution, Int. J. Adhes. Adhes. 30 (2010) 160–165.
[37] D.N. States, K.L. de Vries, Geometric factors impacting adhesive lap joint strength and design, Int. J. Adhes. Adhes. 26 (2012) 89–107.
[38] L.D.R. Grant, R.D. Adams, L.F.M. da Silva, Effect of the temperature on the strength of adhesively bonded single lap and T joints for the automotive industry, Int. J. Adhes. Adhes. 29 (2009) 535–542.
[39] L.F.M. da Silva, N.M.A.J. Ferreira, V. Richter-Trummer, E.A.S. Marques, Effect of grooves on the strength of adhesively bonded joints, Int. J. Adhes. Adhes. 30 (2010) 735–743.
[40] J.M. Sykes, Surface treatment for steel, in: D.M. Brewis (Ed.), Surface Analysis and Pretreatment of Plastics and Metals, Applied Science Publishers, London, 1982, pp. 153–174.
[41] K.L. DeVries, R.D. Adams, Mechanical testing of adhesive joints, in: D.A. Dillard, A.V. Pocius (Eds.), The Mechanics of Adhesion, Adhesion Science and Engineering-1, Elsevier Science B.V, Amsterdam, 2002 (Chapter 6).
[42] N.R.E. Domingues, R.D.S.G. Campilho, R.J.C. Carbas, L.F.M. da Silva, Experimental and numerical failure analysis of aluminium/composite single-L joints, Int. J. Adhes. Adhes. 64 (2016) 86–96.
[43] J.F. Durodola, Functionally graded adhesive joints, Int. J. Adhes. Adhes. 76 (2017) 83–89.
[44] M.K. Apalak, A. Engin, Effect of adhesive free-end geometry on the initiation and propagation of damaged zones in adhesively bonded lap joints, J. Adhes. Sci. Technol. 18 (2004) 529–559.
[45] E.M. Moya-Sanz, I. Ivañez, S.K. Garcia-Castilo, Effect of geometry in the strength of single-lap adhesive joints of composite laminates under uniaxial tensile load, Int. J. Adhes. Adhes. 72 (2017) 23–29.
[46] K.B. Armstrong, Long-term durability in water of aluminium alloy adhesive joints bonded with epoxy adhesives, Int. J. Adhes. Adhes. 17 (1997) 89–105.
[47] M. Bordes, P. Davies, J.-Y. Cognard, L. Sohier, V. Sauvant-Moynot, J. Galy, Prediction of long term strength of adhesively bonded steel/epoxy joints in sea water, Int. J. Adhes. Adhes. 29 (2009) 595–608.
[48] M. Heshmati, R. Haghani, M. Al-Emrani, Effects of moisture on the long-term performance of adhesively bonded FRP/steel joints used in bridges, Compos. B 92 (2016) 447–462.
[49] M. Heshmati, R. Haghani, M. Al-Emrani, Durability of CFRP/steel joints under cyclic wet-dry and freeze-thaw conditions, Compos. B 126 (2017) 211–226.
[50] X. Jiang, X. Qiang, M.H. Kolstein, F.S.K. Bijlaard, Experimental investigation on mechanical behavior of FRP-to-steel adhesively-bonded joint under combined loading—part 2: after hygrothermal ageing, Compos. Struct. 125 (2015) 687–697.
[51] V. Shenoy, I.A. Ashcroft, G.W. Critchlow, A.D. Crocombe, M.M. Abdel Wahab, An evaluation of strength wearout models for the lifetime prediction of adhesive joints subjected to variable amplitude fatigue, Int. J. Adhes. Adhes. 29 (2009) 639–649.
[52] M. Gholami, A.R.M. Sam, J.M. Yatim, M.M. Tahir, A review on steel/CFRP strengthening system focusing environmental performance, Constr. Build. Mater. 47 (2013) 301–310.
[53] G. Jhin, S. Azari, A. Ameli, N.V. Datla, M. Papini, J.K. Spelt, Crack growth rate and crath path in adhesively bonded joints: comparison of creep, fatigue and fracture, Int. J. Adhes. Adhes. 46 (2013) 74–84.

[54] V. Shenoy, I.A. Ashcroft, G.W. Critchlow, A.D. Crocombe, M.M. Abdel Wahab, Strength wearout of adhesively bonded joints under constant amplitude fatigue, Int. J. Fatigue 31 (2009) 820–830.
[55] M.V. Fernández, M.F.S.F. de Moura, L.F.M. da Silva, A.T. Marques, Characterization of composite bonded joints under pure mode II fatigue loading, Compos. Struct. 95 (2013) 222–226.
[56] S. Azari, M. Papini, J.A. Schroeder, J.K. Spelt, Fatigue threshold behaviour of adhesive joints, Int. J. Adhes. Adhes. 30 (2010) 145–159.
[57] M.V. Fernández, M.F.S.F. de Moura, L.F.M. da Silva, A.T. Marques, Composite bonded joints under mode I fatigue loading, Int. J. Adhes. Adhes. 31 (2011) 280–285.
[58] L.S. Penn, H. Wang, Epoxy resins, in: S.T. Peters (Ed.), Handbook of Composites, Chapman and Hall, London, 1998 (Chapter 3).
[59] O. Moussa, A.P. Vassilopulos, T. Keller, Effects of low-temperature curing on physical behaviour of cold-curing epoxy adhesives in bridge construction, Int. J. Adhes. Adhes. 32 (2012) 15–22.
[60] G.L. Jialanella, E.O. Shaffer II, The effect of adhesive modulus on the performance of SMC lap shear joints, J. Adhes. Sci. Technol. 7 (1993) 1171–1181.
[61] L.F.M. da Silva, R.D. Adams, Adhesive joint at high and low temperatures using similar and dissimilar adherends and dual adhesives, Int. J. Adhes. Adhes. 27 (2007) 216–226.
[62] K. Uehara, M. Sakurai, Bonding strength of adhesives and surfaces roughness of joined parts, J. Mater. Process. Technol. 127 (2002) 178–181.
[63] G.W. Critchlow, D.M. Brewis, Review of surface pretreatments for alluminium alloys, Int. J. Adhes. Adhes. 16 (1996) 255–275.
[64] C. Czaderski, E. Martinelli, J. Michels, M. Motavalli, Effect of curing conditions on strength development in an epoxy resin for structural strengthening, Compos. B 43 (2012) 398–410.
[65] A. Rudawska, Pressure during curing and the strength of 2024, 2017A and 1050 aluminium alloy sheet adhesive joints, Adv. Sci. Technol. Res. J. 9 (26) (2015) 96–103.
[66] S. Ebnesajjad, C. Ebnesajjad, Surface Treatment of Materials for Adhesive Bonding, second ed., William Andrew Inc., Norwich, NY, 2013.
[67] S. Ebnesajjad, Adhesives Technology Handbook, second ed., William Andrew Inc., Norwich, NY, 2008.
[68] J.R.J. Wingfield, Treatment of composite surfaces for adhesive bonding, Int. J. Adhes. Adhes. 13 (1993) 151–156.
[69] O. Lunder, F. Lapique, B. Johnsen, K. Nisancioglu, Effect of pre-treatment on the durability of epoxy-bonded AA6060 aluminium joints, Int. J. Adhes. Adhes. 24 (2004) 107–117.
[70] S.G. Prolongo, A. Ureña, Effect of surface pre-treatment on the adhesive strength of epoxy-aluminium joints, Int. J. Adhes. Adhes. 29 (2009) 23–31.
[71] G.W. Critchlow, K.A. Yendall, D. Bahrani, A. Quinn, F. Andrews, Strategies for the replacement of chromic acid anodizing for the structural bonding of aluminium alloys, Int. J. Adhes. Adhes. 26 (2006) 419–453.
[72] N. Brack, A.N. Rider, The influence of mechanical and chemical treatments on the environmental resistance of epoxy adhesive bonds to titanium, Int. J. Adhes. Adhes. 44 (2014) 20–27.
[73] A. Rudawska, Selected aspects of the effect of mechanical treatment on surface roughness and adhesive joint strength of steel sheets, Int. J. Adhes. Adhes. 50 (2014) 235–243.
[74] P. Hass, O. Kläusler, S. Schlegel, et al., Effects of mechanical and chemical surface preparation on adhesively bonded wooden joints, Int. J. Adhes. Adhes. 51 (2014) 95–102.

[75] R.M. Podhajny, Comparing surface treatments, Converting 8 (1990) 46–52.
[76] C. Sperandio, J. Bardon, A. Laachachi, H. Aubiert, D. Ruch, Influence of plasma surface treatment on bond strength behaviour of an adhesively bonded aluminium-epoxy system, Int. J. Adhes. Adhes. 30 (2010) 720–728.
[77] C.A. May, Epoxy Resins, Chemistry and Technology, second ed., Marcel Dekker, New York, 1988.
[78] H.L. Lee, H. Neville, Handbook of Epoxy Resins, McGraw-Hill, New York, 1988.
[79] H. Okuhira, T. Kii, M. Ochi, H. Takeyama, Characterization of epoxy resin hardening with ketimine latent hardeners, J. Adhes. Sci. Technol. 18 (2004) 205–211.
[80] I.-N. Yoon, Y. Lee, D. Kang, J. Min, J. Won, M. Kim, Y.S. Kang, S.-h. Kim, J.-J. Kim, Modification of hydrogenated Bisphenol A epoxy adhesives using nanomaterials, Int. J. Adhes. Adhes. 31 (2011) 119–125.
[81] M. Müller, P. Valasek, A. Rudawska, Mechanical properties of adhesive bonds reinforced with biological fabric, J. Adhes. Sci. Technol. 31 (2017) 327–344.
[82] M. Müller, P. Valasek, A. Rudawska, Effect of active rubber powder on structural two-component epoxy resin and its mechanical properties, J. Adhes. Sci. Technol. 32 (2018) 1531–1547.
[83] J. Lee, L.T. Drzal, Surface characterization and adhesion of carbon fibers to epoxy and polycarbonate, Int. J. Adhes. Adhes. 25 (2005) 389–394.
[84] J. Shields, Adhesive Handbook, third ed., Butterworth & Co (Publishers) Ltd, London, 1984.
[85] T. Li, H. Qin, X. Zhong, Y. Yu, A. Serra, Hyperbranched polyester as additives in filled and unfilled epoxy-novolac system, Polymer 53 (2012) 5864–5872.
[86] P. Rachtanapun, P. Heiden, Thermoplastic polymers as modifiers for urea-formaledehyde (UF) wood adhesives. I. Procedures for the preparation and characterization of thermoplastic-modified UF suspensions, J. Appl. Polym. Sci. 87 (2003) 890–897.
[87] A. Tomuta, F. Ferrando, A. Serra, X. Ramis, New aromatic-aliphatic hyperbranched polyesters with vinylic end groups of different length as modifiers of epoxy/anhydride thermosets, React. Funct. Polym. 72 (2012) 556–563.
[88] J.L. Down, The effect of modifiers on the stability of a vinyl acetate/ethylene copolymer dispersion, Stud. Conserv. 61 (2016) 26–45.
[89] A. Rudawska, I. Haniecka, M. Jaszek, M. Osińska-Jaroszuk, The influence of biochemical modification on the properties of adhesive compounds. Polymers 9 (2017) 442, https://doi.org/10.3390/polym9010009.
[90] C.V. Cagle, Handbook of Adhesive Bonding, McGraw-Hill, New York, 1973.
[91] R. Ciardiello, G. Belingardi, B. Martorana, D. Fondacaro, V. Brunella, A study of physical and mechanical properties of a nano-modified thermoplastic adhesive in normal and accelerated ageing conditions, in: ECCM16 European Conference on Composite Materials, Munich, 26–30th June, 2016.
[92] L. Tongxiang, G. Wenli, Y. Yiughui, T. Chunhe, Electroless palting of silver on graphite powders and the study of its conductive adhesive, Int. J. Adhes. Adhes. 28 (2007) 55–58.
[93] E.M. Petrie, Handbook of Adhesives and Sealants, McGraw Hill, New York, 1999.
[94] I. Skeist, Handbook of Adhesives, third ed., Chapmann & Hall, New York, 1990.
[95] S.R. Hartshorn (Ed.), Structural Adhesives, Chemistry and Technology, Plenum Press, New York/London, 1986.
[96] S. Ebnesajjad, A.H. Landrock, Adhesives Technology Handbook, William Andrew, London, 1985.
[97] R. Ciardiello, A. Tridello, V. Brunella, B. Martorana, D.S. Paolino, G. Belingardi, Impact response of adhesive reversible joints made of thermoplastic nanomodified adhesive. J. Adhes. (2017), https://doi.org/10.1080/00218464.2017.1354763.

[98] P. Walker, Organosilanes as adhesion promoters, J. Adhes. Sci. Technol. 5 (1991) 279–305.
[99] M. Mohseni, M. Mirabedini, M. Hashemi, G.E. Thompson, Adhesion performance of an epoxy clear coat on aluminum alloy in the presence of vinyl and amino-silane primers, Prog. Org. Coat. 57 (2006) 307–313.
[100] J. Jang, I. Jang, H. Kim, Effect of the mole ratios of silane-modified polyvinylimidazoles on adhesion promotion of the polyimide/copper system, J. Adhes. Sci. Technol. 12 (1998) 323–337.
[101] L.F.M. da Silva, A. Öchsner, R.D. Adams, Handbook of Adhesion Technology, first ed., Springer, Berlin, Heidelberg, 2011 (Chapter 1).
[102] A. Rudawska, Instrumentation in Bonding Technology, University Publisher, Lublin, 2016 (in polish).
[103] A. Pizzi, K.L. Mittal (Eds.), Handbook of Adhesive Technology, third ed., CRC Press, Boca Raton, FL, 2008.
[104] A.V. Pocius, Adhesion and Adhesives Technology: An Introduction, Carl Hanser, Munich, 2002.
[105] L.-H. Lee (Ed.), Adhesive bonding, first ed., Springer Science+Business Media, New York, 1991.
[106] A. Rudawska (Ed.), Adhesives, Applications and Properties, Intech, Rijeka, 2016.
[107] L. Zhai, G. Ling, J. Li, Y. Wang, The effect of nanoparticles on the adhesion of epoxy adhesive, Mater. Lett. 60 (2006) 3031–3033.
[108] M. May, H.M. Wang, R. Akid, Effects of the addition of inorganic nanoparticles on the adhesive strength of a hybrid sol–gel epoxy system, Int. J. Adhes. Adhes. 30 (2010) 505–512.
[109] H. Brockmann, M. Haufe, J.O. Schulenburg, Mechanism of the curing reaction of model epoxy compounds with monuron, Int. J. Adhes. Adhes. 20 (2000) 333–340.
[110] K. Mimura, H. Ito, Characteristics of epoxy resin cured with in situ polymerized curing agent, Polymer 43 (2002) 7556–7559.
[111] Y. Meng, X.-H. Zhang, B.-Y. Du, B.-X. Zhou, X. Zhou, G.-R. Qi, Thermosets with core-shell nanodomain by incorporation of core crosslinked star polymer into epoxy resin, Polymer 52 (2011) 391–399.
[112] X. Fernández-Francos, X. Ramis, Structural analysis of the curing of epoxy thermosets crosslinked with hyperbranched poly(ethykeneimine)s, Eur. Polym. J. 70 (2015) 286–305.
[113] M. Flores, X. Fernández-Francos, X. Ramis, A. Serra, Novel epoxy-anhydride thermosets modified with a hyperbranched polyester as toughness enhancer. I. Kinetics study, Thermochim. Acta 544 (2012) 17–26.
[114] M. Sadeghinia, K.M.B. Jansen, L.J. Ernst, Characterization and modeling the thermo-mechanical cure-dependent properties of epoxy molding compound, Int. J. Adhes. Adhes. 32 (2012) 82–88.
[115] K.D.V.P. Yarlagadda, S.-H. Hsu, Experimental studies on comparison of microwave curing and thermal curing of epoxy resin used for alternative mould materials, J. Mater. Process. Technol. 155–156 (2004) 1532–1538.
[116] Y.C. Nho, P.H. Kang, J.S. Park, The characteristic of epoxy resin cured by γ-ray and E-beam, Radiat. Phys. Chem. 71 (2004) 241–244.
[117] S. Alessi, A. Parlato, C. Dispenza, M. De Maria, G. Spadaro, The influence of the processing temperature on gamma curing of epoxy resin for the production of advances composites, Radiat. Phys. Chem. 76 (2007) 1347–1350.
[118] S. Du, Z.-S. Guo, B. Zhang, Z. Wu, Cure kinetics of epoxy resin used for advanced composites, Polym. Int. 53 (2004) 1343–1347.

[119] W.A. Dukes, A.J. Kinloch, Non-destructive testing of bonded joints: An adhesion science viewpoint, Non-Destructive Test. 7 (6) (1974) 324–326.
[120] C.J. Hellier, Handbook of Nondestructive Evaluation, Eddy Current Testing, McGraw-Hill, New York, 2001.
[121] R.D. Adams, A.M. Allen, P. Cawley, A new method for nondestructively examining bonded structures, in: Proceedings of the International Conference on Post Failure Analysis of Fibre Reinforced Composites, 1–8 December, Dayton, USA, 1985.
[122] R.D. Adams, Non-destructive testing of adhesively-bonded joints, Weld. Metal Fabr. (1989) 303–311. Aug/Sept.
[123] R.D. Adams, The nondestructive evaluation of bonded structures, Constr. Build. Mater. 4 (1990) 3–8.
[124] R.D. Adams, W. Drinkwater, Nondestructive testing of adhesively-bonded joints, NDT E Int. 30 (1997) 93–98.
[125] S.E. Hanneman, V.K. Kinra, C. Zhu, A new technique for ultrasonic nondestructive evaluation of adhesive joints: part II. Experiment, Exp. Mech. 32 (1992) 332–339.
[126] E. Maeva, I. Severina, S. Bondarenko, G. Chapman, B. O'Neill, F. Severin, R.G. Maev, Acoustical methods for the investigation of adhesively bonded structures: a review, Can. J. Phys. 82 (2004) 891–1025.
[127] R.D. Adams, P. Cawley, Vibration techniques in nondestructive testing, in: R.S. Sharpe (Ed.), Research Techniques in Nondestructive Testing, vol. 8, Academic Press, 1985, pp. 303–360.
[128] I.G. Scott, C.M. Scala, A review of non-destructive testing of composite materials, NDT E Int. 15 (1982) 75–86.
[129] S. Kawamoto, R.S. Williams, Acoustic emission and acousto-ultrasonic techniques for wood and wood-based composites—A review, General Technical Report, FPL-GTR-134, U.S. Department of Agriculture, Forest Service, Forest Products Laboratory, Madison, WI, 2002.
[130] E. Motukisi, Measurement of magnetic field characteristics for MPI, in: 18th World Conference on Non destructive Testing, 16–20 April, Durban, South Africa, 2012. Available from: http://www.ndt.net/article/wcndt2012/papers/77_Motukisi.pdf.
[131] H. Ollendorf, T. Schülke, D. Schneider, Testing the adhesion of hard coating including the non-destructive technique of surface acoustic waves, in: K.L. Mittal (Ed.), Adhesion Measurement of Films and Coatings, vol. 2, VSP, Zeist, 2001, pp. 49–77.
[132] D. Croccolo, R. Cuppini, Adhesive defect density estimation applying the acoustic emission technique, Int. J. Adhes. Adhes. 29 (2009) 234–239.
[133] H. Fagot, F. Albe, P. Smigielski, J.L. Arnaud, Holo-graphic nondestructive testing of materials using pulsed lasers, Report no. CO 223/80, Institut Franco-Allemand de Recherches de Saint-Louis, 1980.
[134] R.D. Adams, P. Cawley, A review of defect types and NDT techniques for composites and bonded joints, Mater. Sci. Technol. 5 (1989) 413–425.
[135] R.L. Vijaya Kumar, M.R. Bhat, C.R.L. Murthy, Some studies on evaluation of degradation in composite adhesive joints using ultrasonic techniques, Ultrasonics 53 (2013) 1150–1162.
[136] S. Yang, L. Gu, R.F. Gibson, Nondestructive detection of weak joints in adhesively bonded composite structures, Compos. Struct. 51 (2001) 63–71.
[137] B. Ren, C.J. Lissenden, Ultrasonic guided wave inspection of adhesive bonds between composite laminates, Int. J. Adhes. Adhes. 45 (2013) 59–68.
[138] C. Jeenjitkaew, F.J. Guild, The analysis of kissing bonds in adhesive joints, Int. J. Adhes. Adhes. 75 (2017) 101–107.

[139] D. Palumbo, R. Tamborrino, U. Galietti, P. Aversa, A. Tatì, V.A.M. Luprano, Ultrasonic analysis and lock-in thermography for debonding evaluation of composite adhesive joints, NDT E Int. 78 (2016) 1–9.
[140] R. Tamborrino, D. Palumbo, U. Galietti, P. Aversa, S. Chiozzi, V.A.M. Luprano, Assessment of the effect of defects on mechanical properties of adhesive bonded joints by using non destructive methods, Compos. B 91 (2016) 337–345.
[141] M. Michaloudaki, E. Lehmann, D. Kosteas, Neutron imaging as a tool for the non-destructive evaluation of adhesive joints in aluminium, Int. J. Adhes. Adhes. 25 (2005) 257–267.
[142] S.A. Titov, R.G. Maev, A.N. Bogachenkov, Pulse-echo NDT of adhesively bonded joints in automotive assemblies, Ultrasonics 48 (2008) 537–546.
[143] O.-Y. Kwon, S.-H. Lee, Acousto-ultrasonic evaluation of adhesively bonded CFRP-aluminum joints, NDT E Int. 32 (1999) 153–160.
[144] ASM, Metals Handbook, Vol. 17, Nondestructive Inspection and Quality Control, Eddy Current Inspection, ninth ed., ASM International, Novelty, OH, 1898164–194.
[145] D.W. Schindel, D.S. Forsyth, D.A. Hutchins, A. Fahr, Air-coupled ultrasonic NDE of bonded aluminum lap joints, Ultrasonics 35 (1997) 1–6.

Further reading

[146] P.R. Underhill, A.N. Rider, D.L. DuQuesnay, The effect of warm surface treatments on the fatigue life in shear of aluminium joints, Int. J. Adhes. Adhes. 26 (2006) 199–205.
[147] R.D. Adams, J.A. Harris, The influence of local geometry on the strength of adhesive joints, Int. J. Adhes. Adhes. 7 (1987) 69–80.
[148] A. Rudawska, D. Stančeková, J. Brzęczek, The visual method in quality assessment of single-lap adhesive joints, Adv. Sci. Technol. Res. J. 10 (2016) 109–118.
[149] C. Bockenheimer, D. Fata, W. Possart, M. Rothenfusser, U. Netzelmann, H. Schaefer, The method of non-linear ultrasound as a tool for a non-destructive inspection of structural epoxy-metal bonds—a resume, Int. J. Adhes. Adhes. 22 (2002) 227–233.
[150] D.J. Hagemaier, Low frequency eddy current testing of aircraft structures, in: S.S. Udpa, P.O. Moore (Eds.), Nondestructive Testing Handbook, third ed., American Society for Nondestructive Testing, Columbus, OH, 2004, pp. 481–485.
[151] S. Tanary, Characterization of adhesively bonded joints using acousto-ultrasonic (thesis), Department of Mechanical Engineering, University of Ottawa, Canada, 1998.

Surface treatment methods

3

3.1 Importance of the surface preparation process

The adherend surface preparation precedes any other operation involved in the technological process of bonding [1–4]. This stage plays a fundamental role in fostering adhesion by modifying the surface of adherends. Pretreated surfaces acquire properties ensuring that the produced joint exhibits required purpose-specific properties (usually desired strength). The main aim of surface preparation is to promote formation of strong physical and chemical bonds in the adhesive joint by ensuring the following [1, 5–8]:

- removal of various contamination off the surface of adherends, including: grease, dust, corrosion products, adsorbed oxide layers, microorganisms, bacteria, etc.,
- removal of any types of undesirable coatings (in the case of regeneration or repeated adhesive joining of elements) or protective coating (e.g. anticorrosion protection),
- favourable stereometric adherend surface texture, and
- proper activation of adherend surfaces, related to the surface energy characteristics of substrates.

There are several factors considered in the selection of surface treatment of substrates for bonding [9–20]:

- the type and properties of substrate material,
- the surface state,
- the type of contamination, the force of adhesion to surface ratio, and the distribution of adhesion forces,
- the shape of adherends,
- the dimensions of adhesively bonded surfaces,
- the positioning of the adherends (location of and access to the joint),
- the type of production,
- the assembly method,
- the workshop conditions,
- the requirements, and
- the instrumentation at disposal.

Surface preparation involves a range of operations, varying in terms of technology and their number and order (Fig. 2.7). Surface treatment methods may be grouped into preliminary treatment, fundamental treatment (basic), and additional surface treatment methods (Fig. 3.1), whose application depends on, e.g. the type of substrate and the type of adhesive.

Preliminary surface treatment methods are expected to aid and facilitate the proper treatment of adherends for bonding by means of: edge deburring, cleaning, or degreasing. In certain cases, pretreatment is effectively the sole surface treatment applied to adherends, which is often the case with degreasing [1, 2, 19, 21].

Surface Treatment in Bonding Technology. https://doi.org/10.1016/B978-0-12-817010-6.00003-5

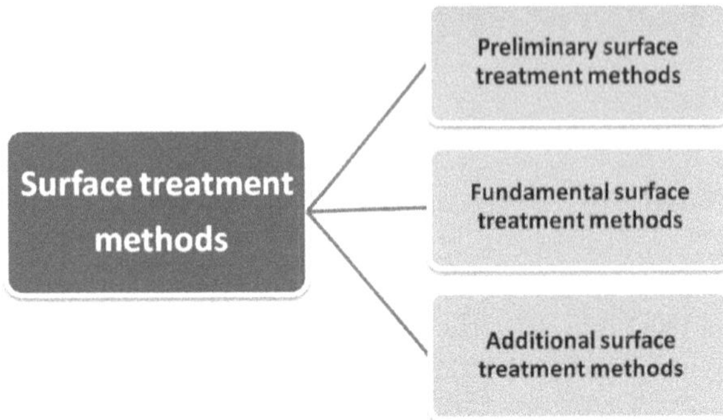

Fig. 3.1 Surface treatment methods.

Fundamental surface treatment methods are implemented with an aim of producing rough surface texture, hence promoting its physicochemical activity as a consequence of increasing the strength of adhesive bonds between the adherends and the adhesive mass [6, 22–29].

Apart from mechanical, chemical, and electrochemical treatments, constituting the fundamental surface treatment methods, there is a group of special methods, including: corona discharge treatment [30–32], plasma treatment [33–42], laser treatment [43–46], flame treatment [32, 47], electromagnetic-radiation-based methods [48, 49], ozone oxidation treatment, and others [50–53]. Additional treatments ensure that the technological process of surface treatment is carried out correctly, and that the obtained surface exhibits properties promoting the final strength of adhesive joints.

3.2 Characteristics of surface treatment methods

3.2.1 Preliminary methods of surface treatment

The surface treatment methods classified in the preliminary surface preparation category are the following: mechanical treatment, degreasing, edge deburring, and rinsing (Fig. 3.2).

Each method is characterised by a set of different properties and technological parameters. The final selection of treatments, agents or parameters of treatment to be applied on adherends is based on such factors as the type of substrate or the workshop conditions. The paragraphs to follow are devoted to presenting different preliminary treatment methods in preparation of surfaces for bonding; *nota bene* rinsing is described later in this chapter.

Mechanical treatment

Surfaces to be subjected to adhesive bonding that have previously been in service frequently demand preliminary treatment in order to remove rust-type contamination. To that end, it is the mechanical treatment that offers superior results, particularly when

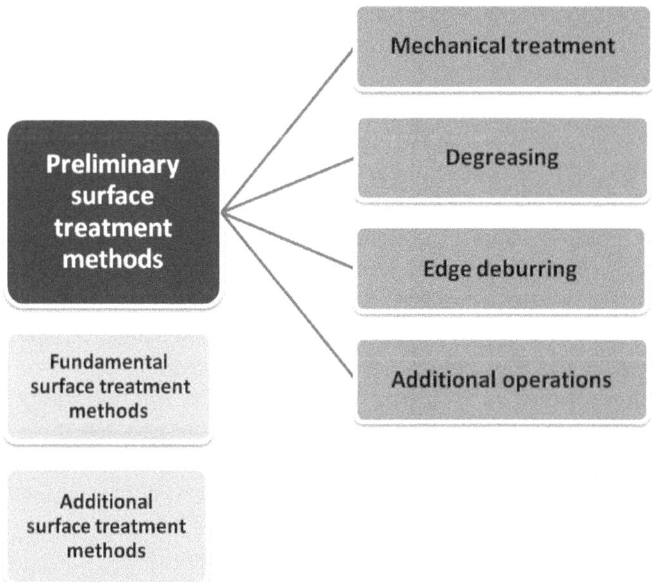

Fig. 3.2 Preliminary surface treatment methods.

the actual method is hammering. Hammering is a rough treatment method consisting in removal of loosely connected corrosion products, such as rust, scale, or paint coats, off the surface of adherends. The operation may be realised manually, or with the use of a pneumatic hammer, and usually precedes abrasive blasting treatment.

Degreasing

Degreasing, also referred to as cleansing, frequently precedes other surface treatments designed to prepare the surface of substrates for bonding [1, 2, 4, 52]. With regards to the surface, it will usually be coated with a layer of grease (e.g. for corrosion protection), a layer of dust, fine debris products (of other different treatments), and organic or nonorganic substances (e.g. layers of oxides, atmospheric pollution, and fats). If the contamination is not fully removed off the surface of adherends, the strength of adhesive bonds between the adhesive and adherends, and therefore the strength of the whole adhesive joint assembly may be compromised. This is the primary reason why degreasing is expected to remove all contamination off the adherend surface [1, 2, 54]. The degreasing surface treatment method is explored in further detail in Chapter 4.

Edge deburring

A smooth edge of elements to be bonded is one of the predicaments of good adhesion. Not only are sharp edges unfavourable from the viewpoint of ergonomics, but roughness in the edge area may furthermore inhibit adhesion by preventing intimate contact between adherends, as well as it can hamper surface treatment. The removal of burr and unevenness of the surface, as well as rounding sharp edges of adherends are mainly conducted by means of manual mechanical treatment with files, or by means of milling cutters with different-sized cutting tools, various grinding wheels, and countersinks.

Machining by hand is quite labour-intensive, and is therefore applied in preparation of small-sized adherends and in piece or job-lot production. Manual treatment of large-sized adherend surfaces would not be viable, and is therefore to a certain extent automated by use of, e.g. grinder, or a power drill. The problem of sharp edges of adherends in job-lot production is resolved by means of polishing or deburring machines with or without the use of abrasive or tumbling media. The technology of this operation consists in that the mixture of abrasive/polishing media and parts is tumbled at a predetermined intensity in a special container, thus creating friction between the elements that provides the surface finish [4].

Rotary wire brushing is a popular surface treatment solution for deburring and smoothing rough edges. The rotary wire brush is a tool composed of wire bristle, where each wire becomes a milling micro-tooth. The wires are in a radial arrangement providing a large number of self-sharpening cutting edges. Wire tools are mounted on milling machines, grinding machines, or turning lathes [2, 4, 54].

3.2.2 Fundamental methods of surface treatment

The pretreated surface may be subsequently subjected to fundamental surface treatment, whose aim is to improve the surface texture and increase the physicochemical activity of the treated surface, leading as a result to the growth in the strength of adhesive bonds between the adherend and the adhesive. Fundamental surface treatment methods include, along the somewhat standard mechanical, chemical, and electrochemical treatments, special methods, such as: corona discharge treatment, plasma treatment, laser treatment, flame treatment, electromagnetic-radiation-based methods, and ozone oxidation treatment (Fig. 3.3). The characteristics of surface treatment fundamental methods of adherends may be found in the next chapters.

The mechanical treatment method produces surfaces of maximum roughness, which increases the surface energy [5, 10, 26, 54–58]. The disadvantage of the method is that it does not act for the activation of the surface. By improving the surface roughness the total surface area and the adhesive contact area are larger, hence providing superior conditions for the interlocking of the adhesive. The method is implemented when the surface after degreasing remains covered with oxide residue. Typical mechanical treatment methods include: abrasive blasting, coated abrasive tool treatment, grit blasting, sanding, peening, brushing, scraping, and polishing [4, 59–63].

Machining by means of coated abrasive tools is one of the most practical methods of mechanical surface treatment. The major advantages of the coated abrasive tools are: cost-efficiency, convenience in use, availability, adaptability to different conditions, producing a nonoriented structure.

The selection of coated abrasive tools for particular applications is mainly dependent on the grit size of the tool. The bigger the grit size, the more efficiently the material is removed from the surface, including the physisorption layer. Coated abrasive tool treatment, however, does not ensure developing proper surface texture.

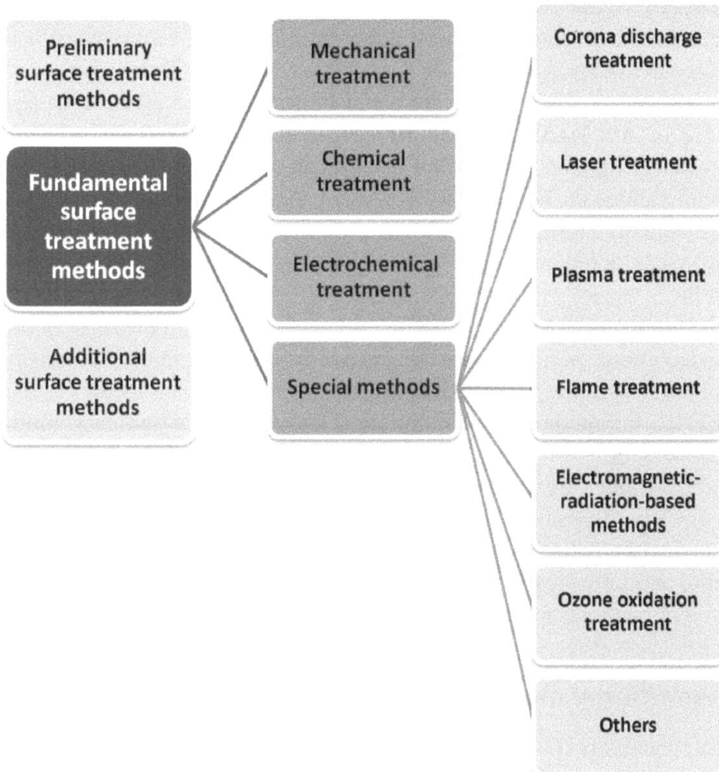

Fig. 3.3 Fundamental surface treatment methods.

Treatment by means of coated abrasive tools is expected to impart adhesive properties to the surface, as a result of [4, 52, 60]:

- producing a rough surface of maximum effective surface,
- mechanical effect, and
- removal of contaminants from the surface of adherends.

The chemical treatment acts towards promoting adhesion by etching the surface of adherends in the bath, whose temperature, time, and composition are strictly specified with a view to improving the strength of adhesive joints. This method provides excellent corrosion protection for metal substrates and imparts roughness to the adherend surface while activating the surface layer. In addition, chemically treated surfaces exhibit high physicochemical activity with a selected adhesive [10, 64].

Degreasing tends to initiate and to conclude the surface cleansing process. On account of its effects, the operation is employed to remove various contaminations from the surface of adherends, e.g. [3, 21]:

- layers of grease or oil which served as a temporary corrosion protection,
- organic and nonorganic compounds, such as oxides coatings, fats or atmospheric contaminants, and
- the layer of dust and small-sized surface treatment residue.

Degreasing in acidic or alkaline bath is designed to remove oxides from the surface of adherends. Such contaminants may reduce the strength of adhesive bonds or hinder other surface treatment operations on the substrate material. The properties of a given substrate material decide whether it should be subjected to alkaline or acidic bath treatment. It is quite significant that this surface treatment ought not to be omitted in the case when the bonded assemblies are expected to be exposed to extremely high or low temperatures during their service life. Otherwise, particularly when potential bath treatment could exert a negative effect on the properties of the assembled adhesive joint, it is recommended that this step should not be introduced in the surface treatment [4].

The electrochemical treatment is predominantly applied in the form of the anodic oxidation technique performed in the chemical bath of specific composition. The finish and the properties of the oxide coating depends on such factors as the type of entry material, the method of oxidation and the selected preliminary surface treatment [65–68].

Special treatments are applied on adherends to produce the desired texture of their surface, which will promote its physicochemical activity and increase the adhesive bonds between the adhesive and the adherend. Mechanical, chemical, electrochemical, plasma, and laser treatments are the main representatives among the special surface treatment methods applied nowadays.

3.2.3 Additional methods of surface treatment

Additional treatments contribute to the correct execution of the technological process of surface treatment, and are a crucial factor for the joint to develop the expected strength characteristics (Fig. 3.4).

Priming

Grounds are applied on materials in the last stage of surface preparation for bonding. The operation consists in coating the adherend surface with primers or activators, which, being built from chemically active functional groups, react with both the adhesive and the adherend surfaces. Application of grounds on both adhered surfaces contributes to increasing the assembled joint strength. However, the use of activator on one of the adherends results in accelerating the cross-linking process [1, 5, 50, 65].

Primers are selected to given applications as a result of analysis of the adherend-adhesive system, and also depending on such technological parameters as:

- the time between the application of primer and adhesive and
- the form of ground.

A number of substances are found to be used as primers, including: chlorinated rubber, zirconians of metal, silanes, titanates, nitrogen-heterocyclic, acrylic, nitrocellulose, polyurethane compounds, as well as phenol formaldehyde or natural resins. If applied, these substances tend to constitute a weighty contribution to the cost of adhesive joining and aid to the complexity of the method. Further details may be found in Chapter 8.

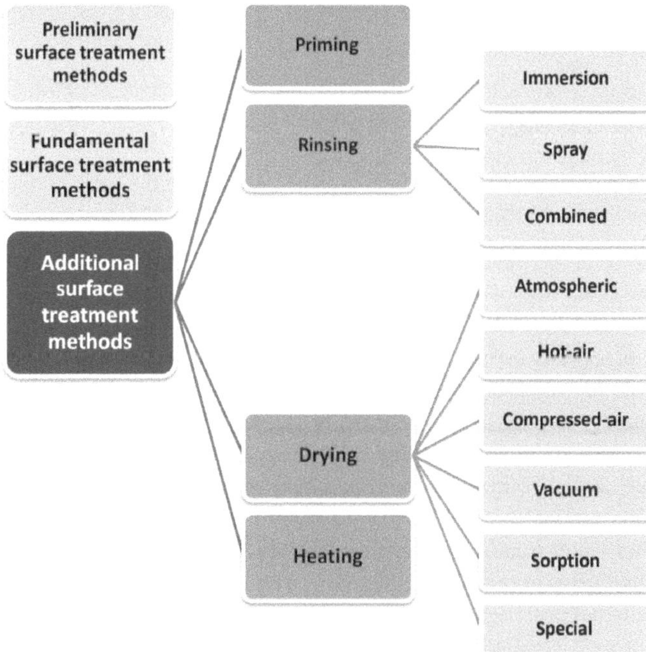

Fig. 3.4 Additional surface treatment methods.

Rinsing

Rinsing operation is conducted on the surface of adherends and typically follows etching in acidic or alkaline bath and a specific treatment operation. The operation aims to remove any chemical substance residue or contamination produced during or by other surface treatment operations off the surface of adherends. Rinsing can be carried out with water (e.g. after chemical etching): warm, cold, distilled, running, demineralised, deionised, or by means of degreasing solvents (following mechanical treatment) [4, 69, 70].

Typical requirements of the properties of water applied in rinsing treatment of substrates to be bonded are shown in Table 3.1 [4].

Rinsing is a process of diluting contamination on the surface of objects. Hence one of the determinants of the process' efficiency is quick and thorough mixing of water in the washer. Other aspects vital to the effectiveness of rinsing include [69, 70]:

- the rinsing method,
- the water flow type,

Table 3.1 Required rinse water properties

Properties	Characteristics
Resistivity	$5\times10^4\,\Omega$cm at 30°C
Total alkalinity	10 ppm as $CaCO_3$
Chloride content	Below 15 ppm
pH	6.5–7.5

- the rinse time,
- the rinse temperature,
- the water properties,
- the addition and type of wetting agents,
- the amount of rinse water,
- the water volume in the washer,
- the immersion method: the arrangement of rinsed elements in the bath, the type of hangers and the removal of elements from the rinse bath, etc.

The literature in the field [4, 17, 24, 62] specifies various technological conditions for the correct execution of rinsing for different materials (the type of water and temperature, the rinse time).

The rinse process is carried out at different water temperatures, from hot 60(80)–90°C, to warm (30–50°C) and cold (i.e. at ambient temperature, 20–25°C). Cold water rinse is frequently applied in the prerinsing stage [4]. In general, hot water rinse is performed on specimens immediately after degreasing in order to remove any soap residue off the surface, otherwise the soap could begin to gel, simultaneously becoming difficult to remove (in the case of cold water treatment). Cold water rinse always follows the hot water treatment (which usually lasts for approximately 15 s), or may be implemented prior to or after etching [69]. The work [17] presents research whose results indicate that prolonged rinse in warm or hot water exhibits an adverse effect on the strength properties of adhesive joints.

Rinsing methods

Rinsing may constitute an intermediate or a final operation [4, 69, 70]. Intermediate rinse takes place within the technological process of surface preparation for bonding, and is applied particularly following acidic or alkaline bath. Final rinse serves as surface protection and consists in removing any chemical substances applied onto the surface in the technological process. Furthermore, rinsing should account for the techno-economic aspects relating to water saving and recycling requirements or sewage disposal; hence, in terms of water supply, rinsing can be classified into closed-loop and open-loop systems.

Rinsing technique and efficiency to a considerable extent determine the quality of treated surfaces. In a number of cases, this may be the key factor in selecting the surface preparation method.

Among rinsing methods we may distinguish the following types, shown in Table 3.2 [4, 69]:

- immersion rinse,
- spray rinse,
- chemical rinse, and
- combination rinse.

Continuous flow immersion rinse is the most widely applied rinsing method. The objects are immersed in a washer, which is fed with an unvarying or changing volume of continuous fresh water. Multiple-tank washer units are usually found in a cascade configuration and are typically supplied in water in a serial arrangement.

Table 3.2 Types of rinsing methods

Rinsing method	Type of rinse	Characteristics
Immersion	One-time or multiple-time immersion	Rinsing carried out by one-time or multiple-time immersion of the object in the washer
	Single-step or multiple-step immersion: – Parallel water feed – Cascade water flow	Rinsing conducted in one or several cascading washers
	Static	Rinsing in static water; water changed periodically
	Continuous: – Continuous and unregulated flow of stream – Continuous and regulated flow of stream	Rinsing in washers equipped with a continuous water flow systems
Spray	With spray nozzles In water mist (aerosol)	Rinsing over/inside an empty washer, rinsing over a filled washer
Chemical		Immersion in specific solutions
Combined		Various rinse methods are combined, e.g. spray under immersion

In *static immersion rinse* systems, water is changed completely or in part at certain intervals. The concentration of contaminants removed from the rinsed objects increases until the water is changed after a specified number of rinsing procedures. Static immersion rinse systems are predominantly found in the form of a single-tank, but also multiple-tank configurations, and are typically supplied in water in a serial arrangement. Due to the necessity of periodic water change, this factor should be accounted for in a given application when selecting the washer tank size, as this is what constitutes an important inconvenience of this rinsing solution.

In *spray rinse* the contaminants are removed from the surface of treated objects in a hydrodynamic rinse procedure, executed by means of a strong jet of fresh water discharged from various-type nozzles. Spray rinse is an extremely economical process characterised by remarkably low rinse water consumption, which is relative to the rinse efficiency factor k (the volume of water involved in rinsing to the total volume of water discharged by nozzles).

Spray rinse is realised in two variants: spray rinse with high-pressure water in an empty washer tank or spray rinse over a full washer tank. Spray rinse in an empty washer tank enables repeated rinsing of objects in the same washer as an operation following various bath treatments. In this system the spray nozzles are typically located inside the washer tank on both sides of the dipping platform elevator, so that the sprayed water reaches the entire rinsed object together with the platform. The water is discharged in a number of patterns, e.g. flat fan or cone, and at desired angles to the rinsed objects. Moreover, the spray pattern and angle are selected according to the shape, the orientation and the geometry of the objects. This method is predominantly applied in rinsing objects of irregular geometry, and is frequently combined with single-step or multiple-step immersion. According to recommendations, the treatment should be carried out with soft or demineralised water with a view to maintaining spray nozzles in the working order.

Spray rinse in water mist (*aerosol*) consists in discharging fine water mist by means of compressed air over the washer with elevated temperature bath. As a result, the water layer deposited on the surface of treated objects becomes thinned and by receding back to the water tank is able to compensate for the bath water loss resulting from the evaporation.

In *chemical rinse* objects are initially immersed in a suitable deactivating solution, and subsequently in water. This procedure enables removal of bath components as a result of chemical reactions.

Drying

Drying operation is most frequently implemented following rinse and is expected to remove excess of water from the treated surface, whether in preparation for application of adhesive or additional treatment operations [2, 4, 69, 70]. Typical drying methods are shown in Table 3.3.

Drying is performed in: the ambient air, the air stream (also heated to 40–50°C), the inert gas atmosphere (of e.g. nitrogen or argon), or the chamber drier (Fig. 3.5).

The selection of drying solution depends to the largest extent on the type of dried material [69, 70]; for instance, subjected to slow drying in the fresh air, iron alloys become exposed to corrosion inhibiting the adhesive joint strength. Corrosion may be eliminated if the drying process is conducted in inert gas atmosphere.

Atmospheric drying

The method consists in drying objects in the atmospheric air heated to a sufficient temperature to cause the water on the object surface to evaporate. Prior to drying, it is advised to rinse treat the object in hot water. An atmospheric drying exhibit higher

Table 3.3 Drying methods

Drying method	Characteristics
Atmospheric drying	Working medium—atmospheric air
Hot-air drying	Working medium—hot atmospheric air
Compressed-air drying	Working medium—oil-free atmospheric air
Vacuum drying	Working medium—vacuum
Sorption drying	Working medium—hot atmospheric air and sawdust
Drying with organic dewatering fluids	Working medium—organic compounds that displace water on the surface of immersed metal objects
Special drying	Working medium—infrared radiation or hot air with centrifugal force

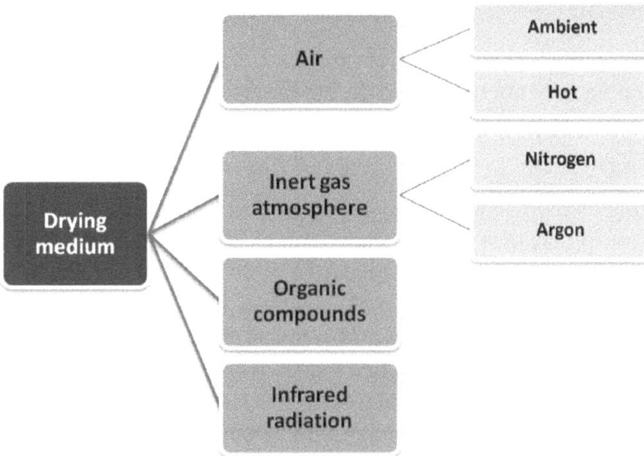

Fig. 3.5 Drying mediums.

effectiveness and takes shorter time under the condition that the dried object's mass-to-surface ratio is higher than a specified limit. Should the ratio be lower than the limit value, the treated object could cool down before the wetting water evaporates. The work [69] approximates minimal thicknesses at which the atmospheric drying shows good efficacy, i.e. when the level of evaporated water is approximately $40\,cm^3/m^2$. The determined minimal thickness of a steel object (steel tape) was equal to 1.8 mm, whereas for aluminium 2.8 mm [69]. In certain circumstances the process can be facilitated by additional removal of water off the surface of a treated object by the application of various surfactants, e.g. $0.1–0.2\,g/dm^3$ of soap [69], which decreases the surface tension of water during rinse.

Hot-air drying

Drying in hot air is typically found in treatment of small-sized objects. The process is performed by means of hot-air circulation driers, powered by electricity, steam or gas with temperature regulation. In the case of hot air medium, in order to remove $1\,dm^3$ of water from the unheated surface we must utilise between 2 and 4 kg of steam or equivalent heating medium [70]. Drying efficacy may be increased by implementing hot water treatment prior to drying.

Compressed-air drying

Owing to its high effectiveness, drying with compressed air is applied whenever time is of the essence and treated objects are of complicated shapes. In addition, the method prevents from development of spots on metal surfaces (particularly of the highly reactive ones, such as copper, zinc, and their alloys). During compressed-air drying the working medium should be dry and free of contamination—particularly free from any type of oil substances.

Vacuum drying

The method is applied when the dried objects are of intricate shape and when the elevated temperature may not be applied during drying.

Sorption drying

Sorption drying involves sawdust and is typically carried out at elevated temperatures. The application of sawdust enables removing all the water off the object's surface and in addition acts as a mechanical medium removing spots or damp patches. The sawdust must exhibit high hardness and essentially be acid-free. Both conditions are fulfilled by maple, beech, or poplar wood—typical sawdust materials used in sorption drying. The method is usually performed manually: the sawdust is placed inside metal crates and heated from beneath. In the mechanical variant of the method, the system is supplemented with mechanical apparatus and elements, such as inclined drums with heating screens and mechanical sawdust separator.

Among the numerous disadvantages of the method there are high labour consumption, at times prolonged drying time, the need for applying separators after drying, the risk of damaging the dried object surface, dustiness. Nevertheless, the method is still relatively popular in many applications.

Drying with organic dewatering fluids

This method employs organic compounds, i.e. dewatering fluids, capable of displacing water on the surface of immersed metals and is recently becoming increasingly popular [69, 70]. The displaced water falls in drops to the bottom of the vessel to become a separate liquid phase, which does not mix with the organic fluid. The dewatering fluid operates on the principle of rapid increase of the wetting angle of water on the surface of metal immersed in the dewatering bath. Therefore, the water cannot wet the surface of the object and is displaced, while the organic substance that exhibits good metal wetting properties replaces it.

Organic dewatering fluids are typically composed of organic solvents in the form of aliphatic carbohydrates of different viscosity, composed of different compounds, such as volatile or nonvolatile surfactants, and if necessary solvents like ethers or glycols, acting as solubility promoters for the dewatering fluids. Volatile polar surfactants evaporate from the surface of dried objects with the solvent to leave the surface completely dry and clean. Nonvolatile surfactants remain on surfaces in the form of coating (acting as, e.g. a corrosion protection layer), and the thickness, type, and properties of the layer depend on the type of the polar component used.

Drying with organic dewatering fluids has a number of significant advantages, such as [69, 70]:

- capacity for removing the entire water layer off the surface of objects, including mineral salts, emulsified oils, and other substances,
- effective removal of water from various holes and blind holes, slots, cavities, which are often of intricate shapes and in hardly accessible places,
- stain-free drying of objects,

- simultaneous deposition of special property layers, e.g. corrosion protection layer, and
- short immersion time, which shortens the total treatment time.

However, the method involves certain drawbacks including:

- relatively high price of organic dewatering fluids,
- high loss of drying agents during removal from bath,
- the risk of fire of flammable substances, and
- the need to use specially designed devices (e.g. inclined bottom)

Alternative drying methods

There are several alternative methods of drying, e.g. [4, 69, 70]:

- infrared drying,
- drying employing centrifugal force (with various rotary driers), and
- combined drying—employing various drying methods,

The selection of drying solution for a given application is determined by such factors as: the type of material, the dimensions and the shape of dried surfaces, the type of production, the workshop, and the technical conditions. It is therefore, highly important that when designing the technological process all factors that could potentially affect the choice of the drying method should be considered.

References

[1] R.D. Adams (Ed.), Adhesive Bonding. Science, Technology and Applications, Woodhead Publishing Limited, Cambridge, 2010.
[2] W. Brockmann, P.L. Geiß, J. Klingen, B. Schröder, Adhesive Bonding. Materials, Applications and Technology, Wiley-VCH Press, Weinheim, 2009.
[3] R.D. Adams, J. Comyn, W.C. Wake, Structural Adhesive Joints in Engineering, second ed., Chapman & Hall, London, 1997.
[4] C.V. Cagle, Handbook of Adhesive Bonding, McGraw-Hill, New York, 1973.
[5] S. Ebnesajjad, C. Ebnesajjad, Surface Treatment of Materials for Adhesive Bonding, second ed., William Andrew Inc., Norwich, NY, 2013.
[6] R.E. Litchfield, G.W. Critchlow, S. Wilson, Surface cleaning technologies for removal of crosslinked epoxide resin, Int. J. Adhes. Adhes. 26 (2006) 295–303.
[7] G.A. Wade, W.J. Cantwell, R.C. Pond, Plasma surface modification of glass fibre-reinforced nylon-6,6 thermoplastic composites for improved adhesive bonding, Interface Sci. 8 (2000) 363–373.
[8] P.H. Winfield, N. Becerra, P. Kaczmar, Investigation of primer/preservative basecoats for timber using vacuum impregnation, Int. J. Adhes. Adhes. 29 (2009) 702–709.
[9] R.D. Bhatt, V. Radhakrishnan, Evaluation of some machining process as adherend surface treatments in bonding aluminium, J. Adhes. Sci. Technol. 3 (1989) 383–396.
[10] N. Brack, A.N. Rider, The influence of mechanical and chemical treatments on the environmental resistance of epoxy adhesive bonds to titanium, Int. J. Adhes. Adhes. 44 (2014) 20–27.
[11] G.W. Critchlow, K.A. Yendall, D. Bahrani, A. Quinn, F. Andrews, Strategies for the replacement of chromic acid anodizing for the structural bonding of aluminium alloys, Int. J. Adhes. Adhes. 26 (2006) 419–453.

[12] G.W. Critchlow, D.M. Brewis, Review of surface pretreatments for alluminium alloys, Int. J. Adhes. Adhes. 16 (1996) 255–275.
[13] L. Domingues, J.C.S. Fernandes, M. Da Cunha Belo, M.G.S. Ferreira, L. Guerra-Rosa, Anodising of Al 2024-T3 in a modified sulphuric acid/boric acid bath for aeronautical applications, Corros. Sci. 45 (2003) 149–160.
[14] L.D.R. Grant, R.D. Adams, L.F.M. da Silva, Effect of the temperature on the strength of adhesively bonded single lap and T joints for the automotive industry, Int. J. Adhes. Adhes. 29 (2009) 535–542.
[15] R.M. Podhajny, Comparing surface treatments, Converting 8 (1990) 46–52.
[16] J.M. Sykes, Surface treatment for steel, in: D.M. Brewis (Ed.), Surface Analysis and Pretreatment of Plastics and Metals, Applied Science Publishers, London, 1982, pp. 153–174.
[17] P.R. Underhill, A.N. Rider, D.L. DuQuesnay, The effect of warm surface treatments on the fatigue life in shear of aluminium joints, Int. J. Adhes. Adhes. 26 (2006) 199–205.
[18] J.R.J. Wingfield, Treatment of composite surfaces for adhesive bonding, Int. J. Adhes. Adhes. 13 (1993) 151–156.
[19] R.F. Wegman, J. Van Twisk, Surface Preparation Techniques for Adhesive Bonding, second ed., Elsevier, Oxford, 2013.
[20] P. Molitor, V. Barron, T. Young, Surface treatment of titanium for adhesive bonding to polymer composites: a review, Int. J. Adhes. Adhes. 21 (2001) 129–136.
[21] A. Rudawska, J. Nalepa, M. Müller, The effect of degreasing on adhesive joint strength, Adv. Sci. Technol. Res. J. (2017) 75–81.
[22] A.M. Guedes Pinto, A.G. Magalhães, F. Gomes da Silva, A.P. Monteiro Baptista, Shear strength of adhesively bonded polyolefins with minimal surface preparation, Int. J. Adhes. Adhes. 28 (2008) 452–456.
[23] B. Bhushan, B.K. Gupta, Handbook of Tribology: Materials, Coatings, and Surface Treatments, McGraw-Hill, New York, 1991.
[24] J. Comyn, Surface treatment and analysis for adhesive bonding, Int. J. Adhes. Adhes. 10 (1990) 161–165.
[25] M. Sedlacek, B. Podgornik, J. Vizitin, Influence of surface preparation on roughness parameters, friction and wear, Wear 266 (2009) 482–487.
[26] M. Shahid, S.A. Hashim, Effect of surface roughness on the strength of cleavage joint, Int. J. Adhes. Adhes. 22 (2002) 235–244.
[27] L. Sorrentino, L. Carrino, Influence of process parameters of oxygen cold plasma treatment on wettability ageing time of 2024 aluminium alloy, Int. J. Adhes. Adhes. 29 (2009) 136–143.
[28] M. Mohseni, M. Mirabedini, M. Hashemi, G.E. Thompson, Adhesion performance of an epoxy clear coat on aluminum alloy in the presence of vinyl and amino-silane primers, Prog. Org. Coat. 57 (2006) 307–313.
[29] G.A. Wade, W.J. Cantwell, Adhesive bonding and wettability of plasma treated, glass fiber-reinforced nylon-6,6 composites, J. Mater. Sci. Lett. 19 (2000) 1829–1832.
[30] J. Gołębiewski, M. Żenkiewicz, Influence of corona treatment energy and storage time on the changes of the contact angle of activated polysteryrene film, Polimery 48 (2003) 134–137.
[31] S. Bhowmik, P.K. Ghosh, S. Ray, K. Barthwal, Surface modification of high density polyethylene and polypropylene by DC glow discharge and adhesive bonding to steel, J. Adhes. Sci. Technol. 12 (1998) 1181–1204.
[32] J. Custódio, J. Broughton, H. Cruz, P. Winfield, Activation of timber surface by flame and corona treatments to improve adhesion, Int. J. Adhes. Adhes. 29 (2009) 167–172.

[33] M.C. Kim, S.H. Yang, J.-H. Boo, J.G. Han, Surface treatment of metals using an atmospheric pressure plasma jet and their surface characteristics, Surf. Coat. Technol. 174–175 (2003) 839–844.
[34] W. Leahy, V. Barron, M. Buggy, T. Young, A. Mas, F. Schue, T. McCabe, M. Bridge, Plasma surface treatment of aerospace materials for enhanced adhesive bonding, J. Adhes. 77 (2001) 215–249.
[35] G.Z. Xiao, Effects of solvent on the surface properties of oxygen plasma-treatment polyethylene and polypropylene films, J. Adhes. Sci. Technol. 11 (1997) 655–663.
[36] M.-H. Jung, H.-S. Choi, Surface treatment and characterization of ITO thin films using atmospheric pressure plasma for organic light emitting diodes, J. Colloid Interface Sci. 310 (2007) 550–558.
[37] Y. Kusano, H. Mortensen, B. Stenum, S. Goutianos, S. Mitra, Atmospheric pressure plasma treatment of glassy carbon for adhesion improvement, Int. J. Adhes. Adhes. 27 (2007) 402–408.
[38] Z.Z. You, J.Y. Dong, Effect of oxygen plasma treatment on the surface properties often-doped indium oxide substrates for polymer LEDs, J. Colloid Interface Sci. 300 (2006) 697–703.
[39] G. Rozovskis, J. Vinkevičius, Plasma surface modification of polyimide for improving adhesion to elekctroless copper coatings, J. Adhes. Sci. Technol. 10 (1996) 399–406.
[40] G.S. Sheu, S.S. Shyu, Surface modification of Kevlar 149 fibers by gas plasma treatment. Part II. Improved interfacial adhesion to epoxy resin, J. Adhes. Sci. Technol. 8 (1994) 1027–1042.
[41] C. Sperandio, J. Bardon, A. Laachachi, H. Aubiert, D. Ruch, Influence of plasma surface treatment on bond strength behaviour of an adhesively bonded aluminium-epoxy system, Int. J. Adhes. Adhes. 30 (2010) 720–728.
[42] N. Encinas, J. Abenojar, M.A. Martínez, Development of improved polypropylene adhesive bonding by abrasion and atmospheric plasma surface modifications, Int. J. Adhes. Adhes. 33 (2012) 1–6.
[43] Q. Bénard, M. Fois, M. Grisel, P. Laurens, Surface treatment of carbon/epoxy composites with an excimer laser beam, Int. J. Adhes. Adhes. 26 (2006) 534–549.
[44] E.G. Babraj, D. Starikov, J. Evans, G.A. Shafeev, A. Bensaoula, Enhancement of adhesive joint strength by laser surface modification, Int. J. Adhes. Adhes. 27 (2007) 268–276.
[45] G. Rotella, M. Alfano, T. Schiefer, I. Jansen, Evaluation of mechanical and laser surface pre-treatments on the strength of adhesive bonded steel joints for automotive industry, J. Adhes. Sci. Technol. 30 (2016) 747–758.
[46] R. Rechner, I. Jansen, E. Beyer, Influence on the strength and aging resistance of aluminium joints by laser pre-treatment and surface modification, Int. J. Adhes. Adhes. 30 (2010) 595–601.
[47] C.D. Volpe, A. Deimichei, T. Riccò, A multiliquid approach to the surface free energy determination of flame-treated surfaces of rubber-toughened polypropylene, J. Adhes. Sci. Technol. 12 (1998) 1141–1180.
[48] M. Żenkiewicz, Effect of electron radiation dose on oxidation of surface layers of polyethylene, polypropylene and poly(ethylene terephthalate) films, J. Adhes. Sci. Technol. 18 (2004) 607–616.
[49] M. Żenkiewicz, P. Rytlewski, J. Czupryńska, J. Polański, T. Karasiewicz, W. Engelgard, Contact angle and surface free energy of electron-beam irradiated polymer composites, Polimery 53 (2008) 446–451.
[50] P. He, M. Huang, S. Fisher, et al., Effects of primer and annealing treatment on the shear strength between anodized Ti6Al4V and epoxy, Int. J. Adhes. Adhes. 57 (2015) 49–56.

[51] J. Aakkula, O. Saarela, Silane based field level surface treatment methods for aluminium, titanium and steel bonding, Int. J. Adhes. Adhes. 48 (2014) 268–279.

[52] S. Ebnesajjad, Adhesives Technology Handbook, second ed., William Andrew Inc., Norwich, NY, 2008.

[53] W. Polini, L. Sorrentino, Adhesion of a protective coating on a surface of aluminium alloy treated by air cold plasma, Int. J. Adhes. Adhes. 27 (2007) 1–8.

[54] A. Rudawska, Selected aspects of the effect of mechanical treatment on surface roughness and adhesive joint strength of steel sheets, Int. J. Adhes. Adhes. 50 (2014) 235–243.

[55] P. Hass, O. Kläusler, S. Schlegel, et al., Effects of mechanical and chemical surface preparation on adhesively bonded wooden joints, Int. J. Adhes. Adhes. 51 (2014) 95–102.

[56] L.F.M. da Silva, R.J.C. Carbas, G.W. Critchlow, M.A.V. Figueiredo, K. Brown, Effect of material, geometry, surface treatment and environment on the shear strength of single lap joints, Int. J. Adhes. Adhes. 29 (2009) 621–632.

[57] L.F.M. da Silva, N.M.A.J. Ferreira, V. Richter-Trummer, E.A.S. Marques, Effect of grooves on the strength of adhesively bonded joints, Int. J. Adhes. Adhes. 30 (2010) 735–743.

[58] D.N. States, K.L. de Vries, Geometric factors impacting adhesive lap joint strength and design, Int. J. Adhes. Adhes. 26 (2012) 89–107.

[59] L. Kumosa, D. Armentrout, D. Kumosa, The effect of sandblasting on the initiation of stress corrosion cracking in unidirectional E-glass/polymer composites used in high voltage composite (non-ceramic) insulators, Compos. Sci. Technol. 62 (2002) 1999–2015.

[60] A.F. Harris, A. Beevers, The effect of grit-blasting on the surface properties for adhesion, Int. J. Adhes. Adhes. 19 (1999) 445–452.

[61] O. Lunder, F. Lapique, B. Johnsen, K. Nisancioglu, Effect of pre-treatment on the durability of epoxy-bonded AA6060 aluminium joints, Int. J. Adhes. Adhes. 24 (2004) 107–117.

[62] X.P. Jiang, X.Y. Wang, J.X. Li, D.Y. Li, C.-S. Man, M.J. Shepard, T. Zhai, Enhancement of fatigue and corrosion properties of pure Ti by sandblasting, Mater. Sci. Eng. A 429 (2006) 30–35.

[63] C. Ageorges, L. Ye, Fusion Bonding of Polymer Composites, Springer, Great Britain, 2002.

[64] S.G. Prolongo, A. Ureña, Effect of surface pre-treatment on the adhesive strength of epoxy-aluminium joints, Int. J. Adhes. Adhes. 29 (2009) 23–31.

[65] C. Fotea, C. D'Silva, Adhesion enhancement of chromium tanned heavy-duty leather (Salz leather) under extreme conditions using photoreagents as surface primers, Int. J. Adhes. Adhes. 25 (2005) 442–449.

[66] B.B. Johnsen, F. Lapique, A. Bjørgum, The durability of bonded aluminium joints: a comparison of AC and DC anodizing pretreatments, Int. J. Adhes. Adhes. 24 (2004) 153–161.

[67] A.C. Kennedy, R. Kohler, P. Poole, A sodium hydroxide anodize surface pretreatment for the adhesive bonding of titanium alloys, Int. J. Adhes. Adhes. 3 (1983) 133–139.

[68] A. Rudawska, I. Danczak, Adhesive properties and adhesive joint strength of aluminium alloys following surface pre-treatment, Recent Pat. Mech. Eng. 6 (2013) 152–165.

[69] L.H. Durney (Ed.), Electroplating Engineering Handbook, fourth ed., Van Nostrand Reinhold, New York, 1984.

[70] A.K. Graham, H.L. Pinkerkton, Electroplating Engineering Handbook, Reinhold Publishing Corporation, New York, 1971.

Degreasing

4

4.1 Characteristics of contaminations and hydrogen bonding classification of solvents

In order to promote adhesion, the surface of adherends must be properly prepared for joining [1–4]. Surface treatments are designed to remove all type of contaminations found on the surface of the substrate, making it possible for the adhesive to wet the actual surface rather than its apparent surface. It should activate and increase the surface free energy of adherends and the active bonding area to improve the adhesion of the adhesives [5–14].

Most surfaces of materials may be contaminated with different substances (Fig. 4.1) hindering adhesion, e.g. [1, 2, 12]:

- a layer of lubricating oil (usually anticorrosive) and greases on metals,
- weak oxide layers on metals and atmospheric contaminants, fats,
- dust, or debris (products of different machining treatment) and other particulate contamination,
- additives and low molecular weight material on the surfaces of plastics, and
- mould-release agents such as silicones, fluorocarbons, and waxes.

All the metals possess an oxide surface layer, but not all are coherent [1, 15, 16]. Some tend to flake (e.g. copper oxide) or powder (e.g. rust). Neglecting to remove contaminants off the surface may lead to weakening adhesive bonds between the adhesive and adherend and therefore contribute to diminishing adhesive joint strength. The aim of surface treatment is to remove these loose layers and then possibly to etch away the existing oxide and to replace it with one which gives good performance, both in the form of high initial joint strength and resistance to high humidity [1]. It is for these reasons that degreasing operation is aimed at removing any contaminants from the surface of adherends.

A contaminant, or an organic residue, is an unwanted material which usually occurs on parts as a thin film or monomolecular layer [17]. Contaminants have a molecular structure; therefore they have properties that are combinations of polarity, nonpolarity, and hydrogen bonding. These characteristics are reflected in the solubility parameter value of the material, which, in turn is based on the cohesive energy of a molecule. Cohesive energy is, in effect, a measure of the properties of the material. Organic contaminations have low surface energy values, as compared to metals, and they can prevent the adhesion of protective coatings which have higher surface energy value. There is no universal solvent to remove all contaminants. Each type of contaminant must be removed by the use of certain types of solvent. In this case the important issue is solubility parameter principles which can contribute to effective removal of contaminants from the surfaces of materials.

Surface Treatment in Bonding Technology. https://doi.org/10.1016/B978-0-12-817010-6.00004-7

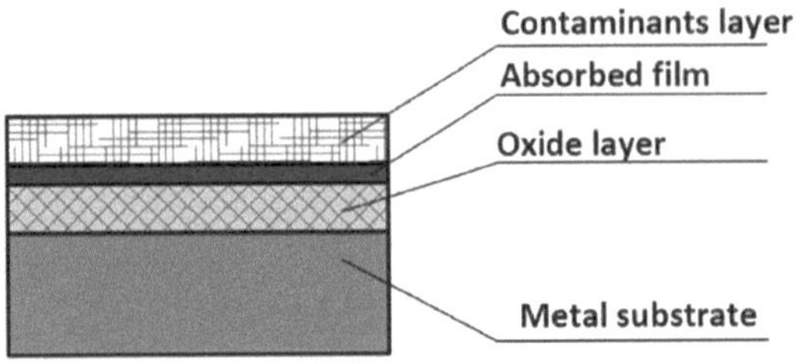

Fig. 4.1 Scheme of typical surface layers on a metal substrate.

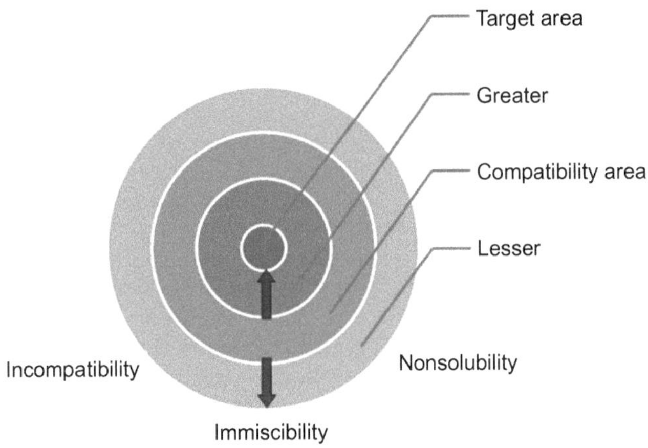

Fig. 4.2 Solubility parameter concept.

The effective removal of a contaminant from a substrate is dependent on the compatibility between contaminant and solvent used. Solubility parameter technology involves a study of the compatibility of materials. Similar molecules will dissolve more easily in one other than will dissimilar molecules. It is the fundamental principle of solubility parameter technology. The solubility parameter of a material to be solved is the target area (Fig. 4.2, prepared on the base of [17]).

The selection of a solvent having a solubility parameter reasonably close to that of target area will increase the solubility. However, if a solvent that has a solubility parameter too far away from the target area is chosen, the molecules will not match or fit together, and there will be little diffusion, molecular interaction, or solubility.

Accuracy in the selection of solvents for contaminant removal can be enhanced by utilising the knowledge that molecules have different levels of activity which significantly influence their compatibility. For this aim, the classification of solvents into three groups having weak, moderate, and strong hydrogen bonding is very useful (Table 4.1).

Moreover there are five considerations which govern the choice of a solvent [1]:

- toxicity,
- flammability,

Table 4.1 Hydrogen bonding classification of solvents (examples) [17]

Hydrogen bonding potential		
Weak	**Moderate**	**Strong**
Heptane	Ethylhexyl acetate	Isopropyl alcohol
Ethylhexyl chloride	Butyl acetate	
Methylchloroethylene	Cellosolve acetate	
Toluene	Acetone	
Trichloroethylene	Cyclohexanone	
Vinyl trichloride		

- efficiency for degreasing,
- environment, and
- cost.

There is also the factor of convenience in that it is desirable to have only one or perhaps two different solvents available for degreasing, whereas the optimum efficiency, if this were the sole criterion, might require a number to be available [1].

These classifications tend to group of solvents according to their electron charge capability. The strong hydrogen bonders are alcohols which have high solubility parameters values. The moderate hydrogen bonders are esters and ketones. The weak hydrogen bonding solvents are the hydrocarbons (heptanes, toluene, and the chlorinated compounds); they have low solubility parameters values. Some examples of solubility parameters of solvents used for the removal of various contaminants are presented in Table 4.2.

According to the test presented by Jackson [17], trichloroethylene is shown to be the best solvent for the removal of machine oil. The grease is more difficult to remove from the surface than machine oil and the longer cleaning time would be required. Isopropyl alcohol can be used for the removal of detergents.

Table 4.2 Solubility parameters of solvents (examples) [17]

Solvent	**Solubility parameter**
Heptane	7.50
2-Ethylhexyl chloride	8.13
2-Ethylhexyl acetate	8.29
Methylchloroethylene (1,1,1-trichloroethane)	8.44
N-Butyl acetate	8.69
Toluene	8.93
1,1,2-Trichloroethylene	9.16
Cellosolve acetate (ether acetate)	9.35
Acetone	9.62
Vinyl trichloride (1,1,2-trichloroeethane)	9.88
Cyclohexanone	10.42
Isopropyl alcohol	11.45

Acetone is often used because it has good balance of degreasing capability and comparatively low environmental concern, although fire-protection measures must be taken due to liability to ignite. This applies especially when degreasing processing cannot be carried out in a hermetically closed circuit [18].

4.2 Characteristics of degreasing methods

In surface preparation this is degreasing that frequently constitutes the first operation in the process. The degreasing can be applied either by itself or in combination with other methods, e.g. mechanical or chemical [6, 18–20]. Degreasing (organic or inorganic) can improve the characteristics of the bond, but even better results can be achieved by mechanical removal of the outer, inactive contaminated layer. Specially formulated chemical etching can produce a receptive and active substrate surface for the adhesive that results in both high initial strength and good durability [2, 21]. Surfaces of adherends can be prepared by one of the following procedures presented in Fig. 4.3 (based on Refs. [18, 22]).

The procedures were listed in order of increasing effectiveness, because when maximum strength is required the degreasing operation should be joined with mechanical or/and chemical (electrochemical) pretreatment.

After degreasing, care must be taken to avoid contaminating the pretreated surface prior to bonding. Contamination may be caused by finger marking, or by cloths which are not perfectly clean, or by using substandard degreasing or chemical solutions. Whatever the surface treatment procedure used, it is good practice to bond the adherends as soon as possible after completion of the pretreatment when the adhesive surface properties are at their best [22]. The surfaces are most "active"—i.e. their surface properties are at the optimum level for bonding [19].

Several degreasing methods may be distinguished; their detailed classification is shown in Fig. 4.4, and includes [1, 18, 19, 23–26]:

- chemical degreasing:
- electrochemical degreasing,

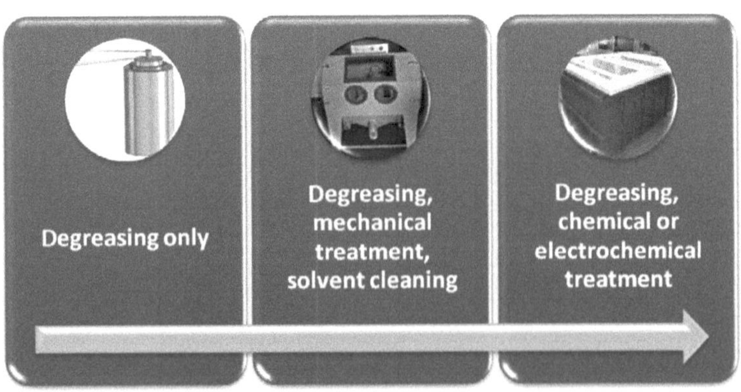

Fig. 4.3 Procedures surface treatment with degreasing.

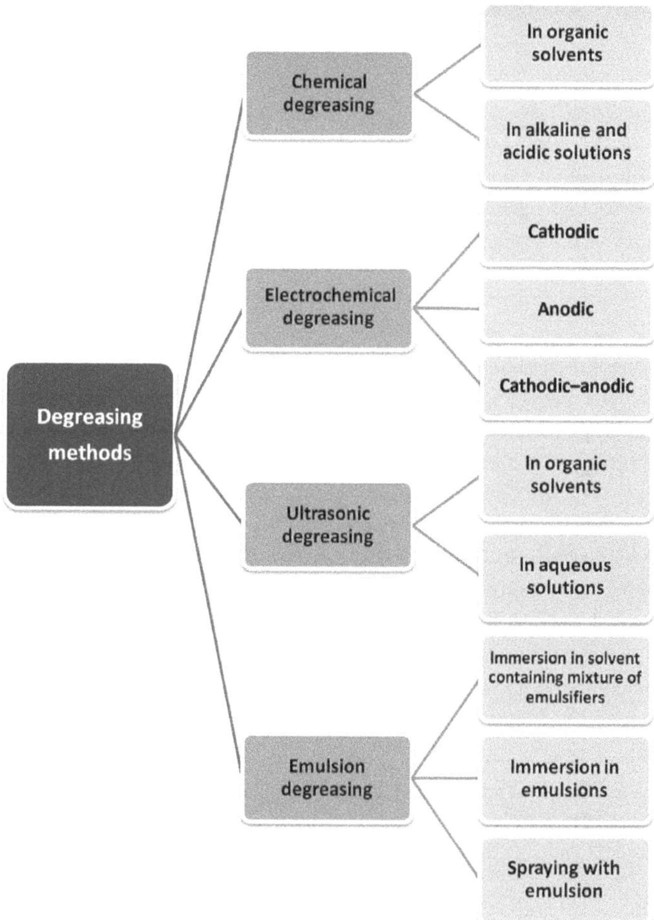

Fig. 4.4 Degreasing methods of surface treatment.

- ultrasonic degreasing (physicochemical method), and
- emulsion degreasing (physical method).

In addition, degreasing may be divided into:

- initial degreasing (rough) and
- fine degreasing (precise).

Several techniques (methods) of degreasing agent applications may be used during degreasing process [1, 2, 9, 17, 19, 26], which were shown in Fig. 4.5. The choice of degreasing techniques is very important because the process that is used for the application of solvent also influences the degree of contaminant removal. Spray applications are effective only on exposed surfaces where the force of solvent can be directed but complex parts cannot be effective degreasing by this technique. Vapour degreasing is a more effective method, since the part is exposed only to clean vapours.

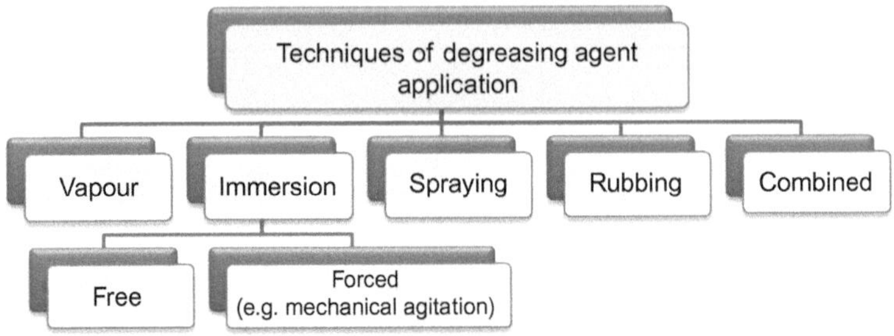

Fig. 4.5 Techniques of degreasing agent applications.

Although ultrasonic degreasing is a very effective method, there is always the possibility of drag-out occurring, whereby contaminants in the cleaning solution coat the part as it is being removed from the tank. Additionally, vapour and ultrasonic degreasing can be combined for greater effectiveness than can be obtained for either of the individual processes. Silicone grease, for example, is one of the difficult materials that remove from substrate surfaces. However, a multicycle cleaning process consisting of vapour degreasing and ultrasonic cleaning, again followed by vapour degreasing, can be used to obtain the maximum adhesion potential for the surface [17].

Selection of a proper degreasing method and the technique of degreasing agent applications will depend on a number of factors. Among factors which play a decisive role in the selection of a degreasing and technique method and also an agent, it may distinguish [2, 4, 17, 27–31]:

- cleaning effectiveness for a given material,
- dimensions and shape of adherends,
- type of substrate material,
- type of contaminants on adherend surface,
- type of production, and
- accessibility of technological instrumentation, etc.

All cleaning methods are improved by additional agitation, which may take form of scrubbing a large adherends or using stirring or ultrasonic agitation in the case of bath. An indication of the wide variation in degreasing performance as a function of degreasing medium and the level of agitation is provide by the date of Table 4.3 [6, 31].

The most widely applied degreasing agents include: acetone, petrol, benzene, ethanol, trichloroethylene, tetrachloroethylene, toluene, methyl ethyl ketone, etc. [1, 2, 7, 19, 31].

Table 4.3 Efficiency of degreasing as a function of process

Degreasing method	% Cleaning efficiency
Pressure washing with detergent solution	14
Mechanical agitation in petroleum solvent	30
Vapour degreasing in tricholoroethylene	35
Wire brushing in detergent solution	92
Ultrasonic agitation in detergent solution	100

Table 4.4 Properties of selected basic components of degreasers

Name of component	Chemical formula	Selected properties
Sodium hydroxide Potassium hydroxide	NaOH KOH	– Highly saponifiable – Poor emulsification properties, – Difficult to remove from the surface of objects – Not suitable for cleaning highly reactive metals and alloys, i.e. Al, Zn, Sn, Pb
Sodium carbonate Potassium carbonate	Na_2CO_3 K_2CO_3	– Saponification properties worse than of hydroxides – Good wettability and emulsification properties – Relatively easy to remove from the surface of objects – Aggressive
Sodium phosphate	$Na_3PO_4 \cdot 12H_2O$	– Saponifiable and emulsifiable – Easy to remove from the surface of objects – Mild degreasing agent – Suitable for treatment of metals reactive with NaOH or Na_2CO_3
Sodium polyphosphate	Compound-dependent formulae, e.g. sodium triphosphate ($Na_5P_3O_{10}$)	– Saponifiable and emulsifiable – Washability depends on the type of compound – Additive to degreasing baths ($2\,g/dm^3$)
Sodium silicate Sodium silicate	Na_2SiO_3 Na_4SiO_4	– Saponifiable and emulsifiable – Easy to remove from the surface of elements – Suitable for treatment of metals reactive with NaOH or Na_2CO_3

Degreasing operation is carried out in a bath washer, a chamber washer, or with the use of cloths [7, 18, 19, 26]. Basic elements selected ingredients of chemical or electrochemical degreasing are shown in Table 4.4 [32].

4.2.1 Chemical degreasing

Chemical degreasing is performed in alkaline solutions, which are usually composed of [2, 32]:

- sodium/potassium hydroxide,
- sodium/potassium carbonate,
- phosphates,
- silicates,
- borates, and
- gluconates.

The concentration of given components will vary depending on the base material and type of contamination. Modern degreasing baths are enriched with special degreasing-enhancing organic additives, which are surface-active agents exhibiting wetting, foaming, emulsification, and floatation properties. A particle of a surface-active agent (surfactant) is composed of hydrophobic and hydrophilic portions. The former moiety is apolar, i.e. insoluble in water and easily soluble in apolar liquids, such as oils, whereas the latter is polar, i.e. water-soluble and fat-insoluble. These substances are amphiphiles, i.e. compounds whose moieties are soluble in different solvents. Surfactants reduce the surface tension of liquids, which in turn facilitates wetting of solids by liquids. Moreover, they enable forming a mixture of two immiscible liquids (e.g. water and oil) to produce a system of colloids—emulsion (Fig. 4.6, prepared on the base of Ref. [32]).

Depending on their application several types of surfactants are distinguished:

- emulsifying agents (stabilising emulsion),
- wetting agents,
- foam producing agents, and
- detergents (cleaning agents).

Surfactants are classified into four groups according to their chemical constitution: anionic, cationic, nonionic, and amphoteric. In degreasing baths the most widely applied surface-active agents are anionic (containing the acid hydrophilic group) and nonionic substances, which do not dissociate in water to ions, and whose water-soluble character leads to the emergence of organic groups of high affinity for water.

Chemical degreasing in organic solvents

Chemical degreasing in organic solvents is frequently carried out as an initial operation, preceding other surface treatment methods, and is applied to very large surfaces with difficult-to-remove contaminants. This operation is furthermore applied as an ultra-fine cleaning solution in applications of high level of cleanliness requirements.

Organic solvents are effective degreasing agents in both the vapour and liquid phase. This advantage has lead the design of degreasing tanks in which solvent is

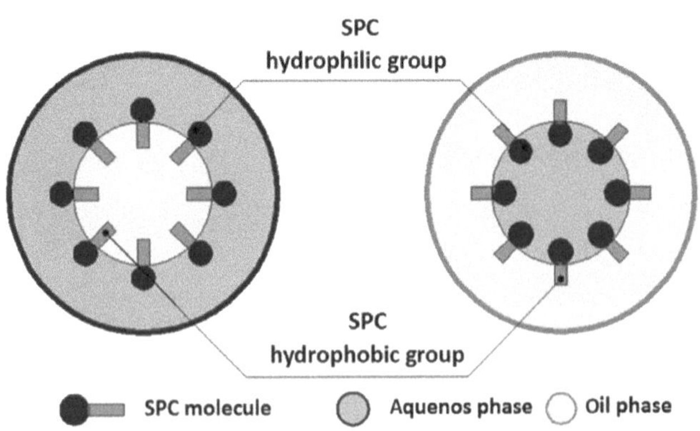

Fig. 4.6 Schematic representation of emulsion formation—(colloid system of oil and water).

heated to provide a vapour blanket above the liquid and adherends to be cleaned are held for a short time in the vapour phase prior to immersion in the liquid [31].

Degreasing in organic solvents consists in removing oil and oil-like contaminants (oils, lubricants, etc.) by means of physical dissolution.

Major advantages of the method listed in literature are:

- removing contaminants difficult to remove in a degreasing bath, such as lubricants, mineral oils, waxes, etc.,
- waste-free degreasing, and
- long-life as cleaning agents, as they do not require as frequent changes as degreasing baths.

Organic solvents applied in degreasing represent predominantly three groups of chemical compounds:

- chlorinated hydrocarbons (organochlorides),
- aromatic hydrocarbons (arenes), and
- aliphatic hydrocarbons.

The choice of solvent has been revised dramatically in the light of environmental and safety concerns and some years ago, chlorinated hydrocarbons (e.g. trichloroethane, trichloroethylene, and perchloroethylene) were weakly recommended as a degreasing solvent for metals. For both environmental and health and safety reasons, this practice has been largely superseded and a range of proprietary solvents has been successfully introduced into the marketplace with improved environmental and safety properties: environment and the impact of adhesive technology [31].

Aliphatic hydrocarbons are the most environmentally friendly solvents as the maximum workplace concentration level for this substance is relatively high, similarly as in air quality standards, and the pollution charges for the use and altering the environment are comparably low.

Chemical degreasing in alkaline and acidic solutions

Alkaline and acidic solutions are employed in a degreasing bath in order to remove the layer of oxides or scale off the surface of metal substrates. Such contaminants may reduce the strength of adhesive bonds or hinder other surface treatment operations on substrate material [32].

Chemical degreasing in alkaline solutions consists in saponification of animal and plant fats and in emulsification of mineral oils by alkalis. However, since alkalis are incapable of ensuring sufficient emulsification of oil, small amounts of emulsifying agents are added to the solution.

Chemical cleaning is commonly performed by immersion of elements (e.g. sheets) in steel tanks fitted with heating and mixing systems. The temperature of cleaning bath is 70–90°C, and the immersion time is several minutes. If degreasing is to be followed by etching or phosphating, the workpiece should be rinsed in cold and warm water beforehand. If the degreased object is to be painted, all alkalis must be removed from its surface (as an inaccuracy will eventually decrease the life of the applied paint coating). Cleaning may be also conducted in acidic solution.

4.2.2 Electrochemical degreasing

Electrochemical degreasing is regarded as one of the most efficient cleaning solutions for metal substrate applications, however, it requires strict adherence to composition of solutions and their working parameters. The chemical degreasing action occurring in this process is intensified by anodic or cathodic cycle, when either oxygen or hydrogen precipitates, and consists in emulsification of fats and oils by means of oxygen or hydrogen gas bubbles produced in electrolysis [32].

In practice it is anodic treatment (or cathodic/anodic with an extended final anodic cycle) that is preferred over cathodic process despite higher efficiency of the latter. This preference results from a major disadvantage of cathodic treatment inhered in the mechanics of the process: the amount of gas produced is twice as big compared to anodic treatment, however, as a consequence the degreased material becomes hydrated, which in turn changes the mechanical properties of the surface and may cause contaminants from the degreasing bath to deposit on the workpiece.

The essential composition of the electrochemical degreasing bath is largely similar to that in chemical degreasing. Time, temperature, and current density specifications are adjusted to specific conditions, i.e. surface material and the type and degree of surface contamination. In electrochemical degreasing, increasing the concentration of bath ingredients, current density, or temperature as well as extending the bath time beyond the necessary minimum will accelerate the side effects of degreasing, e.g. surface etching. It is therefore imperative that adequate current density is maintained throughout the process, as deviations from specifications may lead to poor degreasing (insufficient current density), surface etching or blacking (as a consequence of excessive density). Degreasing time is most frequently adjusted experimentally, and is a resultant of the other parameters.

4.2.3 Ultrasonic degreasing (physicochemical method)

Physicochemical method of degreasing, i.e. ultrasonic degreasing, depends on the propagation of ultrasonic vibration (of usually 20–50 kHz in industrial conditions) in solutions. The process offers numerous advantages, including: acceleration of degreasing, fine degreasing of the surface of treated elements, elimination of manual tasks, and of inflammable or poisonous solutions.

Ultrasonic degreasing has numerous applications in interoperational surface treatment, at the end of the production cycle, as well as in maintenance of instrumentation and tooling. Degreasing by means of ultrasonic method is suitable for treating small-sized objects of complex geometry. Although treatment time is relatively short, high consumption of different agent substances determines that the method is considerably cost consuming.

Ultrasonic degreasing process is regulated by a series of factors:

- ultrasonic field strength and frequency in the degreasing solution,
- type of cleaning solution,
- type of contaminant and required degree of degreasing, and
- material of degreased elements.

The effect of ultrasonic agitation on the cleaning process is impressive and the cleaning efficiency can be improved from 10% in a still solution at ambient temperature to 85% [31].

The process itself utilises cavitation action in the cleaning solution, which is capable of breaking mechanical forces that keep contaminants on the surface of objects, and accelerating the process of dissolving the contaminants. Effectiveness of cavitation action depends heavily on the physical properties of the cleaning solution.

Factors relevant to cavitation action in physicochemical degreasing are listed below:

- high surface tension of liquid intensifies cavitation action, however, in order for cavitation to occur a significantly high ultrasonic power must be produced,
- low surface tension of liquid promotes wettability of degreased surfaces; such solutions find application in cleaning objects of complex geometry,
- low density of liquid decreases intensity of cavitation action,
- high density of liquid increases acoustical resistance,
- gassing the liquid decreases intensity of cavitation,
- low pressure of saturated vapour intensifies cavitation action,
- temperature of liquid has no significant impact on intensity of cavitation action.

Optimal temperature for aqueous solutions should be approximately 55°C, for alcohol-based solvents 10–20°C, whereas for kerosene-based solvents 20–30°C.

Among solutions applied in ultrasonic degreasing we can distinguish:

- organic solvents and
- aqueous solutions.

In general, organic solvents exhibit good capability of dissolving different target contaminants. Nevertheless, prior to application several factors should be inspected, e.g. the level of ultrasonic wave attenuation, toxicity, as well as recoverability and corrosion hazard of degreased objects, bearing in mind that ultrasonic cleaning instrumentation for organic solvents is markedly more expensive than aqueous solution equipment.

The principal advantage of aqueous solutions is a more intensive cavitation action compared with other cleaning baths. Selection of a suitable type of solution to a given application, i.e. the type of surface material and surface contaminant, and adjusting its concentration will ensure that the target-high degreasing effectiveness of the process is obtained.

There are several types of aqueous solutions:

- basic detergents, cleaning mixes based on sodium hydroxide, sodium phosphate, water glass, etc.,
- acid aqueous solutions of tartaric acid, citric acid, and other organic and nonorganic acids, and
- neutral mixtures and emulsions (organic solvent and emulsifying agent and stabilising agent and water).

Basic aqueous solutions are characterised by the pH value, which specifies their cleaning properties and effectiveness of degreasing (Table 4.5).

Table 4.5 Types and characteristics of basic solutions [33]

Type of acidic solution	Properties
pH 10.5–11.2	The least aggressive. Suitable for aluminium and its alloys, brass and high pressure zinc die casts. As a result of the pH level, corrosion occurrence after treatment is possible
pH 11.2–12.4	Applied in cleaning aluminium and its alloys, zinc alloys and steel material mildly contaminated with petroleum-based oils or abrasive compounds
pH 12.4–13.8	Contains sodium compounds and water glass. Removes mineral and plant oils, old paint and tarnish. The most active representative of basic solutions (pH 13.3) removes corrosion and oxidation products, and boiler scale

All types of basic solutions are supplemented with sodium phosphate, which acts as a water softening agent. The concentration of solutions should be in the range of 0.2%–2.0%.

Acidic aqueous solutions are applied in ultrasonic degreasing in order to remove products of corrosion or etching, and scales resulting from forging and drawing processes. Taking into account the impact of ultrasonic energy on intensification of mildly acidic solutions, the degreasing tanks should be properly preserved prior to conducting such a treatment. The type and concentration of acids added to the solution will depend on the type of degreased objects. The working temperature of acidic solutions is 20°C, and the concentration levels should be between 5% and 25% for poorly resistant materials and 20% and 50% for more resistant materials.

Neutral aqueous solutions are characterised by a polarised molecular structure, which decreases the surface tension of water, oil emulsification and dispergation of solid contaminants. The main area of application of such solutions is cleaning medical instrumentation and equipment, as well as electronic assemblies and subassemblies.

Ultrasonic cleaning technology consists of the following operations presented in Fig. 4.7.

Ultrasonic degreasing equipment comes in different sizes, depending on the target application, the dimensions of treated objects and production requirements. The capacity of ultrasonic cleaning baths ranges from 0.5 dm^3 to the order of several cubic metres. Ultrasonic transducers are installed in the bottom or on the walls of the tank. Currently installed high-efficiency piezoelectric transducers and transistor oscillators provide the power of 100–1200 W.

4.2.4 Emulsion degreasing (physical method)

Physical methods of degreasing (emulsion degreasing) may be carried out in several ways: degreasing by means of immersion of objects in a solvent containing a mixture

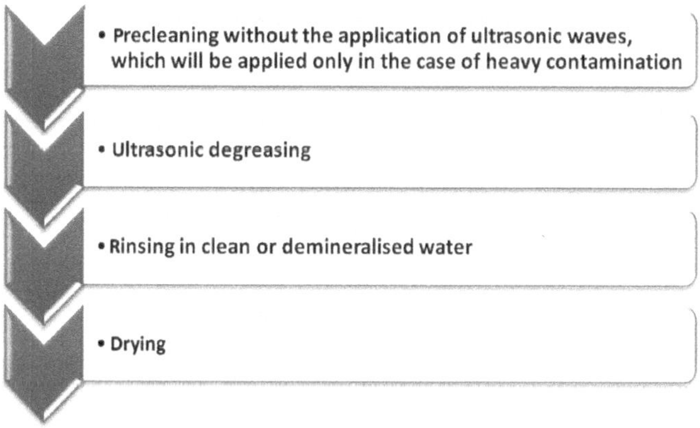

Fig. 4.7 Operations of ultrasonic cleaning technology.

of emulsifiers, or by immersion or spraying contaminated surfaces with a mixture of active composition of emulsifiers and organic solvent emulsified in water.

The procedure of the first application method requires rinsing the degreased surface with cold or hot water [2, 13]. Solvent enhanced with emulsifier removes fat contaminants, whereas water removes traces of chemical compounds and heat or mechanical treatment products. The method is mainly found in such applications as removing lubricants, oils, abrasive compounds, sediments, soot, or dust.

The other method is weaker by comparison, however, sufficiently efficient to ensure contamination-free surface. Its major areas of application are removing: oils from metal surfaces, oil residues of cooling emulsion, swarf, dust, and other loosely-bound mechanical debris.

Emulsion degreasing makes use of organic solvent solution (e.g. trichloroethylene, tetrachloroethylene), water and small amounts of wetting agents in order to reduce the surface tension at the interface. The selected type of degreasing agent will determine the temperature of the process (ambient or elevated); *nota bene*, elevated temperature increases the effectiveness of degreasing. In the final stage, the degreased object is rinsed thoroughly with running water.

Commonly used metal surface degreasing solutions usually include: organic solvents (e.g. acetone, petrol, ethanol, toluene, trichloroethylene, tetrachloroethylene), alkaline solutions (e.g. sodium resinate, sodium hydroxide, and an array of instant solutions such as Alfenol or Sulfapol), degreasing emulsions (ready-made concentrates, such as Emulsol) [32].

Alkaline cleaners present an attractive alternative to solvent cleaners, particularly in the case of very heavy grease deposits, without the problems of unpleasant, or indeed toxic vapours. They are used hot, the principle of an alkaline cleaner being the saponification of the grease layer by the alkali to produce carboxylate salts. Once this has occurred, the reaction products are removed from the surface by emulsification, peptisation, and subsequent dissolution [8, 31].

4.3 Experimental test

4.3.1 Methods

Adherends

The experiments involved producing single-lap adhesive joints of hot-dip galvanised steel sheets DX51+Z275–DIN EN 10142 (thickness of zinc coated is 20 μm) with a thickness ranging from 0.64 ± 0.06 mm. The properties of this steel are listed in Table 4.6.

Cold or hot dip or galvanised sheets have earned a considerable attention in such sectors as building, automotive, and ship-building industries. They are found in such applications as roofing, rails, frames, tanks, and other structures owing to their cost efficiency and good properties. In order to select protective coating suitable for a given application an engineer must consider the properties of a given coating solution, which results from the technology of their production. Coatings obtained by means of immersion or electrolytic zinc coating (cold galvanised) show differences in appearance, thickness, and the structural composition of thus produced coating [33].

The major difference between dip zinc coating and electrolytic zinc coating is that the layer of coating in the former exhibits phase structure, i.e. is composed of several phases of different structure and physicochemical properties, such as different microhardness or electrochemical potential.

Adhesive joint characteristics

Fig. 4.8 shows the schematic design of the tested adhesive joint.

The real dimensions of the produced adhesive joints measured with an electronic slide caliper were as follows: $L = 99.60 \pm 0.12$ mm, $l_a = 14.00 \pm 4$ mm, $w = 20 \pm 0.06$ mm, $t_s = 0.64 \pm 0.06$ mm, $t_a = 0.18 \pm 0.02$ mm, $t_{zn} = 0.020$ mm.

Adhesive properties

Galvanised steel sheets were bonded with a construction adhesive, Loctite 9466 A&B (Tables 4.7 and 4.8). This bi-component epoxy adhesive exhibits high strength and adhesion, low density, average viscosity, and it does not conduct electric current. Prior to adhesive bonding, the adhesive components (resin and curing agent) were mixed in 2:1 volumetric ratio until a uniform mixture. After that, the mixture was spread on one adherend surface. The working life of the produced adhesive mixture is about 60 min.

Table 4.6 The selected mechanical properties of adherends [33]

Properties	Value
Tensile strength R_m min	270 MPa
Tensile strength R_m max	500 MPa
Elongation min A_{80mm}	22%

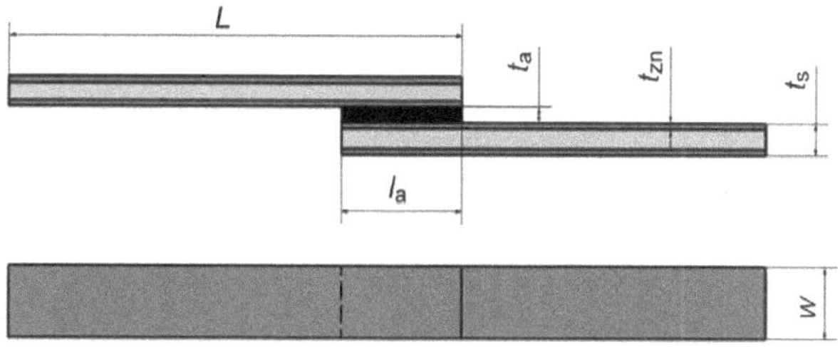

Fig. 4.8 Single-lap adhesive joint of hot-dip galvanised steel sheets.

Table 4.7 Selected properties of loctite 9466 before curing [34]

Property	Resin	Curing agent
Chemical type	Epoxy	Amino
Specific gravity	1.00 N/m^3	1.00 N/m^3
Brookfield viscosity in 25°C spindle 7 at 20 rev/min	42,400 Pa s	5500 Pa s

Table 4.8 Selected properties of Loctite 9466 after curing [34]

Property	Value
Volumetric ratio (epoxy/curing agent)	2:1
Weight ratio (epoxy/curing agent)	100:50
Brookfield viscosity in 25°C	30 Pa s
Lifetime of mixed adhesive	60 min
Working temperature range	−55°C to 120°C

Adhesives Loctite 9466 and 9484 require 24 h in room temperature in order to produce high strength. The process, however, can be accelerated by applying higher temperature.

Adhesives used in the tests are delivered in dual packages (cartridges), facilitating dosing and preparation. Mixing and dosing was performed with special dispensing device—a double syringe and a static mixer. After mixing the adhesive was applied on one of the adhered surfaces.

Epoxy adhesives are among the most effective and widely used adhesive polymers due to high intermolecular forces, i.e. adhesive forces ensuring accurate contact between the joined surfaces. In addition, they are resistant to chemical action.

Experimental details

The surfaces of galvanised metal sheets were prepared using five types of degreasing agents: acetone, extraction naphtha, Loctite 7063, Ultramyt, and Wiko, the properties of which are given in Table 4.9. The degreasing was run by three methods: rubbing, spraying, and immersion. The data are given in Table 4.10, and the schematic designs of these methods are illustrated in Table 4.11.

Table 4.9 Degreasing agents

Degreasing agent	Composition
Acetone (C_3H_6O)	Hydrogen-treated light petrol (petroleum); benzene <0.05%, toluene ≥3% or *n*-hexane ≥3%, <5%
Ultramyt	Toluene, polyoxyethylene ether of synthetic fatty alcohols
Loctite 7063	Hydrogen-treated light petrol (petroleum), <0.1% benzane, ethanol, methylal, carbon dioxide
Wiko	Hydrocarbons, acetone, isobutane , propane, carbon dioxide

Based on A. Rudawska, J. Nalepa, M. Müller, The effect of degreasing on adhesive joint strength, Adv. Sci. Technol. Res. J. 11 (2017) 75–81.

Table 4.10 Degreasing methods used in the experiment [34]

	Degreasing method		
Degreasing agent	**Rubbing (RU)**	**Spraying (SP)**	**Immersion (IM)**
Acetone	●	●	●
Extraction naphtha	●	●	●
Ultramyt	●	●	●
Loctite 7063		●	
Wiko		●	

Table 4.11 Schematic design of the applied degreasing methods [24]

Method	Designation	Schematic design
Rubbing	RU	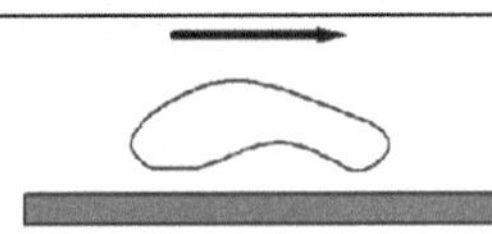
Spraying	SP	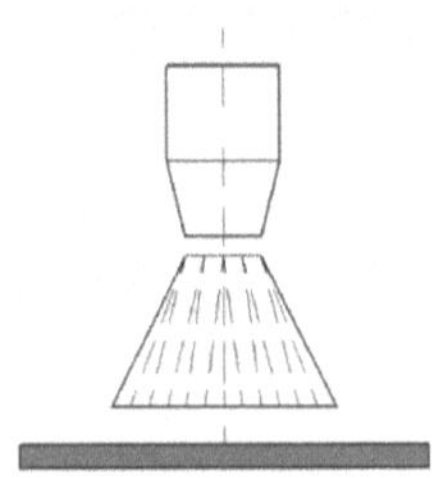
Immersion	IM	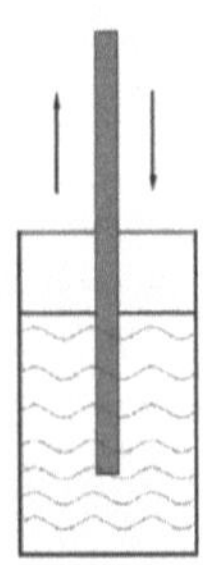

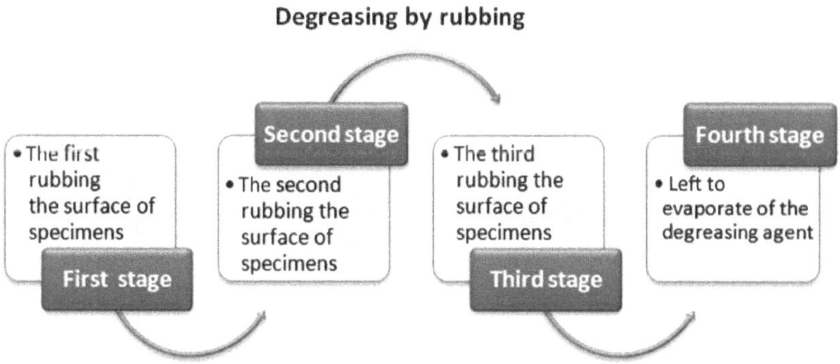

Fig. 4.9 Stages of degreasing by rubbing (RU).

Degreasing by rubbing (RU) consisted in rubbing the specimens three times with a paper towel soaked with a degreasing agent. After the third rub, the degreasing agent was left to evaporate (ca. 2 min) (Fig. 4.9). This method was not used when degreasing with Loctite 7063 and Wiko due to the fact that these agents come in containers with spray applicators. In contrast, other degreasing agents come in glass containers, which enable degreasing by different methods.

Degreasing by spraying (SP) was run in several stages presented in Fig. 4.10. The final operation—evaporation the degreasing agent is about 2 min. This degreasing method was performed using degreasing agents which come in special containers with sprinklers, i.e. acetone, extraction naphtha, and Ultramyt. Loctite 7063 and Wiko came in original packaging provided with spray applicators.

Immersion (IM) consists of immersing the specimen for 2 min in a glass container with a degreasing agent; after that, the specimen was taken out from the container and dried for ca. 3 min. This method was applied for acetone, extraction naphtha, and Ultramyt, but it was not applied for other degreasing agents due to the above-mentioned limitations.

The degreasing process was performed by the above methods in a temperature ranging from 24°C to 26°C and humidity between 28% and 38%.

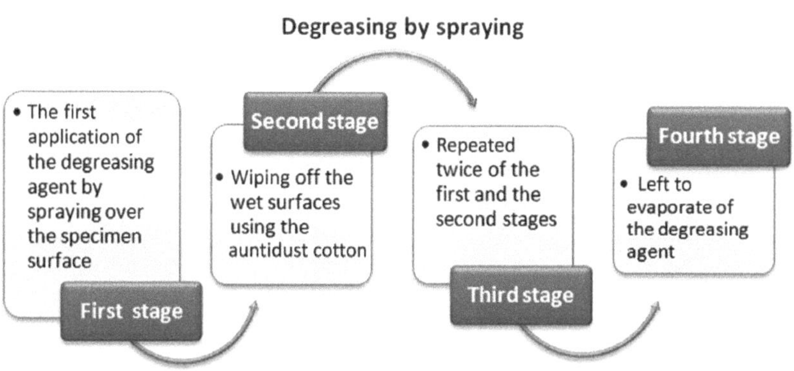

Fig. 4.10 Stages of degreasing by spraying (SP).

Following the degreasing, adhesive joints were made using the Loctite 9466 adhesive. The conditions of the adhesive bonding process for galvanised metal sheets were the same as in the degreasing of metal sheet surface.

First, the adhesive was applied to the adherend's surface; after that, the adherends were fixed and subjected to a load of 0.07 MPa. The adhesive was exposed to single-stage cold curing at a temperature ranging from 24°C to 26°C and humidity between 34% and 38%, and the curing time was set to 72 h. After curing, the specimens of adhesive joints were subjected to strength testing on the Zwick/Roell Z150 testing machine (Fig. 4.11).

The shear strength testing was performed in compliance with the EN DIN 1465 standard [35] and the test speed was 5 mm/min. The tests were performed on 8–10 test pieces per each batch. The strength tests were performed at an ambient temperature of 25 ± 1°C and humidity was 36 ± 2%. The t-Expert program was used to visualisation the strength results.

4.3.2 Results

Adhesive joint strength vs degreasing agent

The diagrams in Figs 4.12–4.14 show the results of strength tests of adhesive joints after surface degreasing with acetone, Ultramyt, extraction naphtha by rubbing, spraying, and immersion.

The diagram in Fig. 4.12 reveals that the shear strength of adhesive joints where the adherend surface was degreased with acetone ranges from 10.78 to 12.53 MPa. The lowest adhesive joint shear strength was observed when the adherend surface was degreased by rubbing, and it amounts to 86% of the highest value of shear strength observed for the specimens degreased by immersion in acetone.

Nonetheless, it seems that the similar values of shear strength produced with the tested degreasing methods point to a fact that the degreasing methods does not have a significant effect on adhesive joint strength; they rather serve for purification to ensure higher adhesion of the adhesive to the adherend surface.

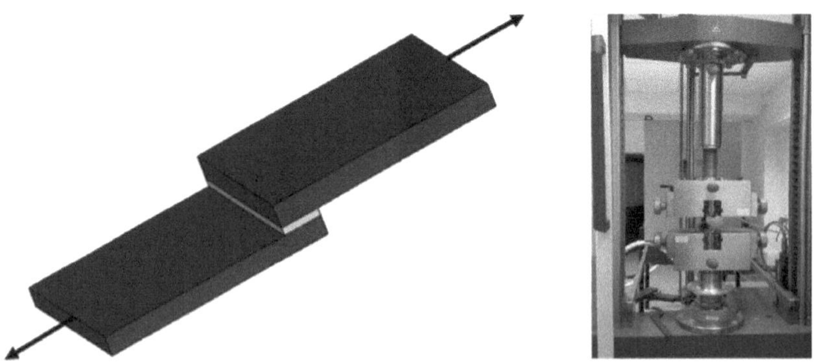

(A) (B)

Fig. 4.11 View of: (A) singlelap shear specimen, (B) shear test according to DIN EN 1465.

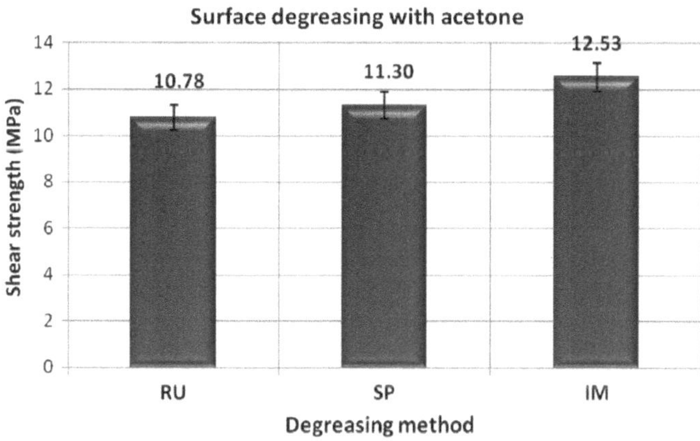

Fig. 4.12 Shear strength results of adhesive joints of galvanised metal sheet after surface degreasing with acetone by rubbing (RU), spraying (SP), and immersion (IM).

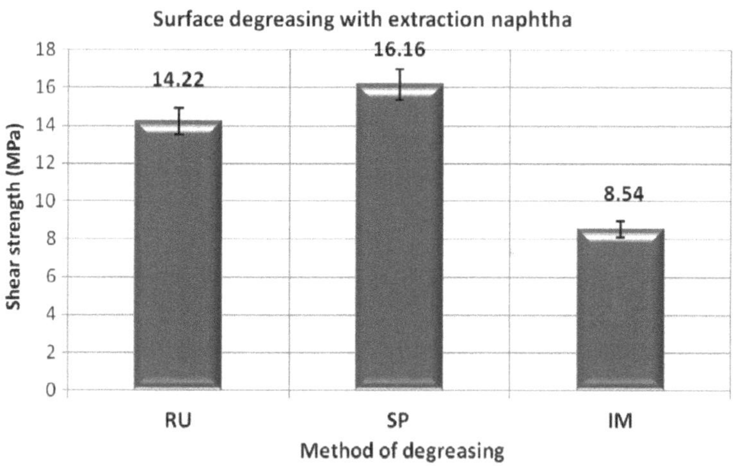

Fig. 4.13 Shear strength results of adhesive joints of galvanised metal sheet after surface degreasing with extraction naphtha by rubbing (RU), spraying (SP), and immersion (IM).

The results reveal that degreasing with extraction naphtha (Fig. 4.13) affects the adhesive joint strength of galvanised metal sheets. The highest shear strength of the tested metal sheet adhesive joints was produced for the specimens degreased by spraying with extraction naphtha, while the lowest—by immersion in extraction naphtha. The strength of the tested adhesive joints of galvanised metal sheet specimens subjected to degreasing by immersion in extraction naphtha before adhesive bonding is 53% of the adhesive joint strength of galvanised metal sheets subjected to surface degreasing by spraying according to the employed experimental method.

The above diagram (Fig. 4.14) demonstrates that the shear strength of the specimens subjected to surface degreasing with Ultramyt ranges between 13.17 and 15.24MPa for the different degreasing methods. Moreover, it is observed that this

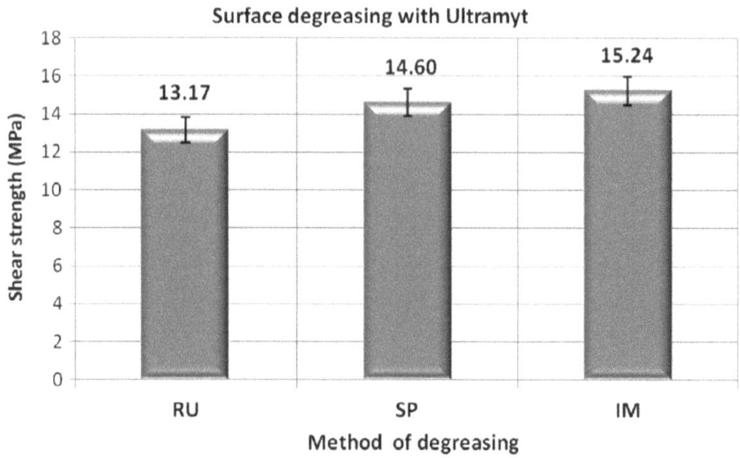

Fig. 4.14 Shear strength results of adhesive joints of galvanised metal sheet after surface degreasing with Ultramyt rubbing (RU), spraying (SP), and immersion (IM).

degreasing agent is effective when applied by immersion. Comparing the results given in Figs 4.12 and 4.14, it can be observed that the lowest shear strength of the tested adhesive joints produced by rubbing with Ultramyt is equal to the highest adhesive joint strength obtained by degreasing by immersion in acetone. In addition, degreasing by immersion using both Ultramyt and acetone has a positive effect on the adhesive joint strength of galvanised metal sheets.

Adhesive joint strength vs degreasing method

The shear strength results following surface degreasing by rubbing, spraying, and immersion using five degreasing agents are given in Figs 4.15–4.17.

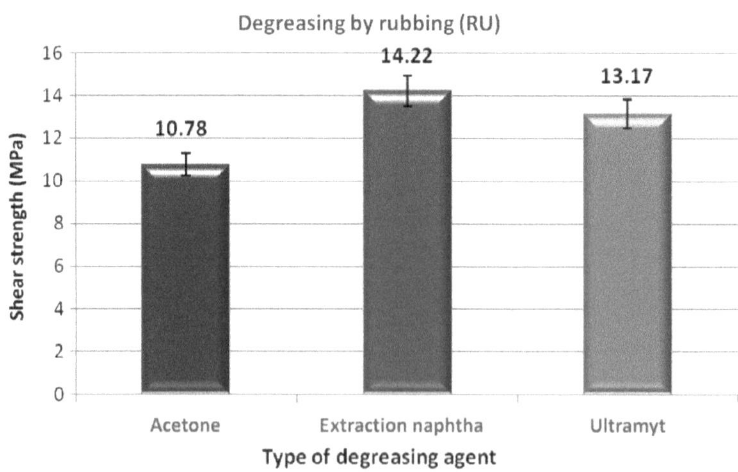

Fig. 4.15 Shear strength results of adhesive joints of galvanised metal sheet after surface degreasing by rubbing (RU).

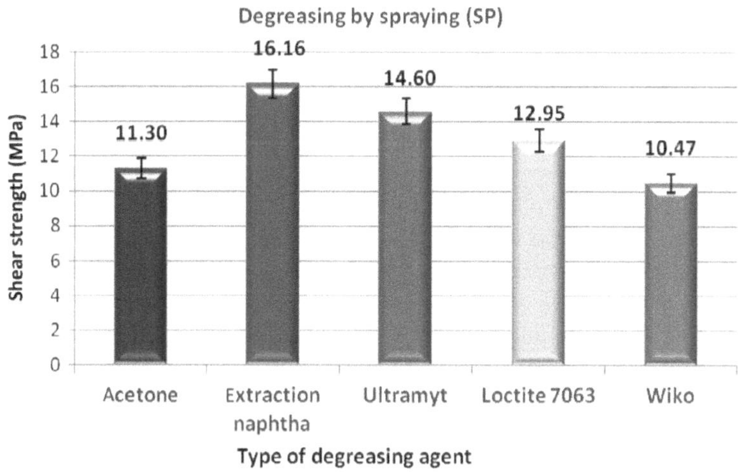

Fig. 4.16 Shear strength results of adhesive joints of galvanised metal sheet after surface degreasing by spraying (SP).

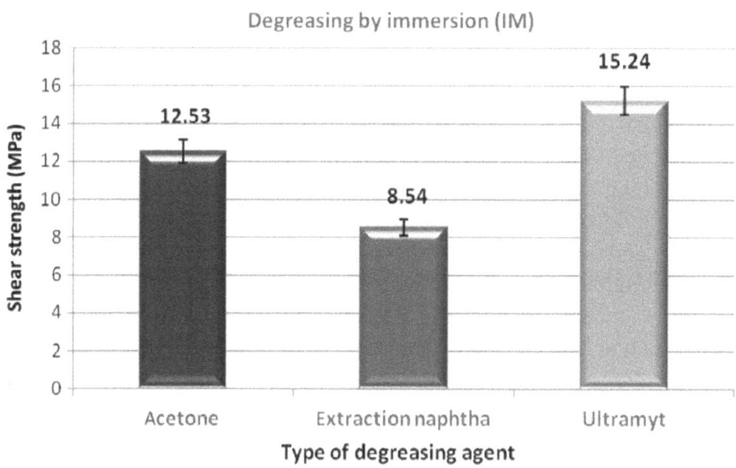

Fig. 4.17 Shear strength results of adhesive joints of galvanised metal sheet after surface degreasing by immersion (IM).

The results (Figs 4.15–4.17) demonstrate that the applied degreasing method using five different degreasing agents has impact on the adhesive joint strength of galvanised metal sheets. It can be observed that there are differences in the values of shear strength depending on the applied degreasing agent and degreasing method.

As for surface degreasing by rubbing (RU) (Fig. 4.15), the highest adhesive joint strength was observed for the specimens degreased with extraction naphtha (14.22 MPa). With degreasing by rubbing with acetone, the adhesive joint strength is 76% of the highest value of the adhesive joint strength of galvanised metal sheets subjected to degreasing with extraction naphtha.

As regards degreasing by spraying (SP), it is also important to emphasise that the choice of a degreasing agent has an impact on adhesive joint strength. Following the application of the tested degreasing agent by spraying, the results demonstrate that the adhesive joint strength differs by 35%. In addition, it is observed that the degreasing of surfaces of galvanised metal sheets using acetone or acetone-containing degreasing agents is the least effective (Fig. 4.16), as the tested adhesive joints have the lowest shear strength following the application of these degreasing agents. These results also confirm that the degreasing agent must be selected depending on the material. Interestingly, surface degreasing by spraying is a widely used method, as many degreasing agents come in containers provided with applicators which facilitate this operation. From an economic point of view, it is also important that this method ensures the lowest consumption of adhesive agents.

The highest differences in adhesive joint strength are observed for the specimens degreased by immersion (IM) (Fig. 4.17). The adhesive joint strength of galvanised metal sheets after surface degreasing by immersion in extraction naphtha is 56% of the adhesive joint strength of the specimens subjected to surface degreasing by immersion in Ultramyt. What is more, when using the immersion method one should take into consideration not only the consumption of the degreasing agent (higher consumption than with other methods) but also the degree of its impurity resulting from subsequently degreased specimens.

In terms of choice of a degreasing method, the results demonstrate that the largest differences in the adhesive joint strength of galvanised metal sheets occur in degreasing by immersion (IM), whereas the smallest differences are observed for degreasing by rubbing (RU).

It can be claimed that the adopted surface treatment method of materials for adhesive bonding should take into account not only the type of degreasing agent but also the degreasing method, as the results reveal that these two factors have a significant effect on the strength of the tested adhesive joints. Importantly, it is necessary to examine the type of adherend because the use of both an unsuitable degreasing agent and a degreasing method can lead to lower strength. Moreover, it is necessary to take account of the available workshop conditions (the possibility of application the immersion method, production type, available devices), operational conditions (the size and accessibility of degreased surfaces), economic conditions (the amount of degreasing agent used directly for surface degreasing), and many others.

4.3.3 Conclusions

The experimental results demonstrate that adhesive joint strength greatly depends on the applied degreasing agent. In addition, adhesive joint strength is affected by the method of application of the degreasing agent, too.

The strength results reveal that the most effective degreasing method for the adhesive bonding of galvanised metal sheets is degreasing with extraction naphtha by spraying. Thereby degreased specimens of adhesive joints have the highest shear strength. Extraction naphtha is also effective when applied by rubbing; however, it is the least effective when applied by immersion (the lowest shear strength). The results of degreasing by spraying demonstrate that Wiko and acetone are the least effective degreasing agents, hence it can be claimed that the degreasing agents containing acetone are not recommended for the degreasing of galvanised metal sheets. The highest shear strength is exhibited by the adhesive joints degreased with toluene and extraction naphtha. The highest and most uniform strength results were observed for the speci mens degreased by immersion in acetone and Ultramyt, even though the results tend to differ to a significant degree. In such a case, the choice of a degreasing agent often depends on two aspects: the economic aspect (a less expensive degreasing agent is applied) and the technological aspect (e.g. a way of application, as the use of agents which come with applicators enables a faster, easier and more economic application of the degreasing agent to the adherend surface).

Notwithstanding the above, a crucial aspect of joining elements for aircraft, automotive, or other machinery designs is to ensure the highest possible tightness and strength of the adhesive joint, which means that a given material should be degreased using the best degreasing agent available.

References

[1] R.D. Adams, J. Comyn, W.C. Wake, Structural Adhesive Joints in Engineering, second ed., Chapman & Hall, London, 1997.

[2] C.V. Cagle, Handbook of Adhesive Bonding, McGraw-Hill, New York, 1973.

[3] L.F.M. da Silva, R.J.C. Carbas, G.W. Critchlow, M.A.V. Figueiredo, K. Brown, Effect of material, geometry, surface treatment and environment on the shear strength of single lap joints, Int. J. Adhes. Adhes. 29 (2009) 621–632.

[4] P. Molitor, V. Barron, T. Young, Surface treatment of titanium for adhesive bonding to polymer composites: a review, Int. J. Adhes. Adhes. 21 (2001) 129–136.

[5] Y. Boutar, S. Naïmi, S. Mexlini, M. Ben Sik Ali, Effect of surface treatment on shear strength of aluminium adhesive single-lap joints for automotive applications, Int. J. Adhes. Adhes. 67 (2016) 38–43.

[6] M.J. Troughton (Ed.), Handbook of Plastic Joining: A Practical Guide, second ed., Wiliam Andrew Inc., Norwich, NY, 2008.

[7] S. Ebnesajjad, Adhesives Technology Handbook, second ed., William Andrew Inc., Norwich, NY, 2008.

[8] S. Ebnesajjad, C. Ebnesajjad, Surface Treatment of Materials for Adhesive Bonding, second ed., William Andrew Inc., Norwich, NY, 2013.

[9] K. Leena, K.K. Athira, S. Bhuvaneswari, S. Suraj, V. Lakshmana Rao, Effect of surface pre-treatment on surface characteristics and adhesive bond strength of aluminium alloy, Int. J. Adhes. Adhes. 70 (2016) 265–270.

[10] A. Rudawska, J. Kuczmaszewski, Surface free energy of zinc coating after finishing treatment, Mater. Sci. Pol. 24 (2006) 975–981.

[11] R.F. Wegman, J. Van Twisk, Surface Preparation Techniques for Adhesive Bonding, second ed., Elsevier, Oxford, 2013.

[12] R.E. Litchfield, G.W. Critchlow, S. Wilson, Surface cleaning technologies for removal of crosslinked epoxide resin, Int. J. Adhes. Adhes. 26 (2006) 295–303.

[13] P.R. Underhill, A.N. Rider, D.L. DuQuesnay, The effect of warm surface treatments on the fatigue life in shear of aluminium joints, Int. J. Adhes. Adhes. 26 (2006) 199–205.

[14] Q. Bénard, M. Fois, M. Grisel, P. Laurens, Surface treatment of carbon/epoxy composites with an excimer laser beam, Int. J. Adhes. Adhes. 26 (2006) 534–549.
[15] B.N. Zand, M. Mahdavian, Evaluation of the effect of vinyltrimethoxysilane on corrosion resistance and adhesion strength of epoxy coated AA1050, Electrochim. Acta 52 (2007) 6438–6442.
[16] R. Rechner, I. Jansen, E. Beyer, Influence on the strength and aging resistance of aluminium joints by laser pre-treatment and surface modification, Int. J. Adhes. Adhes. 30 (2010) 595–601.
[17] L.C. Jackson, Surface cleanliness measurements, contaminant characterization and control, in: Society of Manufacturing Engineers' Substrate Selection and Conditioning Seminar, 3–4 October, Chicago, Illinois, 1978.
[18] W. Brockmann, P.L. Geiß, J. Klingen, B. Schröder, Adhesive Bonding. Materials, Applications and Technology, Wiley-VCH Press, Weinheim, 2009.
[19] J. Bishopp, Surface pretreatment fir structural bonding, in: P. Cognard (Ed.), Adhesives and Sealants, Basic Concepts and High Tech Bonding, vol. 1, Elsevier Science, Amsterdam, 2005, pp. 163–214 (Chapter 4).
[20] S. Correia, V. Anes, L. Reis, Effect of surface treatment on adhesively bonded aluminium-aluminium joints regarding aeronautical structures, Eng. Fail. Anal. 84 (2018) 34–45.
[21] L. Kozma, I. Olefjord, Surface treatment of steel for structural adhesive bonding, Mater. Sci. Technol. 3 (1987) 954–962.
[22] 3M, Surface Preparation and Pretreatment for Structural Adhesives, Technical Bulletin (2014). January, http://www.3M.com/adhesives/4 (Accessed July 20, 2018).
[23] R.M. Podhajny, Comparing surface treatments, Converting 8 (1990) 46–52.
[24] A. Rudawska, J. Nalepa, M. Müller, The effect of degreasing on adhesive joint strength, Adv. Sci. Technol. Res. J. 11 (2017) 75–81.
[25] G.W. Critchlow, D.M. Brewis, Review of surface pretreatments for alluminium alloys, Int. J. Adhes. Adhes. 16 (1996) 255–275.
[26] A. Rider, P. Chalkley, Durability of an off-optimum cured aluminium joint, Int. J. Adhes. Adhes. 24 (2004) 95–106.
[27] N. Brack, A.N. Rider, The influence of mechanical and chemical treatments on the environmental resistance of epoxy adhesive bonds to titanium, Int. J. Adhes. Adhes. 44 (2014) 20–27.
[28] S.G. Prolongo, A. Ureña, Effect of surface pre-treatment on the adhesive strength of epoxy-aluminium joints, Int. J. Adhes. Adhes. 29 (2009) 23–31.
[29] J.A. Bishop, E.K. Sim, G.E. Thompson, G.C. Wood, The adhesively bonded aluminium joint: the effect of pretreatment on durability, J. Adhes. 26 (1988) 237–263.
[30] A. Rudawska, Surface free energy and 7075 aluminium bonded joint strength following degreasing only and without any prior treatment, J. Adhes. Sci. Technol. 26 (2012) 1233–1247.
[31] D.E. Packham (Ed.), Handbook of Adhesion, second ed., John Wiley & Sons, Ltd, Chichester, 2005.
[32] A. Rudawska, Surface Treatment to Bonding of Selected Constructional Materials, University Publisher, Lublin, 2017. (in polish).
[33] DIN EN 10142: Continuously hot-dip zinc coated low carbon steel sheet and strip for cold forming—Technical delivery conditions.
[34] http://www.loctite-kleje.pl/, 2018 (Accessed June 26, 2018).
[35] DIN EN 1465: Adhesives. Determination of tensile lap-shear strength of bonded joints.

Mechanical treatment

5

5.1 Characteristics of mechanical methods

Mechanical pretreatment is the oldest and the most common technique for improving the performance of adhesive metal bonds [1–7]. The primary reason for the application of mechanical treatments is surface cleaning and modification of surface geometry [8–15]. The shape of the outer layer of substrates is one of the features related to the degree of surface finish (i.e. surface texture) [1, 7, 10, 15, 16]. Mechanical treatment enables modification of surface geometry in order to obtain an adequately rough texture (surface roughness); however, it commonly fails to impart its proper activation and can later have an effect on adhesive joint strength [2, 13, 14, 17–24].

Machining is used to remove oxides which cannot be removed from material surface by means of degreasing and to clean material surface from all minor impurities which considerably reduce adhesive joint strength [1, 11, 17, 18, 25]. Machining helps remove thick corrosive layers from material and enables development of its geometrical structure via shaping it as desired [4, 24, 26]. Machining methods for surface treatment include:

- abrasive blasting (sandblasting [8, 26–30], sanding [31–34] and shot blasting [35], grit blasting [13, 14, 26, 36]),
- grinding [18, 37, 38],
- shot-peening [39],
- polishing [10], and
- milling [28, 40, 41].

In addition to the above-mentioned mechanical treatment methods, the following other mechanical treatment can be used during surface treatment of joining materials, e.g. Refs. [1, 17, 18, 21, 39]:

- lapping,
- superfinish,
- wire brushing,
- roller-burnishing, and
- others.

There exists an extensive range of treatment methods for the preparation of surfaces for bonding which combine different types of methods, e.g. mechanical treatment with degreasing or different type of chemical or special treatment [1, 2, 4, 5, 42–44]. Very often each of the above methods of mechanical treatment must be followed by material surface degreasing prior to the application of an adhesive [6, 14, 32, 34, 36, 40, 45–47].

The mechanical treatment methods often using for the preparation of surfaces for bonding are collated in Fig. 5.1. Several of them have been characterised in further subsections.

Surface Treatment in Bonding Technology. https://doi.org/10.1016/B978-0-12-817010-6.00005-9

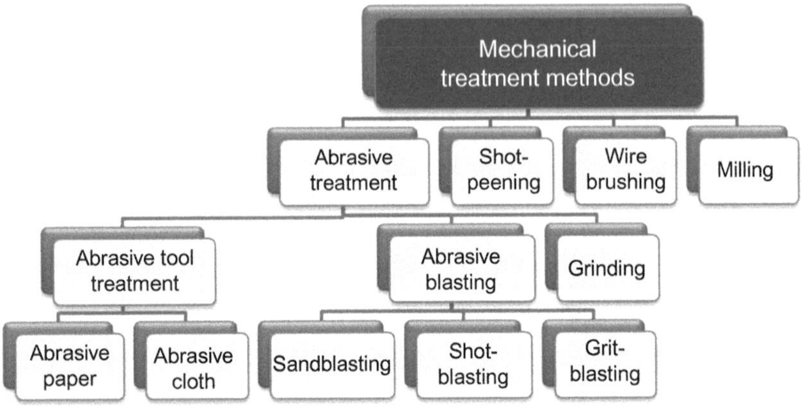

Fig. 5.1 Mechanical treatments methods often using in bonding process.

The various issues of mechanical treatments as surface treatment of various kind of construction materials prior to bonding process have been discussed in numerous studies [2, 6, 7, 13, 30, 31, 40, 47–54].

5.1.1 Steel

Harris and Beevers [10] investigated the effects of grit blasting with different alumina grits on the surface characteristics of mild steel and aluminium alloy. Islam et al. [17] presented the test of using polishing (with 150, 320, 600, 800, and 1200 size grit paper), grit blasting (using sandblasting cabinet and four different grit sized garnet grits) and wire brushing as surface treatment of mild carbon steel. The evaluation of sandblasting treatment of steel on the strength of adhesive bonded steel joints was investigated by Rotella et al. [27]. Experimental and numerical results of the effect of surface roughness (obtained after five variants of grit blasting and diamond polishing) on the cleavage bonded joints strength of steel was presented by Shahid and Hashim [14]. The results of test of mechanical treatment realised by various kinds of abrasive papers treatment of steel, 2024 T0 aluminium alloy- and zinc-coated steel sheets was reported by Rudawska and Łukaszewicz [55].

5.1.2 Aluminium and aluminium alloys

Review of surface pretreatments for aluminium alloys, including mechanical treatments (grit blasting, sandblasting, and mechanical abrasion) was presented by Critchlow and Brewis [26]. Spadaro et al. [38] tested different surface treatments applied to aluminium alloy 2024 to bonding, including mechanical grinding. Da Silva et al. [40] presented the research of different surface patterns and depths created during the milling process, which shows that the surface patterns influence the joint strength of aluminium alloy AA6082 T6. Bockenheimer et al. [3] investigated the effect of mechanical pretreatment by blasting with alumina grit or with glass beads on the surface state of an aluminium adherend (EN-AW 5754) and how in turn these pretreated surfaces influence the formation of the network structure in bonds with an epoxy adhesive system. Bhatt and Radhakrishnan [28] reported detailed studies on some machining processes such as shaping, milling, turning, and blasting used as aluminium alloys surface treatment methods. The effects of grit blasting treatment with different alumina grits on the surface characteristics of mild steel and aluminium alloy was investigated by Harris and Beevers [10]. Rider and Arnott [42] tested the different surface treatment (combination of the grit blasting with other treatment) prior to bonding of 2000, 5000, and 7000 series aluminium alloys.

5.1.3 Titanium and titanium alloys

The review of titanium surface treatment, including mechanical treatment methods, was resented by Molitor et al. [36]. Jiang et al. [8] investigated sandblasting as a surface treatment method, commercially used pure titanium for surface cleaning and corrosion removal of 35A.

Brack and Rider [13] examined the effect of mechanical (abrasion and grit blasting) and chemical treatments of titanium alloy (Ti6Al4V) for bonding to a rubber toughened epoxy adhesive. The effect of mechanical surface treatments on the fatigue performance in titanium, aluminium, and magnesium alloys was presented by Wagner [56].

5.1.4 Polymers and composites

Some information of mechanical surface treatment of polymers and composites are presented in [29, 36, 37, 57, 58]. Kumosa et al. [29] reported the tests of sandblasting (low sandblasting and medium sandblasting) of polymer composites (E-glass/vinyl ester, E-glass/epoxy, and E-glass/modified polyester). The effect of (among others) mechanical surface preparation (grinding with two different tungsten carbide burrs of composite (UDMA Eclipse denture base resin) on the surface roughening and shear bond strength of this composite was presented by Ariff et al. [37]. Various methods of surface treatment (including mechanical treatment like grit blasting with alumina grit blast and abrasion using 180/220-mesh alumina) for both polymer composites and titanium were discussed by Molitor et al. [36].

5.1.5 Wood

The effect of the mechanical (face-milling and sanding) and chemical surface preparation on the adhesively bonded wooden joints was reported by Hass et al. [31].

The usage of these methods most often results from specific accuracy requirements (defined, e.g. by surface roughness parameters) of the joined surfaces.

5.2 Abrasive machining—General information

The basic machining used to prepare the adherend's surface is abrasive machining. It is a type of machining in which the removal of unnecessary material is carried out with

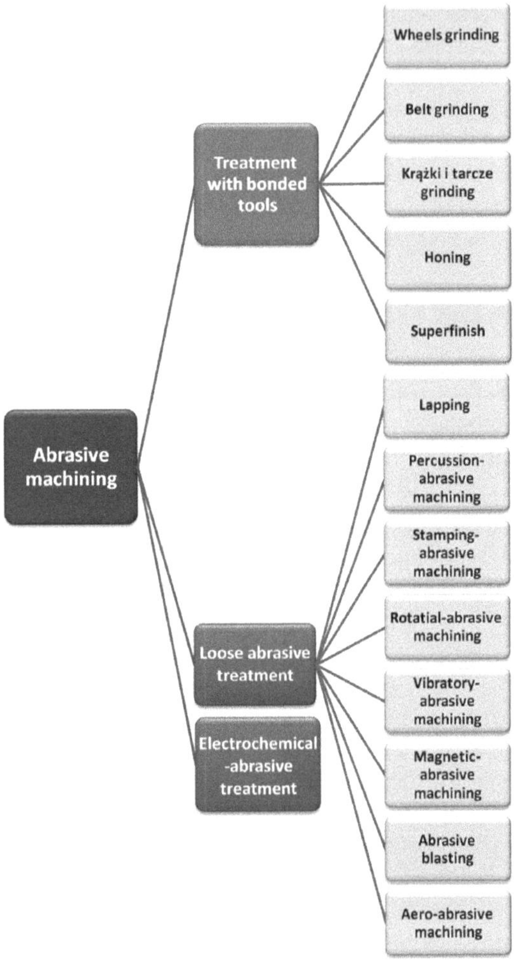

Fig. 5.2 Abrasive machining methods.

abrasive tools or loose grains. During the abrasive machining, apart from the chip removal of the allowance, elastic and plastic deformations of the material also occur [59].

Two groups are distinguished in abrasive machining (Fig. 5.2):

- treatment with bonded tools (grinding wheels, discs, belts, and whetstones) in which abrasive grains are bonded with a binder and
- abrasive treatment using loose grains contained in pastes or liquids.

Typical abrasive materials found in mechanical surface treatment operations are presented in Table 5.1, whereas Table 5.2 shows recommended sandblasting and grinding parameters.

When applying these treatments the geometry of surface irregularities must be closely monitored [6, 10, 14, 31, 40]. Extreme surface roughness can result in excessive stress

Table 5.1 Abrasive materials used in mechanical treatment

Type of abrasive	Mohs scale of hardness	Chemical composition	Applications
Natural abrasives			
Diamond	10	Allotrope of carbon	– Lapping paste – Special-purpose abrasive paste
Corundum	6	60%–90% Al_2O_3	– Abrasive paper – Coated abrasive tapes or discs – Abrasive and polishing paste
Silica	7	SiO_2	– Grinding and polishing of soft materials – Abrasive blasting
Pumice	5–6	AlO_2, Al_2O_3	– Polishing pastes for soft materials
Kaolin	2	$Al_2O_3{\cdot}SiO_2{\cdot}2H_2O$	– Additive in polishing pastes
Synthetic abrasives			
Corundum: Green Black	9.5	98% SiC 95% SiC	– Abrasive paper – Coated abrasive tapes or discs – Polishing pastes – Resin-bonded tumbling media – Abrasive blasting
Electrocorundum: White (A) Regular (B) Black (C)	9	98.5% Al_2O_3 96% Al_2O_3 90% Al_2O_3	– Abrasive paper – Coated abrasive tapes or discs – Polishing paste – Ceramic tumbling media – Abrasive blasting
Alumina (metallurgical)	9	Al_2O_3	– Polishing paste for steel and nonferrous metals
Chromium oxide	8–9	Cr_2O_3	– Multipurpose polishing pastes – Polishing chromium

Source: Own work based on Poradnik galwanotechnika (Handbook of Electropating), Wydawnictwa Naukowo-techniczne, Warszawa, 2002 (in Polish); Narzędzia ścierne i szlifowanie (Abrasive tools and grinding) (in Polish). http://grafit.mchtr.pw.edu.pl/~askalski/PTW-lab/PTW4.pdf/, 2018 (Accessed 22 June 2018).

concentrations, which may as a result negatively affect final joint strength [12]. It has been shown that optimal joint strength characteristics are obtained for adherends exhibiting the surface roughness (maximum peak to valley height) in the range of $R_m = 7–25\,\mu m$ [60].

The state of surface geometry affects its wettability: it has been shown that from the perspective of wetting the most advantageous is the surface in the shape

Table 5.2 Recommended sandblasting and grinding treatments for metals and metal alloys

	Treatment		
	Sandblasting	Grinding	
	Recommended grit size	Recommended grit size	
Metal	**50–100 μm**	**120–240 μm**	**320–400 μm**
Aluminium and its alloys	+	+	–
Copper and its alloys	+	+	–
Iron and its alloys	+	+	–
Precious metals (Ag, Au, Pt)	–	–	+
Nickel and its alloys	+	+	–
Titanium and its alloys	+	+	–
Zinc	–	+	–

+, recommended; –,not recommended.
Source: Own work based on L.H. Durney (Ed.), Electroplating Engineering Handbook, fourth ed., Van Nostrand Reinhold, New York, 1984; S. Malkin, Ch. Guo, Grinding Technology: Theory and Applications of Machining With Abrasives, Industrial Press, New York, 2008.

of densely arranged elongated surface protrusions of small vertical angle [14, 19, 20, 22, 23, 61, 62]. Furthermore, it ought to be remembered that mechanical treatments can contribute to generation of excessive compressive stresses in the surface region of adherends, and consequently cause the adherends to deform. These deformations, in turn, are responsible for the introduction of additional stresses in the bond and as a consequence, the decrease in the strength of adhesive joints even by 10%–50% [60].

5.3 Abrasives tool methods

In general, abrasive tools offer the highest convenience of use among available surface treatment solutions [59, 63–67]. These tools are cost effective, widely available, easy to use in manual operations, and suitable for various conditions and environments [68]. In abrasive treatment it is imperative, particularly from the viewpoint of structural bonding, that the treated surface should exhibit a nonoriented structure [6, 59, 66].

Selecting a suitable tool for performing abrasive treatment operations requires considering the grit size [18, 32, 33, 59, 64–67]. Abrasive tools of high grit sizes show higher efficiency of material removal, including the physisorption layer; however, they do not produce a desired surface texture. Lower grit size tools, on the other hand, produce a more desirable surface texture; however, at the expense of lower rate of physisorption layer removal.

5.4 Abrasive blasting methods

5.4.1 Abrasive blasting

Abrasive blasting is one of the surface treatment methods involving the use of loose abrasive media with high kinetic energy (developing velocity of approximately 100 m/s [59]), which apart from removing the surface leads to deformation of the treated surface [63, 69, 70]. The process is controlled by several parameters, including the blast angle and cutting properties of abrasive media (e.g. edge sharpness, hardness, and impact resilience). These parameters are crucial to the surface finish and effectiveness of the process [59, 63, 71–74].

According to the Polish Standard PN-EN ISO 8504-2, abrasive blasting methods are classified as shown in Fig. 5.3. Applying different criteria, we can also classify abrasive treatments according to grit circulation into, closed-loop and open-loop, or considering the pressure criterion, into three groups:

- low pressure (0.3–0.5 MPa),
- high pressure (1.0–1.2 MPa), and
- hydrodynamic (1.0–200 MPa).

Abrasive blasting can be carried out with or without the use of water medium [59, 63, 67, 71]. Dry abrasive blasting delivers a more coarse surface finish compared to hydro-abrasive blasting; however, with the application of glass media (1 mm in diameter), the effects of dry blasting are comparable to grinding ($R_a \approx 1$ μm) [59]. Hydro-abrasive blasting is performed with the use of a blast of liquid with abrasive media (e.g. electrocorundum and water) and compressed air. The advantage of this method is reducing the dust hazard on the surrounding environment. With regards to surface roughness, in hydro-blasting the peaks of surface irregularities are removed, while the

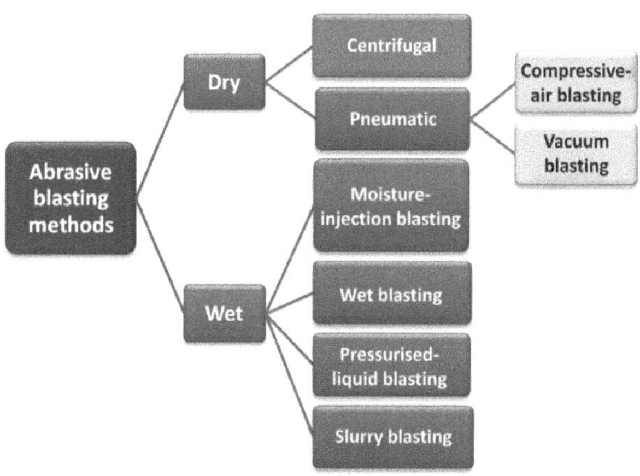

Fig. 5.3 Classification of abrasive blasting methods.
Source: own work based on [Polish Standard PN-EN ISO 8504-2] and Metody przygotowania powierzchni (Surface treatment methods) (in Polish). http://anticorr.pl/metody-przygotowania-powierzchni/, 2018 (Accessed 30 August 2018).

valleys are not deepened, which is an effect similarly observed during dry blasting. The introduction of liquid into the airstream/grit mix enables obtaining the surface finish characterised by lower roughness than following dry abrasive blasting [59, 67]. However, the main disadvantage of the hydro-abrasive blasting method is that it may cause or accelerate corrosion, which is why the water is typically supplemented with corrosion inhibitors, such as soda or sodium nitrite [59, 69, 70].

The final effect of abrasive blasting treatment depends on such factors as: shape of abrasive media, concentration of abrasive media in the slurry, air pressure, and the type of abrasive medium [63–65, 73]. Particular parameters of the process are adjusted to the applied abrasive media geometry; therefore, depending on the process parameters, abrasive blasting treatment is divided into two groups: abrasive and blasting [59, 66, 67, 69].

During abrasive treatment a considerable amount of material is subtracted thus producing surface roughness, expressed by R_a, in the range of 1.25–0.63 μm. In blast-cleaning treatment, which employs micro-powders as the abrasive medium, the roughness of the treated surface is in the range of $R_a = 0.63 \div 0.16$ μm, while the surface exhibits slight deformation [59]. Furthermore, grit size is decisive to the radius of corners and wedges of the treated surface [59, 67]. Surface roughness decreases with the size of the abrasive grain; Table 5.3 shows specifications regarding the correlation between grit size and surface roughness (expressed by R_z).

As mentioned in the preceding paragraphs, the selection of abrasive medium affects the results of abrasive blasting. The type and size of abrasive particles should be adjusted to the type of abrasive blasting treatment (Fig. 5.3), the type of surface contamination, and the type and size of workpiece, its material, and target surface finish.

There is a wide range of materials used as media in abrasive blasting, including [59, 63, 64, 66–74]:

- chilled iron shot, round,
- hardened chilled iron shot,
- steel shot, round,
- cut wire,
- silica sand,

Table 5.3 Medium size in abrasive blasting

Grit size (grain for bonded abrasive tools, abrasive pastes and loose abrasive media)	Surface roughness parameter R_z (μm)	
	Prior to treatment	**Posttreatment**
120	6.3–10	8
180	3.2–6.3	4–5
F320/29	1.6–3.2	1.2
P360/23	0.8–1.6	1.0
P500/13	0.4–0.8	0.3
P1000/5	0.2–0.4	0.2

Source: Own work based on Obróbka skrawaniem (Machining), Wydawnictwa Naukowo-techniczne, Warszawa, 1993 (in Polish).

Table 5.4 Grading of abrasive media applied in blasting

European P-Grade	Grit size (μm)
P 12	2000–1700
P 20	1000–850
P 36	600–500
P 80	212–180
P 120	125–106
P 220	75–63
P 320	47.7–44.7
P 360	42.0–39.0
P 400	36.5–33.5

Source: Own work based on Polish Standard [PN-ISO 8486-1:1998].

- corundum,
- electrocorundum,
- silicon carbide,
- blast furnace slag,
- copper furnace slag,
- glass beads, and
- crushed glass.

Table 5.4 shows typical grit sizes of natural and synthetic abrasive media applied in abrasive blasting treatment.

The surface finish following abrasive blasting is typically described by means of such indicators as: mean roughness depth R_z, waviness, structure, defects, and stresses. These indicators exhibit distinct dependence on: abrasive mix composition, nozzle design, abrasive stream control, and the type and properties of the substrate material.

Abrasive blasting offers relatively high efficiency of machining, and it may be easily mechanised. Furthermore, it enables machining workpieces of irregular geometry, which would otherwise be difficult or impossible to machine using standard solution. The negative aspect of abrasive treatment is the fact that it does not remove the machining marks produced in preceding treatments and that it does not reduce the waviness of the workpiece surface.

5.4.2 Sandblasting

Sandblasting is classified as one of the abrasive treatment methods [8, 26–29, 59, 67, 74]. It is a technological process employing abrasive particles in the stream of compressed air or liquid which cleanses or shapes the surface of the workpiece. By adjusting the applied stream pressure and grit diameter to the particular substrate material, sandblasting is capable of producing a surface of desired roughness properties for further adhesive bonding processes (including adhesive joining or painting) [69, 70]. Surface roughness following sandblasting is comparable to the finish produced by grinding with P36-P50 tools [59].

Sandblasting is often recommended to be applied to surfaces of various structural materials prior to adhesive bonding, these materials including steel, titanium or titanium alloys, aluminum alloys, and polymers (polyester, polyethylene) or composites and some information of description of sandblasting test is presented in [8, 14, 15, 26, 29, 51, 75].

Sandblasting is typically performed with the use of silica sand, however other media may be employed: crushed glass, metal particles, dry ice particles, garnet, coking slag, and glass or ceramic beads [5, 6, 29, 63, 67, 76–83]. A specific surface finish is obtained when employing the crushed walnut shell grit or other hard plant material [59, 63, 71].

Dry-ice blasting is a special sandblasting method known for its high effectiveness and environmental safety [76–81]. The effectiveness of the method is attributed to two factors: thermal shock and pneumatic effect. The impact of dry ice particles crushes the surface contamination, which subsequently suffers thermal shock on account of the considerable temperature disparity (dry ice temperature is −79°C) thus becoming detached from the surface. In addition, the high velocity of the jet stream/ice grit effectively brushes off the loose debris. Immediately after the release of dry ice particles, the ice turns to gas and sublimates entirely to the atmosphere. Due to the nonabrasive nature of the process the substrate surface remains unaffected, whereas the surface contaminants are removed.

The sandblasting machine is typically equipped with a container from which the dry ice is fed to the blasting hose. Simultaneously, the gun is 'propelled' by compressed air, which generates negative pressure, gently leading the dry ice particles from the tank to the gun, while a highly efficient jet system allows the mixture to accelerate to the velocity of approximately 300 m/s [59].

The dry-ice blasting method offers a range of advantages, including the following [76–83]:

- dry ice particles turn to gas CO_2 and sublimate to the atmosphere upon the impact with surface contamination, as a result waste management problem is nonexistent,
- blasting by means of dry ice media is environmentally harmless as it does not require applying any additional chemicals, and the removed surface contamination is the sole by-product of the process,
- the method is entirely safe for the substrate surface due to its nonabrasive character and the fact that the no chemicals inhibiting the corrosion resistance are found in the mix,
- the functionality of machine parts subjected to dry-ice blasting is unaffected, and the majority of elements can be cleaned without the need for disassembly,
- the method is highly time efficient: the surface preparation for treatment is simple and the surface following dry-ice blasting is always contaminant free,
- sandblasting machines are fairly uncomplicated in use and therefore no specialist training or knowledge is required, and
- high functionality of sandblasting machines is attributed to their design (a system of special nozzles and extensions), which guarantees consistent quality in various, even unfavourable, operating conditions.

In addition to the aforementioned benefits, dry-ice blasting is a remarkably 'gentle' method—safe for the substrate surface, which consequently prolongs the service lifetime of the machined elements.

Sandblasting is a process that to a certain extent resembles grinding; however, the surface following sandblasting exhibits lower coarseness. In addition, the treatment itself enables high-precision removal of contamination from places such as cavities—which are inaccessible with traditional methods. The treatment is recommended as a replacement for standard grinding operations on workpieces of intricate geometry. It is moreover advised that the surface following sandblasting should be protected from adverse effect of air [59, 63, 71, 82, 83].

5.4.3 Shot blasting

In essence, shot blasting resembles sandblasting, however, the abrasive medium here is metallic shot, which carves deeper into the substrate surface, contributing to deeper surface roughness and increased adhesion to various substrates [59, 84–89]. The process consists in impacting the workpiece surface with metallic shot with high kinetic energy. As a result, the surface becomes deformed, and the extent of deformation is relative to the value of the kinetic energy of the shot medium (its mass and velocity) [89]. In the first stage of treatment, the oxide layer and other surface contaminants surrender under the impact of shot media against the surface, whereas the subsequent impacts lead to cold working of the surface layer of the substrate. Shot blasting employs both chilled iron (hard and brittle) and steel shot. Due to the risk of surface damage, this treatment is not recommended for workpieces lighter than 0.5 kg and of wall thickness lower than 3 mm (Table 5.5).

Basic parameters of shot blasting include: shot size, material, velocity, coverage (duration of treatment), workpiece geometry, substrate hardness, and target surface roughness. During shot blasting, shot coverage is relative to shot velocity—the higher the velocity the shorter the time of exposure. The effects of shot blasting are: surface cleansing, increase in the fatigue strength, and surface roughness R_a= 20–10 μm (ISO-8501-1). The surfaces treated with shot blasting may be subjected to surface cleanliness and roughness control in accordance with International Standard ISO-8501-1 Sa 2½.

Table 5.5 Shot size and air stream pressure specifications for different workpiece wall thicknesses in shot blasting treatment

Wall thickness (mm)	Shot size (mm)	Air pressure (MPa)
>1.00	0.15–0.30	0.20–0.50
1.00–3.00	0.30–0.50	0.40–0.50
<3.00	0.60–0.80	0.50–0.60

Source: Own work based on Obróbka skrawaniem (Machining), Wydawnictwa Naukowo-techniczne, Warszawa, 1993 (in Polish).

5.5 Grinding

Grinding is classified as an abrasive treatment in which the surface layer of the substrate material is removed by means of bonded abrasive tools [63, 67, 71, 90–95]. The geometry of the grain is individual and irregular—multiple edges and tips—and the distribution of grains on the tool surface is random, as is the grain orientation upon contact with the substrate [59, 63, 90]. Grinding is typically performed with grinding machines, equipped with elastic grinding disks, which are either attached directly to the rotor engine axis or come as a flexible extension shaft attachment. Steel substrates treated with abrasive grinding tools exhibit smooth finish with small grooves carved by abrasive grit. Grinding may also be carried out with mechanically powered belt grinders, in which case the tool may be operated manually, or be a part of a grinding unit or of a mechanised line [59, 63, 71].

Belt grinding, which is carried out with an abrasive belt, provides surface finish that is comparable or even surpasses the effect of grinding wheel; however, it occurs at the expense of dimensional accuracy [59]. This treatment is performed by means of belt grinders of adjustable belt speed; therefore, Table 5.6 shows recommended grinding conditions for typical substrates.

Among the factors affecting the surface roughness of machined surfaces following grinding, the grit size of the grinding tool appears to play a particularly important role. Lower grit size translates to lower surface roughness (expressed by, e.g. R_a—Table 5.7); however, the surface finish also depends on the type of substrate and the order of roughing and finishing operations [59, 63].

Grinding operations are divided into roughing and finishing, which differ, e.g. in terms of grit size [59, 63, 71, 90, 91, 94, 95].

5.6 Shot-peening

Shot-peening is a cold-working treatment, during which the surface of the workpiece is impacted by shot—small, hard, and round particles (such as steel shot, ball bearings, or glass beads) [39, 96–102]. This treatment is a representative of dynamic

Table 5.6 Recommended belt speeds

Type of substrate	Peripheral velocity of grinding disk or belt velocity in grinding (m/s)
Carbon steel	20–50
Stainless steel	25–35
Aluminium alloys	12–15
Copper and its alloys	15–20
Thermosetting materials	10–12
Thermoplastic materials	5–8

Source: Own work based on Obróbka skrawaniem (Machining), Wydawnictwa Naukowo-techniczne, Warszawa, 1993 (in Polish).

Table 5.7 Correlation between R_a and grit size

Grit size	Surface roughness parameter R_a (μm)
P80	2.5–5.0
P120–P180	1.6
P220 and higher	0.4

Source: Own work based on Obróbka skrawaniem (Machining), Wydawnictwa Naukowo-techniczne, Warszawa, 1993 (in Polish).

cold-working methods, and is carried out to strengthen machine parts and therefore the robustness of the machines [59].

In shot-peening the substrate surface is subjected to the effect of shot blasted at it with the velocity in the range of 40–90 m/s, depending on the specific parameters of treatment [59]. Contrary to other cold-working methods, shot-peening does not lead to decreasing the depth of the substrate surface roughness. It does, however, increase the hardness of the surface layer and produce a compressive residual stress layer, which is why it is an ideal solution applied to increase material fatigue strength. The major advantage of shot-peening is that it is suitable for treating workpieces of intricate geometries, highly irregular shape, and high hardness; secondly, it is a cost effective and highly practical cold-working method for the enumerated applications.

According to shot propelling mechanism, shot-peening machines are classically divided into mechanical (with a propelling rotor) and pneumatic (using compressed air to accelerate the abrasive media). The latter process, pneumatic shot-peening, is controlled by a set of basic modifiable process parameters, i.e. [39, 96–102]:

- air pressure,
- distance from nozzle to workpiece,
- exposure time, and
- shot characteristics (mass and size of shot).

There are several parameters of fundamental influence on the effects of vibration shot-peening process: number of nozzles, impact angle, shot velocity, the kinematics of the process [39, 98, 102]. These parameters, however, are largely nonmodifiable, being dependent on the shot-peening machine design and construction.

5.7 Experimental test—Milling

The information in this subchapter was prepared base on article [41].

5.7.1 Methods

Adherends

The experiments were performed on C45 steel sheets. This nonalloyed constructional steel is used for hardening and tempering. The selected properties of C45 steel sheets are presented in Table 5.8.

Table 5.8 The selected properties of C45 steel [103]

Property	Value
Young modulus, E	205–210 GPa
Poisson's ratio, ν	0.30
Yield point, R_e	430 MPa
Tensile strength, R_m	650–800 MPa
Soft-annealed hardness	250 HB

The machining method involved the use of two milling techniques: conventional milling [using a high-speed steel cutter (HSS)] and high-performance cutting (HPC) using an indexed cutter. The applied types of milling cutters are specified below:

- monolithic high-speed steel cutter NFPa φ20 $z=4$—Variant I
- indexed cutter R390-020B20-11L— Variant II

Machining method

The technological parameters of the tools were set as follows (Tables 5.9 and 5.10): cutting speed $V_c=75$ m/min, number of revolutions $n=1200$ rev/min, depth of cut $a_p=1$ mm. On their one end, the C45 specimens were subjected to milling over a length that was equal to the diameter of the cutter. This means that both in the first and second case the length of milling were 20 mm.

The milling of flat surfaces produces burrs on workpiece edges. The application of different feed rates resulted in burrs of different sizes and thicknesses. Defects of this kind are particularly troublesome if they occur on surfaces to be adhesive-bonded with each other. For this reason, the burrs had to be removed with a metal file to prevent disarrangement of material surface structure.

Surface roughness measurement

Surface roughness parameters were examined with a contact profilograph using the T8000 RC120-400 Hommel–Etamic 3D profilometer. The specimen surfaces after milling were examined in three regions: at the beginning, in the centre, and at the end of the machined surface. The surface roughness parameters were measured in accordance with ISO 4278 [104] and ISO 25178 [105] standards. Although the experiments were performed on 12 specimens per each technological parameter variant (listed in

Table 5.9 Technological parameters of conventional milling (Variant I)

Variants of surface mechanical treatment	Technological parameters	
	f_z (mm/tooth)	V_f (mm/min)
Variant I-A	0.05	240
Variant I-B	0.15	720
Variant I-C	0.30	1440

Table 5.10 Technological parameters of high-performance cutting (HPC)—Variant II

Variants of surface mechanical treatment	Technological parameters	
	f_z (mm/tooth)	V_f (mm/min)
Variant II-A	0.05	120
Variant II-B	0.15	360
Variant II-C	0.30	720

Tables 5.9 and 5.10), and 8–12 measurements of surface roughness parameters per each tested specimens were conducted.

The main parameter of surface roughness is R_a which describes the arithmetic average profile roughness, or—in practice—the spacing between two parallel lines limiting the surface area of material-filled profile regions located above a mean line *m* and the material-free profile regions located below the mean line [104, 106].

Other fundamental parameters for describing roughness include [104]:

- ten-point roughness height—R_z (calculated by averaging the five highest peaks and five lowest valleys over each sampling length),
- average maximum roughness height—R_{max},
- maximum profile peak height—R_p,
- reduced peak height—R_{pk},
- core roughness depth—R_k,
- reduced valley depth—R_{vk},
- total profile height—R_t,
- mean spacing of profile irregularities—S_m, and
- mean spacing of local peaks of the profile—*S*.

Adhesive joints

The single-lap adhesive joints made of C45 constructional steel sheet were subjected to strength testing. The scheme of the tested adhesive joint with averaged dimensions was shown in Fig. 5.4.

The adhesive joint had the following average dimensions presented in Table 5.11.

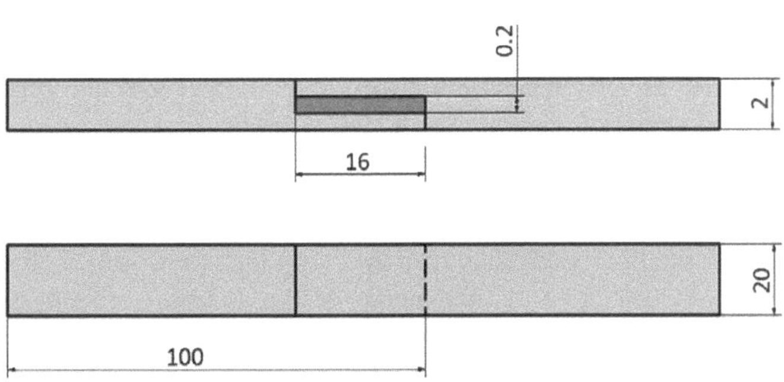

Fig. 5.4 Scheme of the tested single-lap adhesive joint—New.

Table 5.11 Theoretical dimensions of single-lap joints

Theoretical dimensions of single lap-joints (mm)				
Specimen length	**Specimen width**	**Specimen thickness**	**Adhesive layer thickness**	**Lap length**
100	20	2	0.2	16

The tested adhesive joints were bonded using epoxy adhesive compounds made of Epidian 57 epoxy resin and Z-1 curing agent Z-1, and mixed in a ratio of 100–10. The adhesive is denoted as Epidian 57/Z-1/100:10.

Epidian 57 is an epoxy resin which is created by adding to the base epoxy resin, Epidian 5, polyesters comprising hydroxyl groups (products of Organica—Sarzyna Chemical Works, Poland [107]. This type of epoxy resin is predominantly used for preparing the adhesive compounds for cold bonding. Z-1 curing agent is a high-boiling aromatic polyamine with mice odour. It is mainly used for curing small-particle resins. The excess of the curing agent relative to the stoichiometric quantity causes the worsening of properties of the material being cured due to the fact that it does not get completely bonded in the resin. On the other hand, the deficiency of the curing agent leads to lower reactivity and has a negative effect on properties of the material, as it leads to lower mechanical properties and chemical resistance of this material. The adhesive layer prepared from this type of epoxy adhesive compound is characterised by high shear and peel strength.

The epoxy adhesive compound was mixed for 3 min in a plastic container with a mixer to prevent air bubbles from occurring in the mixture. The mixer was then used to apply the adhesive to one of the surfaces of the adherend. The appropriate mass of epoxy resin and curing agent was determined with the ONYX OX-120 analytical balance with an accuracy of 0.0001 g.

Prior to application of the adhesive, the C45 steel sheet specimens were degreased with Loctite 7063 Cleaner Reiniger, a unary solvent formulation for cleaning and degreasing surfaces to be adhesive bonded. The degreasing operation involved applying this agent twice to the substrates and wiping them, and finally, applying this agent and leaving it to evaporate. After 3 min, the adhesive was applied to surfaces of the specimens of the adherend.

After that, the bonded joints were subjected to one-stage curing under a 0.018 MPa pressure for 7 days in room conditions: temperature—$23 \pm 2°C$ and humidity—$27 \pm 3\%$. Then all specimens of adhesive joints were subjected to control with respect to the assessment of their real lap length and joint thickness.

Strength tests

After curing process, the specimens of adhesive joints were subjected to strength tests using the ZWICK/ROELL Z150 materials testing machine according to the DIN EN 1465 standard [108]. The specimens were mounted in the wedge-screw grips of the testing machine. The bonded joints were then subjected to shear strength tensile testing.

5.7.2 Results

The test results were subjected to statistical analysis in order to compare the strength of adhesive joints of C45 steel where surface treatment involved two variants of milling at three different feed rates. During this analysis the extreme values in each measurement series were omitted in the process of results analysis.

Surface roughness parameters

Surface roughness parameters measurements were performed on C45 steel sheet both prior to and after milling using different methods of milling (Variant I and Variant II). The results of these measurements are presented in the tables:

- after degreasing—Table 5.12,
- after conventional milling (Variant I)—Table 5.13, and
- after HPC milling (Variant II)—Table 5.14.

The results (Table 5.12) indicate a slight roughness of the degreased surface compared to the surface after the milling operation.

Examining the data given in the Table 5.13, it can be observed that the R_a parameter at different feed rates does not differ to a significant degree, which points to small variations in roughness of the tested surfaces. The greatest differences can be observed in values of the mean spacing of profile irregularities. The highest mean spacing was produced for the surface machined with the HSS cutter at the feed rate set to f_z=0.30 mm/tooth. This can stem from the fact that the feed value was the highest of all applied rates of feed. The lower the feed rate applied is, the highest the surface roughness of the material being machined is produced.

Table 5.12 Selected surface roughness parameters for C45 steel sheet after degreasing

	Surface roughness parameters (μm)			
Material/surface treatment	R_a	R_z	R_t	R_p
C45 steel sheet/degreasing	0.058	0.292	0.755	0.146

Table 5.13 Selected surface roughness parameters for C45 steel sheet after conventional milling (Variant I)

Conventional milling (Variant I)										
f_z (mm/tooth)	Surface roughness parameters (μm)									
	R_a	R_q	R_z	R_{max}	R_p	R_{pk}	R_k	R_{vk}	R_t	S_m
0.05	3.88	4.97	25.18	34.18	14.73	6.12	12.51	4.85	34.43	146
0.15	3.42	4.33	22.45	28.45	12.71	4.93	10.85	3.92	29.53	140
0.30	3.70	3.70	23.76	29.75	13.41	5.29	11.32	5.94	31.56	179

Table 5.14 Selected surface roughness parameters for C45 steel sheet after HPC milling (Variant II)

HPC milling (Variant II)										
f_z (mm/ tooth)	**Surface roughness parameters (μm)**									
	R_a	R_q	R_z	R_{max}	R_p	R_{pk}	R_k	R_{vk}	R_t	S_m
0.05	1.08	1.53	8.20	12.63	4.62	2.60	2.98	2.49	13.27	390
0.15	1.39	1.81	7.80	13.96	4.50	1.86	4.12	2.70	14.20	191
0.30	3.20	3.76	13.52	15.05	7.71	4.28	7.80	1.86	15.18	418

In the Varian II of machining (HPC milling—Table 5.14), it can be noticed considerable differences in the R_a parameter depending on the feed rate applied. The highest arithmetic mean deviation of the roughness profile is observed after milling at a feed rate of 0.30 mm/tooth. As for the other two machining variants, this parameter does not differ to any significant extent. Similar relationships were observed for other parameters.

Comparison of selected surface roughness parameters after milling methods

The comparison of the selected surface roughness parameters after two variants of milling: conventional (Variant I) and HPC milling (Variant II) was presented in Figs 5.5–5.7.

The highest mean value of R_a (Fig. 5.5) was obtained after conventional milling (Variant I)—in machining with the HSS cutter at the feed rate set to $f_z = 0.05$ mm/tooth, while the lowest R_a was obtained for the same feed rate in HPC milling (Variant II). The R_a parameter describes the arithmetic average deviation of the roughness profile.

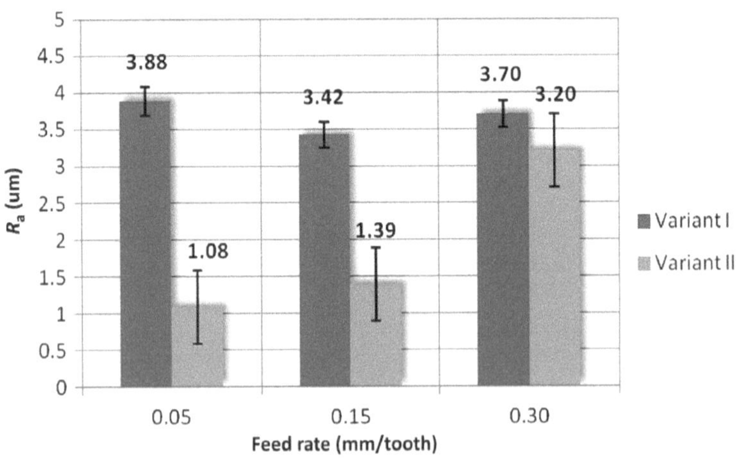

Fig. 5.5 R_a surface roughness parameter after two variants of milling.

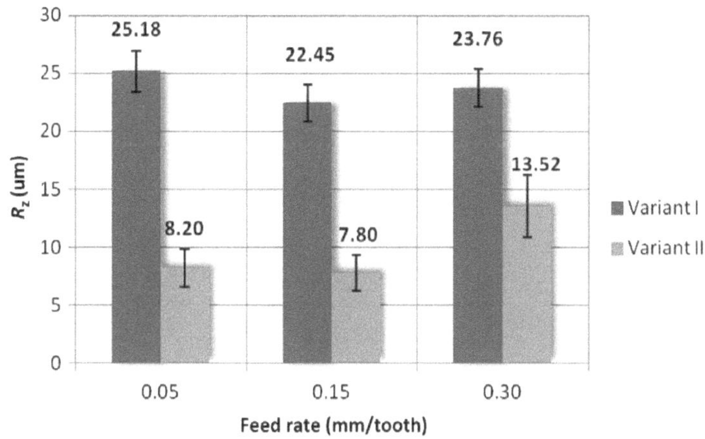

Fig. 5.6 R_z surface roughness parameter after two variants of milling.

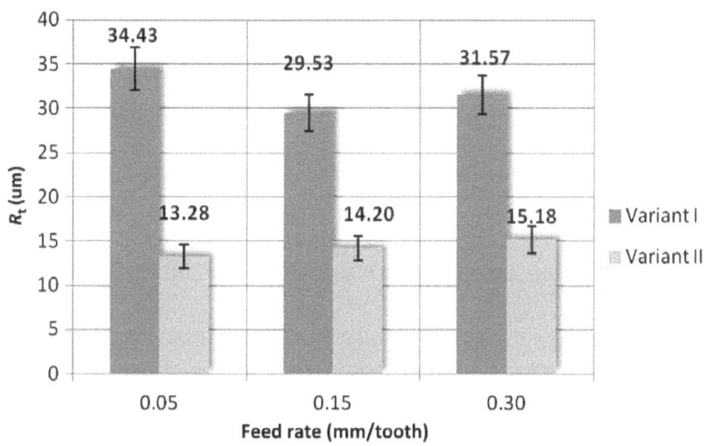

Fig. 5.7 R_t surface roughness parameter after two variants of milling.

The other two charts compare the R_z (Fig. 5.6) and R_t (Fig. 5.7) parameters with respect to two variants of milling with the two end mill types. These parameters denote the arithmetic average of the five highest peaks above the mean line reduced by the five lowest valleys below the mean line and the total profile height, respectively.

The above comparisons (Figs 5.6 and 5.7) demonstrate that the surface roughness parameters obtained after surface machining are directly proportionate. With increasing the feed rate, the parameters increase after milling by indexed cutter and decrease when the machining is done using the HSS cutter. The highest total profile roughness height for each feed variant was obtained after traditional milling with the HSS cutter (Variant I). After HPC milling, the value of R_t is approximately half lower (Variant II).

The other parameters with respect to two variants of milling were presented in Fig. 5.8.

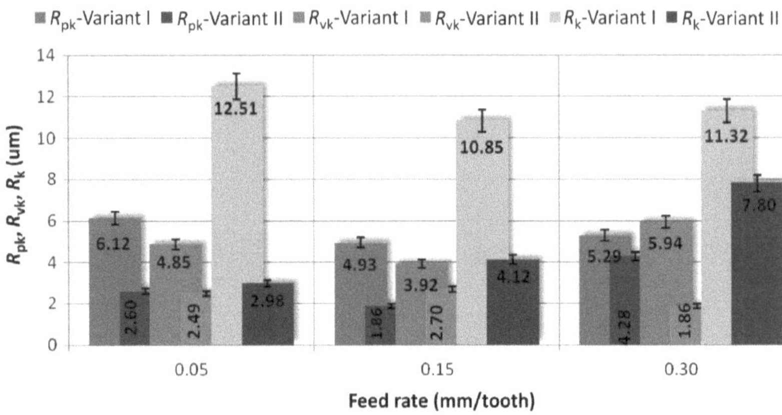

Fig. 5.8 R_{pk}, R_{vk}, and R_k surface roughness parameters after two variants of milling.

The reduced peak height R_{pk} describes the behaviour of sliding and running surfaces in abrasion. A low R_{pk} means high resistance to abrasive wear. The reduced valley depth R_{vk} defines oil retention capacity of the surface; e.g. lubricated surfaces are required to have a high R_{vk}. It might seem then that this parameter could be applied to determine the degree of roughness depth filling by an adhesive in making adhesive joints. However, due to differences in viscosity of adhesives and lubricants, it is the width of valleys that plays a significant role here. In lots of cases of the lowest R_{pk}, R_{vk}, and R_k parameters were obtained at the feed rate set to $f_z = 0.15$ mm/tooth. The highest parameters (Fig. 5.8) were generated after Variant I (traditional milling).

Adhesive joint strength—Conventional milling (Variant I) and HPC milling (Variant II)

A comparison of the shear strengths (mean values) of C45 steel adhesive joints after two variants of milling: conventional milling (Variant I) and HPC milling (Variant II) is given in Fig. 5.9.

Considering the results of the mean values of shear strength at different feed rates and at the constant cutting speed, V_c set to 75 m/min, in milling with the HSS cutter NFPaφ20—conventional milling, the following observations were formulated:

- The highest adhesive joint strength was obtained at the feed rate set to $f_z = 0.05$ mm/tooth (Variant I-A)—the result is almost twice as high as that produced at other feed rates, and it ranges between 13.40 and 15.10 MPa.
- A slightly lower shear strength was noticed at the rate of feed set to $f_z = 0.15$ mm/tooth (Variant I-B), and it ranged from 7.86 to 9.08 MPa.
- The lowest shear strength was obtained when the feed rate was 0.30 mm/tooth (Variant I-C)—it ranged from 6.48 to 7.41 MPa. This value is 47% of the highest obtained shear strength.

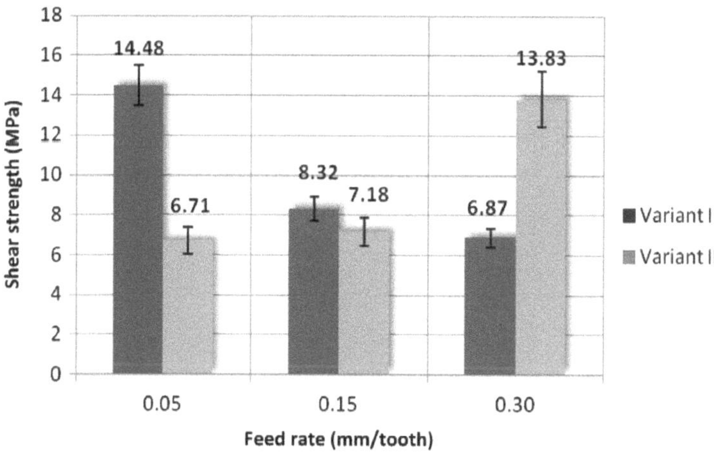

Fig. 5.9 Shear strength of adhesive joints after two variants of milling.

The results of shear strength of steel sheet adhesive joints where surface preparation involved HPC milling at different feed rates (Variant II) indicated that:

- The highest adhesive joint strength was obtained for the adherends milled at a feed rate of 0.30 mm/tooth (Variant II-C). The shear strength of the adhesive joints ranges from 12.10 to 14.70 MPa. This feed rate is the fastest of the feed rates applied in the tests.
- The other two machining variants produced similar strength results. The adhesive joints where surface preparation involved HPC milling at the feed rate f_z set to 0.05 mm/tooth (Variant II-A) have the shear strength between 5.68 and 7.13 MPa. Similar results were obtained in HPC milling where the feed rate was 0.15 mm/tooth (Variant II-B): here, the shear strength ranged from 6.92 to 7.50 MPa.

As can be seen from the above chart (Fig. 5.11), the highest shear strength of steel sheet adhesive joints is produced after conventional milling (Variant I) with the HSS cutter at the feed rate set to f_z=0.05 mm/tooth and it amounts to 14.48 MPa. Similar shear strength equal to 13.83 MPa was obtained for the adhesive joints where surface preparation involved HPC milling (Variant II) with the indexed cutter at a feed rate of 0.30 mm/tooth. The lowest shear strength was recorded for the adhesive joint of steel sheets where the surfaces were machined by the indexed cutter at feed rate f_z=0.05 mm/tooth, and by the high speed steel cutter at the feed rate set to 0.30 mm/tooth; they were 6.71 and 6.87 MPa, respectively.

As it was already mentioned, the above chart shows that the highest strength was obtained after HPC milling (Variant II) at the feed rate set to 0.30 mm/tooth, which is exactly the opposite to the results yielded in machining with the HSS cutter—conventional milling (Variant I).

Statistical analysis of the adhe.sive joint strength results

The statistical analysis results of shear strength of C45 adhesive joints where surface preparation involved milling at three different feed rates using two different types of variants of milling were presented in Table 5.15. The selection of an adequate

Table 5.15 Comparison of statistical analysis results of shear strength of C45 steel adhesive joints after application two variants of milling

Surface treatment method: Variant I/Variant II						
Feed rate (mm/tooth)	**F-test statistic**	**Critical value F**	**Conclusion**	**t-Test statistic**	**Critical value t**	**Conclusion**
$f_z=0.05$	1.15	9.28	$\delta_1^2=\delta_2^2$	−15.340	1.9432	$m_1 \neq m_2$
$f_z=0.15$	4.83	9.28	$\delta_3^2=\delta_4^2$	−3.830	1.9432	$m_3 \neq m_4$
$f_z=0.30$	9.20	9.28	$\delta_5^2=\delta_6^2$	11.240	1.9432	$m_5 \neq m_6$

where: δ, variance; m, mean.

analysis method depends on the truth of preliminary assumptions. The main criterion for selection of a suitable statistical model rests upon the assumption of equality of two population variances. The truth of this assumption was tested using Fisher's test [109, 110].

The results of the Student's t-test reveal that the mean shear strength of the tested adhesive joints is significantly dependent on the applied feed rates. In every case, the null hypothesis about the equality of means had to be rejected in favour of an alternative hypothesis.

Influence of the surface roughness parameters on the adhesive joint strength

The analysis was performed based on the results of strength testing and surface roughness parameters of adherends surface after two variants of milling. The charts given in Figs 5.10–5.12 compare the shear strength results with selected surface roughness parameters.

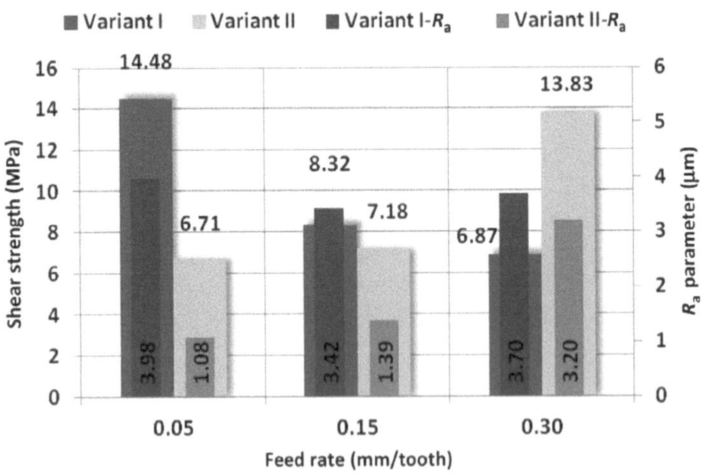

Fig. 5.10 Shear strength of steel sheet adhesive joints vs R_a surface roughness parameter after two variants milling.

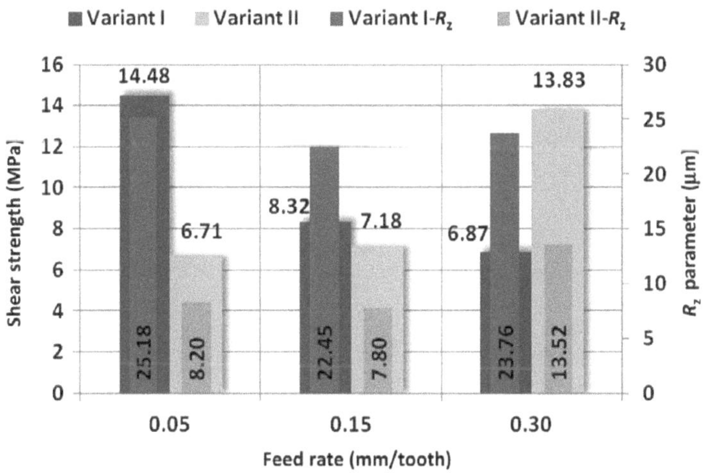

Fig. 5.11 Shear strength of steel sheet adhesive joints vs R_z surface roughness parameter after two variants milling.

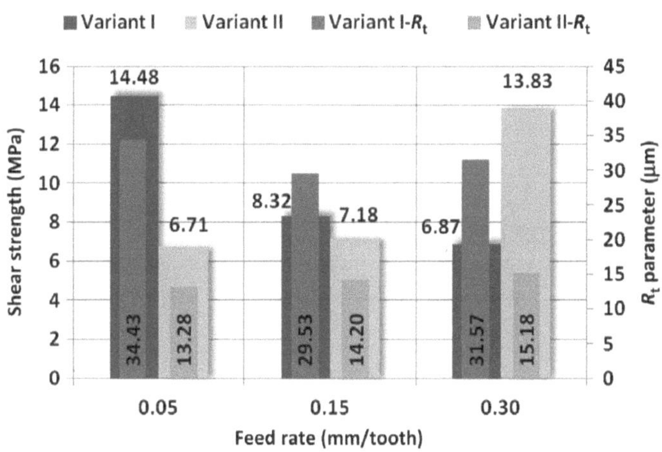

Fig. 5.12 Shear strength of steel sheet adhesive joints vs R_t surface roughness parameter after two variants milling.

The above charts (Figs 5.10–5.12) demonstrate that the shear strength of steel sheet adhesive joints where surface treatment is involved; the higher the use of HPC milling (Variant II) is, higher the roughness parameters are; in this particular case, the shear strength is most proportionate to the values of R_a.

The results of shear strength of steel sheet adhesive joints after surface treatment by conventional milling (Variant I) with the HSS cutter and the surface roughness parameters obtained after this surface treatment reveal that the application of the highest roughness parameters results in the highest shear strength.

5.7.3 Conclusions

The investigation involved examining the effects of the type of milling methods, conventional milling by a high speed steel cutter (Variant I), HPC milling by an indexed cutter (Variant II), and also the feed rate. Based on the results and survey of the available literature, the following conclusions have been drawn:

- For every applied variant of technological parameters, machining with the HSS cutter (conventional milling) produces specimens with more similar surface roughness than HPC milling (Variant II). This, however, did not result in a proportionate increase in strength of adhesive joints.
- Surface treatment for adhesive bonding by HPC milling with a indexed cutter type R390-020B20-11L (Variant II) yields the best shear strength results for the feed rate f_z set to 0.30 mm/tooth, whereas the application of the high speed steel cutter NFPaφ20 (conventional milling) produces the highest shear strength at the feed rate set to $f_z = 0.05$ mm/tooth.
- The highest surface roughness is produced for exactly the same milling variants which exhibited the highest adhesive joint strength, which points to a certain relationship between surface roughness and adhesive joint strength.
- It can be observed that the application of a lower feed rate in conventional milling produces better strength results; in HPC milling, however, the highest the feed rate is, the highest the strength of adhesive joints.

The statistical analysis of the results of mean shear strength of adhesive joints reveals that the application of the same feed rate for different milling methods and cutter types does not produce the same strength. In all cases, the statistical analysis results show significant differences at the adopted significance level $\alpha = 0.05$.

5.8 Experimental test—Sandblasting

The information in this subchapter was prepared based on article [30].

5.8.1 Methods

Materials and specimen

The tests were conducted on samples made of one type of carbon steel sheet: C45 (EN 10083-2/1.0503) [111]. C45 is a medium carbon steel and it is used when greater strength and hardness is desired. Selected mechanical properties were presented in Table 5.15. The main chemical composition of C45 steel is: carbon (C) from 0.42% to 0.50%, manganese (Mn) from 0.50% to 0.80%, sulphur (S) max 0.025%, phosphorus (P) max 0.025%, and silicon (Si) from 0.15% to 0.35% [112].

Surface treatment

The C45 steel specimens were subjected to sandblasting as a surface treatment. The variables applied in this mechanical treatment operation were: three different types of abrasive material and three values of pressure (1×10^5 Pa, 2×10^5 Pa, and 4×10^5 Pa).

General characteristics of the applied abrasive materials are given in Table 5.16. The time of surface treatment in all variants of the process was maintained the same.

The sandblasting process was performed using the German sandblasting machine Clemco HS 200 P-1. The process was made at right angle to the substrate surface (to the horizontal) and at a distance of about 100 mm from the nozzle for the time of about 1 min (Fig. 5.13).

The applied variants of sandblasting are characterised in Table 5.17. For comparative aspects, the analysis of surface geometrical structure, adhesive properties, and

Table 5.16 Characteristic of abrasive materials

Symbol	Abrasive material	Granulation (μm)
A-BA	Brown fused alumina	150–180
C-WA	White fused alumina	1400–1700
D-G	Glass beads	70–110

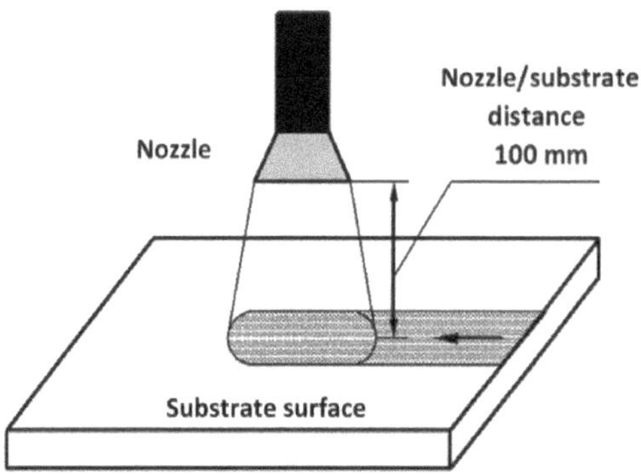

Fig. 5.13 Schematic of distance between nozzle and substrate surface during sandblasting process.

Table 5.17 Surface treatment variants

No.	Surface treatment variant	Pressure (Pa)	Designation
1	Reference samples, without sandblasting	–	O
2	Sandblasting with A-BA abrasive	1×10^5	A-BA-1-1
3		2×10^5	A-BA-2-1
4		4×10^5	A-BA-4-1
5	Sandblasting with C-WA abrasive	1×10^5	C-WA-1-1
6		2×10^5	C-WA-2-1
7		4×10^5	C-WA-4-1
8	Sandblasting with D-G abrasive	1×10^5	D-G-1-1
9		2×10^5	D-G-2-1
10		4×10^5	D-G-4-1

strength was conducted also on reference specimens that were only degreased with an acetone-based degreaser, without sandblasting (designation—0).

Surface roughness parameters

Surface roughness parameters were measured using the Perthometer 2 profilometer and NanoFocus µscan AF2 in compliance with standard ISO 4278 requirements [104]. In the tests, we measured the following parameters: R_a, R_z, R_p, R_v, R_k, R_{pk}, R_{vk}, M_{r1}, and M_{r2}. Characteristics of the surface roughness parameters was presented in Table 5.18.

The material ratio curve provides reliable information about profile shape and even about the maximum valley depth of a profile. This curve can be used to describe variations in properties of a profile depending on the profile's depth. The parameters defined using the information provided by the material ratio curve include: R_k, R_{pk}, R_{vk}, M_{r1}, and M_{r2}. These parameters are usually sufficient to identify surfaces with the same R_a. Not only do they provide information about roughness depth but also roughness shape. Given this, e.g. it seems that the material ratio curve and parameters determined thereby can be used to analyse the suitability of a given surface for adhesive bonding. The above parameters are graphically illustrated in Fig. 5.14.

Contact angle and hysteresis

The contact angle can be examined by the contact angle of measuring liquids. The contact angle is used for determination the wettability, work of adhesion and surface

Table 5.18 Characteristics of surface roughness parameters [30, 104]

Parameters	Characteristics
R_a	Roughness average R_a is the arithmetic mean of the absolute values of the roughness profile ordinates
R_z	Mean roughness depth R_z is the arithmetic mean value of the single roughness depth R_{zi} of consecutive sampling lengths. Single roughness depth R_{zi} is the vertical distance between the highest peak and the deepest valley within a sampling length The overall height of the roughness profile R_z is the sum of the maximum peak height of the roughness profile R_p and the maximum valley depth of the roughness profile R_v within one sampling length
R_p	R_p is the maximum peak height of the roughness profile within one sampling length
R_v	R_v is the maximum valley depth of the roughness profile within one sampling length. Previously, the parameter symbol R_m was used in place of R_v
R_k	Core roughness depth R_k is the depth of the roughness core profile
R_{pk}	Reduced peak height R_{pk} is the mean height of the peaks protruding from the roughness core profile
R_{vk}	Reduced valley depth R_{vk} is the mean depth of the valleys protruding from the roughness core profile
M_{r1}	M_{r1} is the lowest material ratio of the roughness core profile
M_{r2}	M_{r2} is the highest material ratio of the roughness core profile

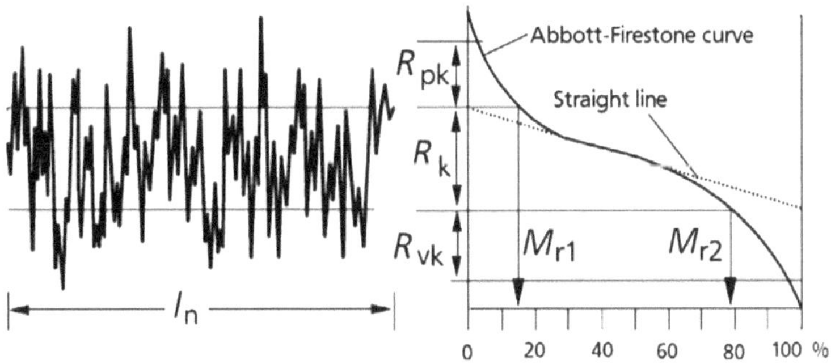

Fig. 5.14 Material ratio curve [30].

free energy. Distilled water (polar liquid) and diiodomethane (apolar liquid) are often used as reference liquids in analyses of the interactions of polar and apolar liquid with solid surfaces [61, 113–116].

Contact angle hysteresis [117–122] has been attributed to solid surface roughness, surface chemical heterogeneity, penetration of liquid into solid surface, swelling, and also surface reorientation of functional group. Surface roughness is not expected to result in contact angle hysteresis for smooth surfaces (roughness surfaces less than approximately 0.5 μm) [37]. Contact angle hysteresis H (1), exists when there is inequality of the advancing angle θ_a and receding contact angle θ_r [116, 118] and the contact angle hysteresis was calculated according to Eq. (5.1).

$$H \equiv \theta_a - \theta_r \tag{5.1}$$

The distilled water was used as a measuring liquid in the experimental test to determine the contact angle and the hysteresis.

Surface free energy

The Owens–Wendt [113, 121, 123] method for calculation of solid surface free energy was applied. The surface free energy of solid γ_S (Eq. 5.2) can be expressed by sum of the dispersive (γ_S^d) and polar (nondispersive) (γ_S^p) components [121–126]

$$\gamma_S = \gamma_S^d + \gamma_S^p \tag{5.2}$$

The measurements were taken using two measuring liquids: polar (distilled water) and apolar (diiodomethane), with known surface free energy and its polar and dispersive components. Values of the surface free energy γ_S and its components as determined with the above measuring liquids are given in Table 5.19.

The contact angle measurements were made with distilled water and diiodomethane taking an average of 8–12 drops with each type of liquid of each variant of treatment. Every sandblasting variant was tested on 3–5 specimens, with the number of measurements for each specimen ranging from 25 to 30 (for: 4–6 drops of each measuring liquid). The experimental details of contact angle measurement were presented

Table 5.19 Values of the surface free energy γ_L and its components determined using measuring liquids [61, 113, 124]

Surface free energy and its components (mJ/m^2)	**Measuring liquids**	
	H_2O	**CH_2J_2**
γ_L	72.8	50.8
γ_L^d	21.8	48.5
γ_L^p	51.0	2.3

in work [30]. Some test conditions of contact angle measurement are derived from the ASTM D5946 standard [127]. The surface free energy of steel was calculated only apparently because the contact angle values were used to SFE calculating not fulfilled the Young equation.

To calculate the surface free energy of steel, in terms of dispersive and polar components, (Eq. 5.2) was used. The components $\gamma_S{}^d$ and $\gamma_S{}^p$ of the examined materials may be calculated from the formulas presented in [123, 124] and in Chapter 7 (Eqs 7.2, 7.3).

5.8.2 Results

Surface roughness

Surface roughness was investigated by measuring selected surface roughness parameters, determining surface roughness profile and geometrical structure of surfaces of the tested materials. Surface roughness parameters formed by sandblasting (Table 5.20) were estimated from 3 locations on each surface of 5 samples of each variant of sandblasting (15 measurements of each variant of sandblasting).

Table 5.20 Selected surface roughness parameters for C45 steel sheet surface after degreasing and sandblasting (ISO 4287)

Variant of surface treatment						
	Surface roughness parameters (μm)					
Designation	R_a	R_q	R_z	R_p	R_v	R_t
O	0.086	0.117	0.703	0.348	0.355	1.510
A-BA-1-1	1.590	2.020	10.100	4.870	5.200	13.700
A-BA-2-1	1.540	1.940	10.010	5.080	5.060	12.300
A-BA-4-1	1.320	1.680	8.290	4.130	4.160	10.700
C-WA-1-1	1.160	1.490	8.030	3.530	4.500	11.300
C-WA-2-1	1.430	1.910	10.500	4.590	5.930	16.700
C-WA-4-1	1.500	1.980	10.200	5.210	5.010	17.900
D-G-1-1	0.827	1.030	5.060	2.660	2.400	6.580
D-G-2-1	0.789	0.959	4.380	2.370	2.020	5.270
D-G-4-1	0.677	0.824	3.700	1.780	1.920	4.810

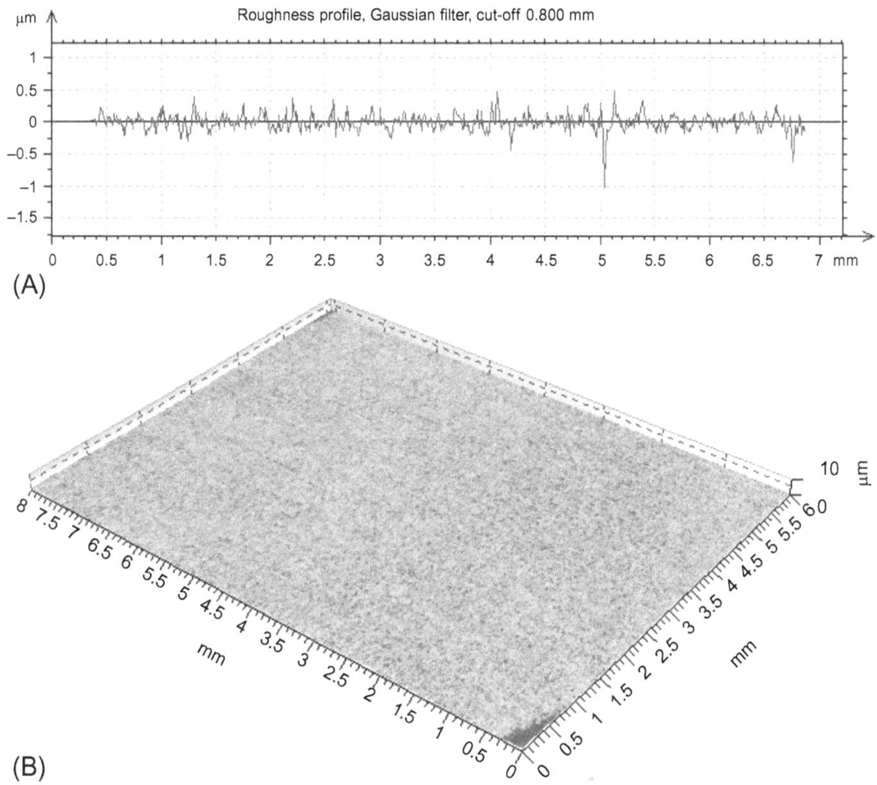

Fig. 5.15 Roughness profile (A) and surface (B) of C45 steel sheet after degreasing [30].

Examples of the roughness profile and view of steel sheet surfaces after degreasing are shown in Fig. 5.15.

The surface of a C45 steel sheet that was subjected to degreasing exhibits slight surface roughness ($R_a=0.086\,\mu m$, $R_z=0.665\,\mu m$) as well as a considerable uniformity of roughness across the entire area of the specimen.

Examples of the roughness profile of steel sheet surfaces following the application of three variants of sandblasting (Table 5.17) with abrasive A-BA (brown fused alumina) are illustrated in Fig. 5.16, while Fig. 5.17 shows the surface treated with sandblasting variant A-BA-2-1.

Examining the obtained roughness profiles (Fig. 5.16) and of the surface itself (Fig. 5.17), it can be observed that the value of pressure in sandblasting with abrasive A does not have any significant effect on the surface roughness profile.

The surface roughness and profiles of surface roughness after the application of the highest applied sandblasting pressure (4×10^5 Pa) and different abrasive materials (Tables 5.16 and 5.17) are compared in Fig. 5.18.

The results given in Figs 5.18 and 5.20 show a certain relationship between the application of a higher sandblasting pressure and surface roughness, depending on the

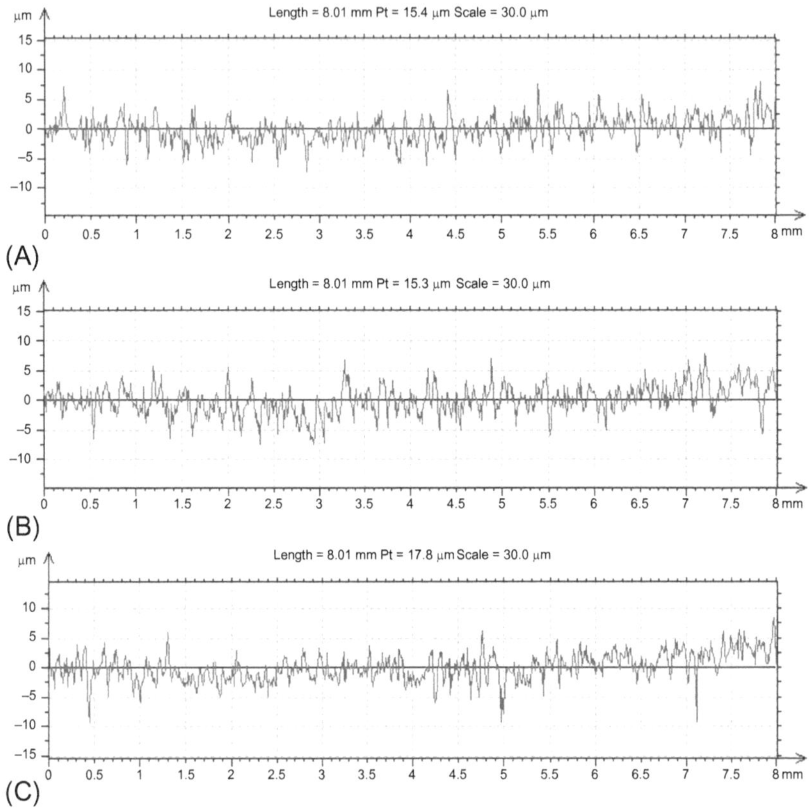

Fig. 5.16 Roughness profile of a C45 steel sheet surface after sandblasting: (A) A-BA-1-1, (B) A-BA-2-1, and (C) A-BA-4-1 [30].

type of abrasive used. The surfaces subjected to sandblasting with abrasive D-G (glass beads) have the smallest roughness (Fig. 5.20C).

The Bresson et al. [75] tested the sandblasting under various conditions: the effect of relative distance between the sandblasting nozzle and the aluminium surface ('far' and 'near'). They observed that increasing the period of sandblasting for 'far' configuration decrease dispersion of the results in terms of roughness measurement, but it does not modify the average significantly. For the 'near' configuration an arithmetic mean roughness of about $2.85 \pm 0.17\,\mu m$ was found and for 'far' configuration—$2.72 \pm 0.16\,\mu m$ was evaluated.

Contact angle and hysteresis

The values of advancing contact angle θ_a, receding contact angle θ_r and contact angle hysteresis H for surface of steel samples after sandblasting variants were listed in Table 5.21.

The hysteresis generally decreases with decreasing of the contact angles. The lowest water contact angle hysteresis H_w was obtained for reference variant O—without sandblasting—where only degreasing treatment was used ($H_w = 3.5$ degrees).

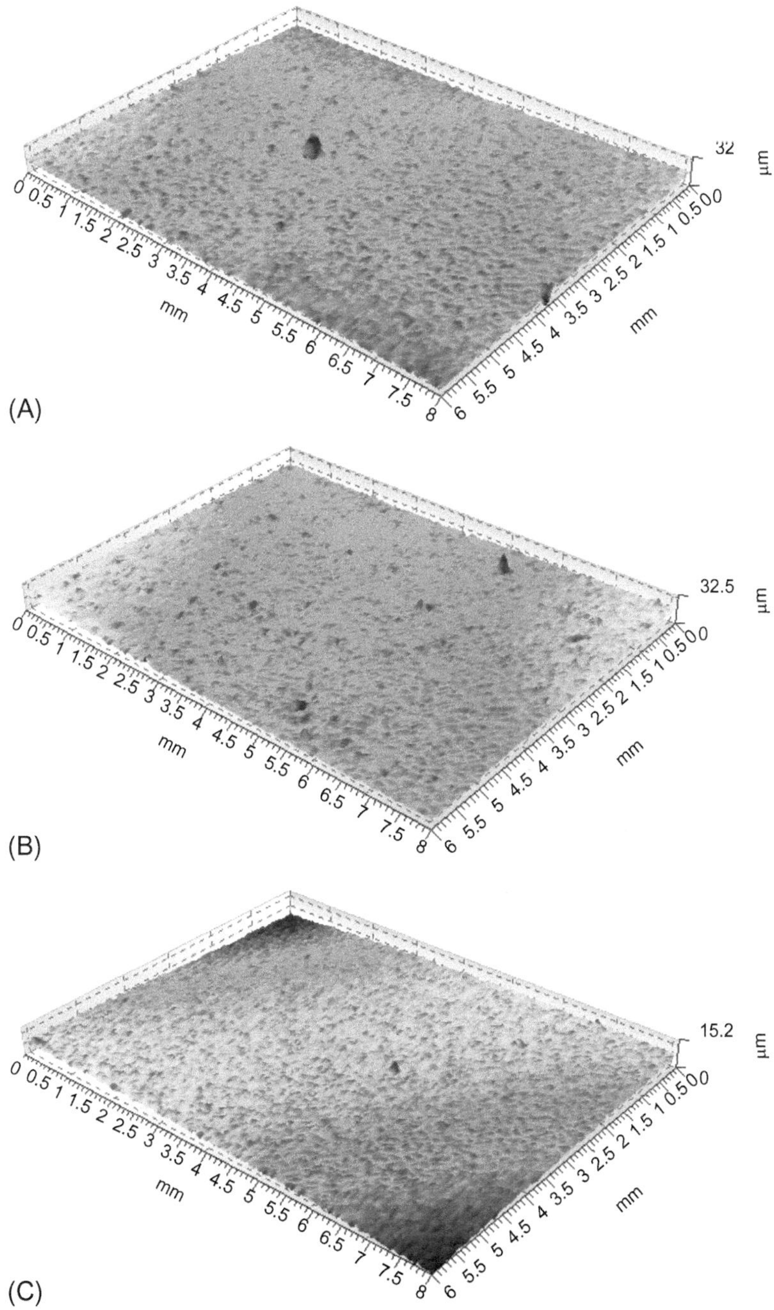

Fig. 5.17 C45 steel sheet surface after sandblasting: (A) A-BA-2-1 [30], (B) C-WA-2-1, and (C) D-G-2-1.

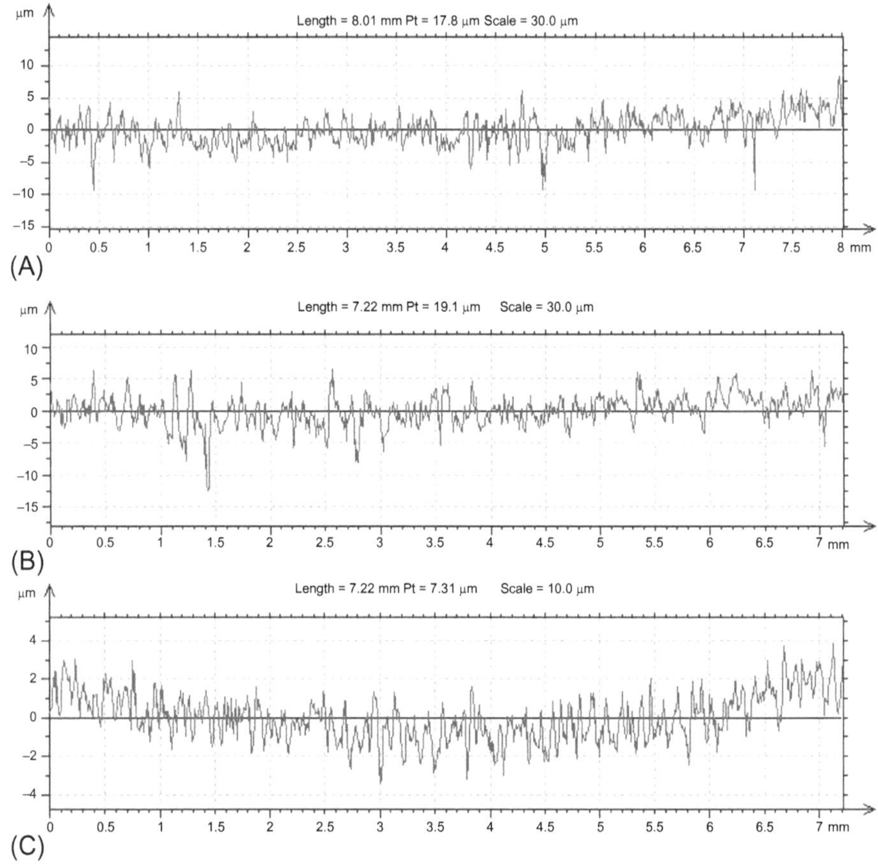

Fig. 5.18 Roughness profile of a C45 steel sheet surface after sandblasting: (A) A-BA-4-1, (B) C-WA-4-1, and (C) D-G-4-1 [30].

Table 5.21 Advancing contact angle θ_a, receding contact angle θ_r and contact angle hysteresis $\boldsymbol{H}_w$ for steel samples after sandblasting variants

Variant of surface treatment	**Measuring liquid—Distilled water**		
	θ_a degrees	**θ_r degrees**	**H_w degrees**
O	84.6	81.1	3.5
A-BA-1-1	105.5	88.9	16.6
A-BA-2-1	111.5	78.2	33.3
A-BA-4-1	107.9	100.0	7.9
C-WA-1-1	96.0	84.4	11.6
C-WA-2-1	96.8	84.1	12.8
C-WA-4-1	89.9	84.1	5.8
D-G-1-1	49.7	42.4	7.3
D-G-2-1	46.4	42.2	4.2
D-G-4-1	41.7	36.7	5.0

After sandblasting the lowest water contact angle hysteresis H_w was obtained in sandblasting variant D-G-2-1 (H_w=4.2 degrees) and in variants of sandblasting with abrasive D. The highest water contact angle hysteresis H_w was observed for the surfaces subjected to sandblasting with abrasive A—brown fused alumina (Table 5.21).

The abrasive material C was characterised by the largest size of particles, while abrasive D (glass beads) had the smallest used to sandblasting particles. Also, abrasives A and B are brown fused alumina with different sizes of abrasive particles. Given the above results, it can be observed that the size of abrasive material particles does not affect the contact angle and contact angle hysteresis H_w. Rather, this effect depends on the type of abrasive material used.

Surface free energy γ_S

Surface free energy and its components of the steel specimens, obtained from the mean contact angle measurement using distilled water and diiodomethane and Owens–Wendt method were presented in Table 5.22.

Analysing the results of surface free energy and its components (Table 5.22), it can be observed that the highest values were obtained in sandblasting with abrasive D (glass beads), while the lowest ones were produced by degreasing. Sandblasting with abrasive D also produced the highest value of the polar component of surface free energy.

It was found that the type of abrasive used is of more significance to surface treatment than pressure. The results of the experiments in Table 5.6 show that, in general, glass beads (abrasive D) display the higher surface free energy values than brown or white fused alumina. Moreover white fused alumina displays the higher surface free energy values than brown fused alumina. The Harris and Beevers [10] also observed during grit blasting that white grit compositions usually cause the higher values of surface free energy than brown. The surface free energy values after grit blasting are from 45 to 70 mJ/m^2 for mild steel and 43–68 mJ/m^2 for aluminium alloy.

Table 5.22 Surface free energy γ_S and its components of a steel sheet surface following different surface treatments [30]

Variant of surface treatment	Surface free energy and its components (mJ/m^2)		
	γ_S	γ_S^d	γ_S^p
O	45.5 ± 1.4	43.5	2.0
A-BA-1-1	49.1 ± 0.5	49.1	0.0
A-BA-2-1	49.2 ± 1.4	49.2	0.0
A-BA-4-1	49.6 ± 0.1	48.7	1.1
C-WA-1-1	50.3 ± 0.8	50.1	0.2
C-WA-2-1	50.2 ± 1.0	50.0	0.2
C-WA-4-1	50.1 ± 0.9	49.4	0.7
D-G-1-1	63.3 ± 2.7	45.3	18.0
D-G-2-1	63.5 ± 1.9	44.0	19.4
D-G-4-1	66.5 ± 2.2	44.7	21.8

The comparison of surface free energy and its components obtained after the application of different surface treatment variants are given in Figs 5.19–5.21.

The results of surface free energy and its components (Figs 5.19–5.21) after the application of all investigated surface treatment variants demonstrate that the lowest γ_S is exhibited by the surfaces that were only degreased at 45.5 mJ/m^2 (Table 5.22, O variant). In contrast, the highest values were obtained in sandblasting with abrasive D. This variant of surface treatment also yields the highest ratios of polar component of the total surface free energy, these ratios ranging from 28% to 33% of the surface free energy. This regularity is observed for every pressure variant applied in sandblasting.

Surface treatment where the same abrasive material was applied produced similar values of γ_S. Nonetheless, it was found that for every type of abrasive material used, the variant wherein the highest pressure was applied (Fig. 5.21) exhibited the highest ratio of the polar component γ_S^p, compared to the operations in which other pressures were applied.

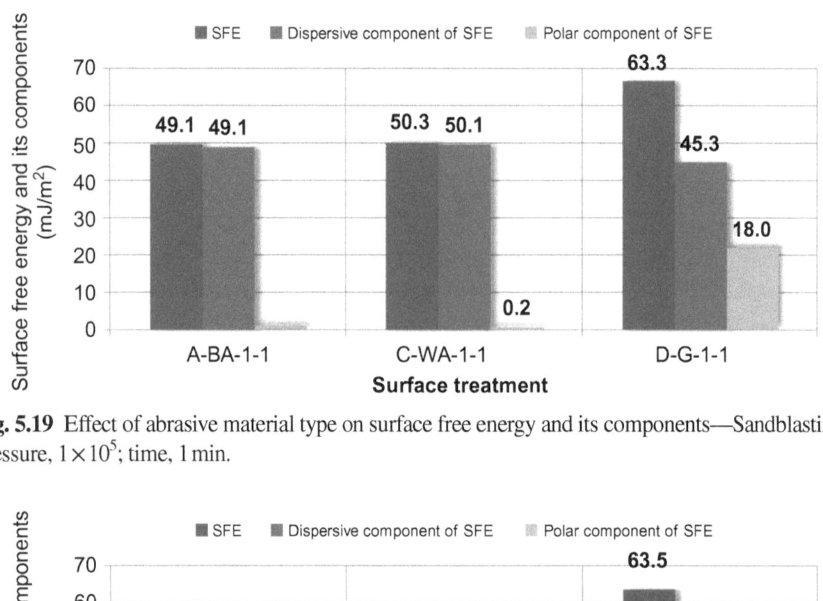

Fig. 5.19 Effect of abrasive material type on surface free energy and its components—Sandblasting: pressure, 1×10^5; time, 1 min.

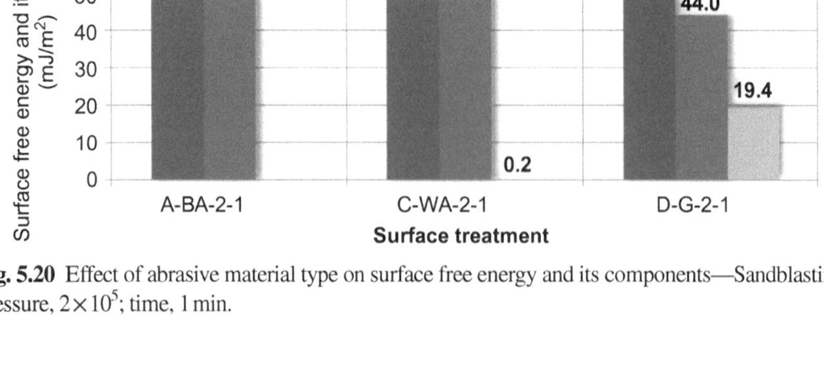

Fig. 5.20 Effect of abrasive material type on surface free energy and its components—Sandblasting: pressure, 2×10^5; time, 1 min.

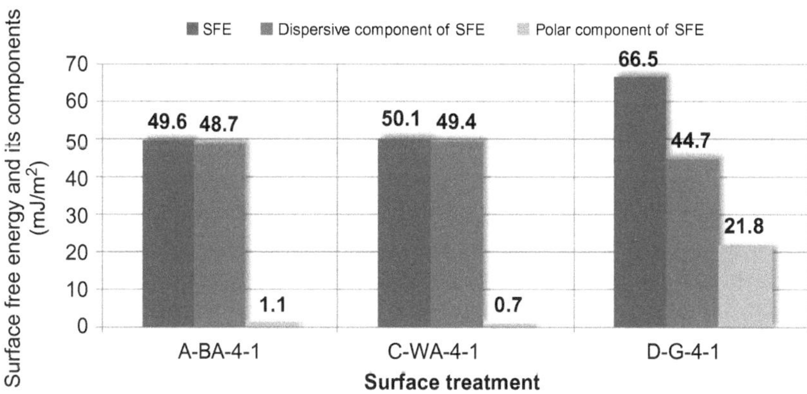

Fig. 5.21 Effect of abrasive material type on surface free energy and its components—Sandblasting: pressure, 4×10^5; time, 1 min.

The Harris and Beevers [10] also observed during grit blasting that, in general, dispersive component of surface free energy is higher than the polar component and the brown and white grit composition produced different values of surface free energy. Also the brown grit composition has a lower surface free energy with higher dispersive and lower polar components than other grit sandblasting medium. In this study (Figs 5.19 and 5.21) the similar observations were obtained.

It can be claimed that this type of abrasive material (glass beads of particular dimensions) helps produce surfaces with considerably different energy parameters than is the case with sandblasting variants based on the use of other types of abrasive material.

Comparison surface free energy and surface roughness parameters

The results of correlation between surface free energy and its components and selected surface roughness parameters (R_a, R_z, R_p, R_v) presented in work [30] show that in the all variants of sandblasting it can be noticed in the following relation: an increase in the surface roughness parameters decreases the polar component of surface free energy. In some cases the closer the coefficient is to either −1, the stronger the correlation between the variables.

Harris and Beevers [10] presented that, in general, smoother grit-blasted surfaces display higher surface energy values. Hitchcock et al. [128] also observed that, within certain limits, roughening a substrate usually causes its wettability to decrease. Some researcher suggested that the peaks, ridges, and asperities from barriers restrict the spreading of the droplet. Shanahan [129] reported that heterogeneities such as surface roughness could produce a positive influence on wetting characteristics of a drop by ‘pulling’ the liquid mass towards the abnormality but the negative effect of peaks and asperities on wetting characteristics was observed. Bikerman [62] suggested that wetting was independent of roughness on grooved and ridged surfaces and implied that as a drop front moves a ridge the true contact remains constant. Packham [20] reported that the extent of contact angle between a liquid and a rough surface depends on the details of the topography. For example, penetration into an idealized pore occurs until the back pressure of trapped air equals the capillary driving pressure.

Several researches have compared the various surface roughness parameters with surface free energy or wetting or contact angle. Although Extrand [130] suggested that surface roughness was relatively unimportant during contact angle measurement. The various surface roughness parameters were used to characterize the surface topography, e.g. to bonding also. Harris and Beevers [10] presented the effect of surface roughness parameter R_q on surface free energy. The Bresson et al. [75] used the arithmetic mean roughness R_a. Shahid and Hasim [14] studied the surface roughness parameters R_a and R_{lo} after polishing and grit blasting. Jiang et al. [8] measured the R_a after sandblasting titanium. Moghadamzadeh et al. [58] used the R_a and R_z (also S_a and S_z) surface roughness parameters to characterize the surface topography after surface treatment. The same parameters were measured by da Silva et al. [40].

5.8.3 Conclusions

Summing up, it can be observed that the application of sandblasting produces surfaces with different roughness parameters and surface properties for adhesion. The above factors play a significant role in the treatment of surfaces which are to be bonded.

In the experiments, we applied three types of abrasive material and variable pressures. The experiments demonstrated that both surface roughness and adhesive properties are more affected by the type of abrasive material used rather than by variations in pressure. It is commonly known that surface free energy and surface characteristic for adhesion are dependent on a complex interaction between surface texture and chemical components. But other factors such a sandblasting media and its particle size influenced on the adhesion characteristics of surface.

The application of abrasive material D-G (glass beads) leads to obtaining the highest values of surface free energy as well as a much higher value of the polar component than is the case with other surface treatment variants. What is more, in sandblasting with abrasive material D-G, the roughness parameters R_a, R_z, R_p, and R_v are the lowest in the tested range. Also, this surface treatment variant reveals a relationship between the polar component and roughness parameters.

The majority of sandblasting variants applied demonstrate that, irrespective of the type of abrasive used, the highest correlation occurs between γ_S^d and R_p, i.e. the maximum peak height of the roughness profile. It was found that an increase in R_p leads to an increase in the dispersive components of surface free energy. It was also observed that an increase in surface roughness leads to a proportional increase in the dispersive component of surface free energy, while the polar component of surface free energy decreases.

It was noted that the various surface roughness parameters could be compared with surface free energy or wetting or contact angle. However, the surface roughness parameters which provide information about roughness depth and also roughness shape can be used to analyse the suitability of a surface for adhesive bonding.

References

[1] R.D. Adams (Ed.), Adhesive Bonding. Science, Technology and Applications, Woodhead Publishing Limited, Cambridge, 2010.

[2] R.D. Adams, J. Comyn, W.C. Wake, Structural Adhesive Joints in Engineering, second ed., Chapman & Hall, London, 1997.

[3] C. Bockenheimer, B. Valeske, W. Possart, Network structure in epoxy aluminium bonds after mechanical treatment, Int. J. Adhes. Adhes. 22 (2002) 349–356.

[4] S. Ebnesajjad, C. Ebnesajjad, Surface Treatment of Materials for Adhesive Bonding, second ed., William Andrew Inc, Norwich, NY, 2013.

[5] S. Ebnesajjad, A.H. Landrock, Adhesives Technology Handbook, third ed., William Andrew, London, 2009.

[6] C.V. Cagle, Handbook of adhesive bonding, McGraw-Hill, New York, 1973.

[7] J. Comyn, Adhesion Science, The Royal Society of Chemistry, Cambridge, 1997.

[8] X.P. Jiang, X.Y. Wang, J.X. Li, D.Y. Li, C.-S. Man, M.J. Shepard, T. Zhai, Enhancement of fatigue and corrosion properties of pure Ti by sandblasting, Mater. Sci. Eng. A 429 (2006) 30–35.

[9] R.D. Adams, J.A. Harris, The influence of local geometry on the strength of adhesive joints, Int. J. Adhes. Adhes. 7 (1987) 69–80.

[10] A.F. Harris, A. Beevers, The effect of grit-blasting on the surface properties for adhesion, Int. J. Adhes. Adhes. 19 (1999) 445–452.

[11] S.G. Prolongo, A. Ureña, Effect of surface pre-treatment on the adhesive strength of epoxy-aluminium joints, Int. J. Adhes. Adhes. 29 (2009) 23–31.

[12] L.F.M. da Silva, R.J.C. Carbas, G.W. Critchlow, M.A.V. Figueiredo, K. Brown, Effect of material, geometry, surface treatment and environment on the shear strength of single lap joints, Int. J. Adhes. Adhes. 29 (2009) 621–632.

[13] N. Brack, A.N. Rider, The influence of mechanical and chemical treatments on the environmental resistance of epoxy adhesive bonds to titanium, Int. J. Adhes. Adhes. 44 (2014) 20–27.

[14] M. Shahid, S.A. Hashim, Effect of surface roughness on the strength of cleavage joint, Int. J. Adhes. Adhes. 22 (2002) 235–244.

[15] Y. Gilbert, G. Verchery, Influence of surface roughness on mechanical properties of joints, in: K.L. Mittal (Ed.), Adhesive joints formation, characteristics, and testing, Plenum Press, New York, 1982, pp. 69–84.

[16] J.M. Arenas Reina, J.J. Narbón Prìeto, C. Alía García, Influence of the surface finish on the shear strength of structural adhesive joints and application criteria in manufacturing process, J. Adhes. 85 (2009) 324–340.

[17] M.S. Islam, L. Tong, P.J. Falzon, Influence of metal surface preparation on its surface profile, contact angle, surface energy and adhesion with glass fibre prepreg, Int. J. Adhes. Adhes. 51 (2014) 32–41.

[18] A. Rudawska, Selected aspects of the effect of mechanical treatment on surface roughness and adhesive joint strength of steel sheets, Int. J. Adhes. Adhes. 50 (2014) 235–243.

[19] R.D. Hazlett, On surface roughness effects in wetting phenomena, in: K.L. Mittal (Ed.), Contact Angle, Wettability and Adhesion, VSP, Zeist, 1993, pp. 173–181.

[20] D.E. Packham, Surface energy, surface topography and adhesion, Int. J. Adhes. Adhes. 23 (2003) 437–448.

[21] J.M. Sykes, Surface treatment for steel, in: D.M. Brewis (Ed.), Surface Analysis and Pretreatment of Plastics and Metals, Applied Science Publishers, London, 1982, pp. 153–174.

[22] K.M. Hay, M.I. Dragila, J. Liburdy, Theoretical model for the wetting of a rough surface, J. Colloid Interface Sci. 325 (2008) 472–477.

[23] R.D. Hazlett, On surface roughness effects in wetting phenomena, J. Adhes. Sci. Technol. 6 (1992) 625–633.

[24] K. Uehara, M. Sakurai, Bonding strength of adhesives and surfaces roughness of joined parts, J. Mater. Process. Technol. 127 (2002) 178–181.

[25] A. Rudawska, Surface free energy and 7075 aluminium bonded joint strength following degreasing only and without any prior treatment, J. Adhes. Sci. Technol. 26 (2012) 1233–1247.

[26] G.W. Critchlow, D.M. Brewis, Review of surface pretreatments for alluminium alloys, Int. J. Adhes. Adhes. 16 (1996) 255–275.

[27] G. Rotella, M. Alfano, T. Schiefer, I. Jansen, Evaluation of mechanical and laser surface pre-treatments on the strength of adhesive bonded steel joints for the automotive industry, J. Adhes. Sci. Technol. 30 (2016) 747–758.

[28] R.D. Bhatt, V. Radhakrishnan, Evaluation of some machining process as adherend surface treatments in bonding aluminium, J. Adhes. Sci. Technol. 3 (1989) 383–396.

[29] L. Kumosa, D. Armentrout, M. Kumosa, The effect of sandblasting on the initiation of stress corrosion cracking in unidirectional E-glass/polymer composites used in high voltage composite (non-ceramic) insulators, Compos. Sci. Technol. 62 (2002) 1999–2015.

[30] A. Rudawska, I. Danczak, M. Müller, P. Valasek, The effect of sandblasting on surface properties for adhesion, Int. J. Adhes. Adhes. 70 (2016) 176–190.

[31] P. Hass, O. Kläusler, S. Schlegel, et al., Effects of mechanical and chemical surface preparation on adhesively bonded wooden joints, Int. J. Adhes. Adhes. 51 (2014) 95–102.

[32] A. Rudawska, I. Miturska, Wpływ operacji odtłuszczania i obróbki mechanicznej na wytrzymałość połączeń klejowych (The influence of degreasing and machining treatment on the adhesive joints strength), Przetwórstwo Tworzyw 2 (2018) 38–46 (in Polish).

[33] A. Rudawska, N. Ściuba, D. Stančeková, Wpływ obróbki mechanicznej na wytrzymałość połączeń klejowych blach ze stali odpornej na korozję (The influence of the mechanical treatmnet on the stainless steel sheets bonded joints strength), Technologia i Automatyzacja Montażu 2 (2016) 55–59 (in Polish).

[34] A. Rudawska, I. Miturska, Statystyczna interpretacja wyników pomiarów wytrzymałości połączeń adhezyjnych wybranych materiałów konstrukcyjnych (Statistical interpretation of measurement results of adhesive joint strength of selected engineering materials), Mechanik 11 (2016) 1620–1623 (in Polish).

[35] G.W. Critchlow, P.W. Webb, C.J. Tremlett, K. Brown, Chemical conversion coatings for structural adhesive bonding of plain carbon steels, Int. J. Adhes. Adhes. 20 (2000) 113–122.

[36] P. Molitor, V. Barron, T. Young, Surface treatment of titanium for adhesive bonding to polymer composites: a review, Int. J. Adhes. Adhes. 21 (2001) 129–136.

[37] F.T.M. Ariff, N. Yunus, M.R. Baig, The effect of mechanical roughening and chemical treatment on shear bond strength of urethane dimethacrylate denture base resin, Int. J. Adhes. Adhes. 31 (2011) 112–118.

[38] C. Spadaro, C. Dispenza, C. Sunseri, The influence of the nature of the surface oxide on the adhesive fracture energy of aluminium- bonded joints as measured by T-peel tests, Int. J. Adhes. Adhes. 28 (2008) 211–221.

[39] K. Zaleski, A study on the properties of surface - active fluids used in burnishing and shot peening processes, Adv. Sci. Technol. Res. 31 (2016) 235–239.

[40] L.F.M. da Silva, N.M.A.J. Ferreira, V. Richter-Trummer, E.A.S. Marques, Effect of grooves on the strength of adhesively bonded joints, Int. J. Adhes. Adhes. 30 (2010) 735–743.
[41] A. Rudawska, M. Reszka, T. Warda, I. Miturska, J. Szabelski, D. Stancekova, A. Skoczylas, Milling as a method of surface pre-treatment of steel for adhesive bonding, J. Adhes. Sci. Technol. 23 (2016) 2619–2636.
[42] A.N. Rider, D.R. Arnott, Boiling water and silane pre-treatment of aluminium alloys for durable adhesive bonding, Int. J. Adhes. Adhes. 20 (2000) 209–220.
[43] M.-H. Kang, J.-H. Choi, J.-H. Kweon, Fatigue life evaluation and crack detection of the adhesive joint with carbon nanotubes, Compos. Struct. 108 (2014) 417–422.
[44] R. Kahramanm, M. Sunar, B. Yilbas, Influence of adhesive thickness and filler content on the mechanical performance of aluminum single-lap joints bonded with aluminium powder filed epoxy adhesive, J. Mater. Process. Technol. 295 (2008) 183–189.
[45] L.F.M. da Silva, M.J.C.Q. Lopes, Joint strength optimization by the mixed-adhesive technique, Int. J. Adhes. Adhes. 29 (2009) 509–514.
[46] A. Rudawska, D. Nowicka, N. Čuboňowǎ, Wytrzymałość połączeń klejowych blach stalowych po obróbce mechanicznej (The bonded joint strength pf steel sheets aft re mechanical treatment), Technologia i Automatyzacja Montażu 1 (2017) 43–47 (in Polish).
[47] Y. Boutar, S. Naïmi, S. Mezlini, M.B.S. Ali, Effect of surface treatment on the shear strength of aluminium adhesive single-lap joints for automotive applications, Int. J. Adhes. Adhes. 67 (2016) 38–43.
[48] S. Hamdan, J.R.G. Evans, The surface treatment and adhesive bonding of polyetheretherketone. Part I. Adhesive joint strength, J. Adhes. Sci. Technol. 1 (1987) 281–289.
[49] J.D. Minford, Handbook of Aluminum Bonding Technology and Data, CRC Press, London, 1993.
[50] S.G. Prolongo, G. del Rosario, A. Ureña, Study of the effect of substrate roughness on adhesive joints by SEM image analysis, J. Adhes. Sci. Technol. 20 (5) (2006) 457–470.
[51] B. Golaz, V. Michaud, S. Lavanchy, J.-A.E. Månson, Design and durability of titanium adhesive joints for marine applications, Int. J. Adhes. Adhes. 45 (2013) 150–157.
[52] J.M. Schuster, C.E. Schvezov, M.R. Resenberger, Analysis of the results of surface free energy measurement of Ti6Al4V by different methods, Procedia Mater. Sci. 8 (2015) 732–741.
[53] J.R.J. Wingfield, Treatment of composite surfaces for adhesive bonding, Int. J. Adhes. Adhes. 13 (1993) 151–156.
[54] W. Brockmann, P.L. Geiß, J. Klingen, B. Schröder, Adhesive bonding. Materials Applications and Technology, Wiley-VCH Press, Weinheim, 2009.
[55] A. Rudawska, M. Łukasiewicz, Wpływ obróbki mechanicznej na wytrzymałość połączeń klejowych wybranych materiałów konstrukcyjnych (The influence of the mechanical treatment on bonding joint strength of chosen construction al materials), Przegląd Spawalnictwa 8 (2008) 27–30 (in polish).
[56] L. Wagner, Mechanical surface treatments on titanium, aluminium and magnesium alloys, Mater. Sci. Eng. A 263 (1999) 2010–2216.
[57] M.J. Troughton, Handbook of Plastics Joining: A Practical Guide, second ed., William Andrew, New York, 2008.
[58] H. Moghadamzadeh, H. Rahimi, M. Asadollahzadeh, A.R. Hemmati, Surface treatment of wood polymer composites for adhesive bonding, Int. J. Adhes. Adhes. 31 (2011) 816–821.
[59] Obróbka skrawaniem (Machining), Wydawnictwa Naukowo-techniczne, Warszawa, 1993 (in Polish).

[60] A. Rudawska, Przygotowanie powierzchni do klejenia wybranych materiałów konstrukcyjnych (Surface treatment of selected contructional materials for bonding), Wydawnictwa Uczelniane, Lublin, 2017 (in Polish).

[61] A.J. Kinloch, Adhesion and Adhesives, Chapman and Hall, New York, 1987.

[62] J.J. Bikerman, Surface roughness and contact angles, J. Phys. Colloid Chem. 54 (1950) 653–658.

[63] S. Malkin, C. Guo, Grinding Technology: Theory and Applications of Machining With Abrasives, Industrial Press, New York, 2008.

[64] V. De Simone, D. Caccavo, G. Lamberti, M. d'Amore, A.A. Barb, Wet-granulation process: phenomenological analysis and process parameters optimization, Powder Technol. 340 (2018) 411–419.

[65] R.J. Zaldivar, H.I. Kim, G.L. Steckel, J.P. Nokes, D.N. Patel, The effect of abrasion surface treatment on the bonding behavior of various carbon fiber-reinforced composites, J. Adhes. Sci. Technol. 26 (2012) 1573–1590.

[66] H. Hamdi, M. Dursapt, H. Zahouani, Characterization of abrasive grain's behavior and wear mechanisms, Wear 12 (2003) 1294–1298.

[67] H.A.-G. El-Hofy, Fundamentals of Machining Processes: Conventional and Nonconventional Processes, CRC Press, 2006.

[68] Narzędzia ścierne i szlifowanie (Abrasive tools and grinding) (in Polish). http://grafit.mchtr.pw.edu.pl/~askalski/PTW-lab/PTW4.pdf/, 2018 (Accessed June 22, 2018).

[69] Poradnik galwanotechnika (Handbook of Electropating), Wydawnictwa Naukowo-techniczne, Warszawa, 2002. (in Polish).

[70] L.H. Durney (Ed.), Electroplating Engineering Handbook, fourth ed., Van Nostrand Reinhold, New York, 1984.

[71] Abrasive blasting, http://en.wikipedia.org/wiki/Abrasive_blasting, 2018 (Accessed April 14, 2018).

[72] Metody przygotowania powierzchni (Surface treatment methods) (in Polish). http://anticorr.pl/metody-przygotowania-powierzchni/, 2018 (Accessed August 30, 2018).

[73] C. Vila, H.R. Siller, C.A. Rodriguez, G.M. Bruscas, J. Serrano, Economical and technological study of surface grinding versus face milling in hardened AISI D3 steel machining operations, Int. J. Prod. Econ. 138 (2012) 273–282.

[74] Y. Taga, K. Kwai, T. Nokubi, New method for divesting cobalt-chromium alloy castings: sandblasting with a mixed abrasive power, J. Prosthet. Dent. 85 (2001) 357–362.

[75] G. Bresson, J. Jumel, M.E.R. Shanahan, P. Serin, Strength of adhesively bonded joints under mixed axial and shear loading, Int. J. Adhes. Adhes. 35 (2012) 27–35.

[76] V. Máša, P. Kuba, Efficient use of compressed air for dry ice blasting, J. Clean. Prod. 111 (2016) 76–84.

[77] G. Spur, E. Uhlmann, F. Elbing, Dry-ice blasting for cleaning: process, optimization and application, Wear 233–235 (1999) 402–411.

[78] E. Uhlmann, R. Hollan, Blasting with solid carbon dioxide—investigation of thermal and mechanical removal mechanisms, Procedia CIRP 26 (2015) 544–547.

[79] Y.-H. Liu, D. Hirama, S. Matsusaka, Particle removal process during application of impinging dry ice jet, Powder Technol. 217 (2012) 607–613.

[80] S.A. Hoening, Cleaning surfaces with dry ice, Compressed Air Mag. 91 (1986) 22–25.

[81] L.R. Milman, J.W. Giancaspro, Environmental evaluation of abrasive blasting with san, water and dry ice, J. Architect. Eng. Construct. 3 (2012) 174–182.

[82] X. Tang, K. Huang, J. Dai, Z. Wu, L. Cai, L. Yang, J. Wei, H. Sun, Influences of surface treatments with abrasive paper and sandblasting on surface morphology. Hydrophilicity, mineralization and osteoblasts behaviors of n-CS/PK composite, Sci. Rep. 7 (2017) 568, https://doi.org/10.1038/s41598-017-00571-4.

[83] I. Grubova, T. Priamushko, M. Surmeneva, O. Korneva, M. Epple, O. Prymak, R. Surmenev, Sand-blasting treatment as a way to improve the adhesion strength of hydroxyapatite coating on titanium implant, J. Phys. Conf. 830 (2017), 012109. https://doi.org/10.1088/1742-6596/830/1/012109.
[84] M. Malli, Review on shot blasting processes, Int. J. Sci. Res. Dev. 2 (2014) 707–709.
[85] R. Wrona, P. Zyzak, E. Ziółkowski, M. Brzeziński, methodology of testing shot blasting machines in industrial conditions, Arch. Foundry Eng. 12 (2012) 97–104.
[86] L. Song, S. Wang, Y. Wang, P. Song, G. Xue, The wear behavior of shots impeller pair in shot blast cleaning, Appl. Mech. Mater. 364 (2013) 37–41.
[87] X. You, Z. Wang, Q. Wang, M. Cao, M. Shen, W. Huang, Effect of shot blasting on fatigue strength of Q345B steel plate with central hole, Metals 7 (2017) 517, https://doi.org/10.3390/met7120517.
[88] N.K.R. Naidu, S.G.S. Raman, Effect of shot blasting on plain fatigue and fretting fatigue behavior of Al-Mg-Si alloy AA6061, Int. J. Fatigue 27 (2005) 323–331.
[89] M. Khairulmani, S.K. Alias, B. Abdullah, J.M. Saida, N. Sulong, M. Mazni, N. Jenal, Single and double shot blasting treatment of 304 stainless steel, Jurnal Teknologi (Science and Engineering) 76 (9) (2015) 49–52.
[90] Y. Novoselov, S. Bratan, V. Bogutsky, Analysis of relation between grinding wheel wear and abrasive grains wear, Procedia Eng. 150 (2016) 809–814.
[91] R. Komanduri, Machining and grinding: a historical review of the classical papers, Appl. Mech. Rev. 46 (1993) 80–132.
[92] V. Gopan, K.L.D. Wins, Quantitative analysis of grinding wheel loading using image processing, Procedia Technol. 25 (2016) 885–891.
[93] T.N. Ivanova, Design and technology support of the grinding process for heavily-machined steel sheets, Procedia Eng. 150 (2016) 782–788.
[94] S. Malkin, Grinding of metals: theory and applications, J. Appl. Metalwork. 3 (1984) 95–109.
[95] T.N. Ivanova, Investigation of the surface layer structure at grinding, Metal Treat. Equip. Tools 1 (2004) 22–24.
[96] S. Lakhwinder, R.A. Khan, M.L. Aggarwal, Effect of shot peening on hardening and surface roughness of nitrogen austenitic stainless steel, Int. J. Eng. Sci. Technol. 2 (2010) 818–826.
[97] B. Hashemi, The wear and corrosion resistance of shot peened-nitrided 316L austenitic stainless steel, Mater. Des. 32 (2011) 3287–3292.
[98] C. Nouguier-Lehon, M. Zarwel, C. Diviani, D. Hertz, H. Zahouani, T. Hoc, Surface impact analysis in shot peening process, Wear 302 (2013) 1058–1063.
[99] Y. Ochi, K. Masaki, T. Matsumura, T. Sekino, Effect of shot-peening treatment on high cycle fatigue property of ductile cast iron, Int. J. Fatigue 23 (2001) 441–448.
[100] A. Zammit, M. Mhaede, M. Grech, S. Abela, L. Wagner, Influence of shot peening on the fatigue life of Cu–Ni austempered ductile iron, Mater. Sci. Eng. A 545 (2012) 78–85.
[101] A. Zammit, S. Abela, L. Wagner, M. Mhaede, M. Grech, Tribological behavior of shot peened Cu–Ni austempered ductile iron, Wear 302 (2013) 829–836.
[102] S. Bagherifard, I. Fernandez-Pariente, R. Ghelichi, M. Guagliano, Effect of severe shot peening on microstructure and fatigue strength of cast iron, Int. J. Fatigue 65 (2014) 64–70.
[103] Metal data sheet—C45 steel properties, http://www.ovako.com/en/Contact-Ovako/, 2018 (Accessed April 12, 2018).
[104] ISO 4278 Geometrical Product Specifications (GPS)—Surface texture: Profile Method—Terms, definitions and surface texture parameters.
[105] ISO 25178 Geometric Product Specifications (GPS)—Surface texture: areal.

[106] PN-EN ISO 4288 Geometrical Product Specifications (GPS)—Surface texture: profile method—rules and procedures for the assessment of surface texture.
[107] Epoxy resins, http://www.ciechzywice.pl/pl/, 2018 (Accessed May 17, 2018).
[108] DIN EN 1465. Adhesives. Determination of tensile lap-shear strength of bonded joints.
[109] W. Krysicki, J. Bartos, K. Królikowska, M. Wasilewski, Rachunek Prawdopobieństwa i statystyka matematyczna w zadaniach. Cz. II Statystyka matematyczna (Probability and Mathematical Statistics Tasks. Vol. II. Mathematical Statistics), PWN, Warszawa, 1997 (in Polish).
[110] E.L. Lehmann, J.P. Romano, Testing Statistical Hypotheses, Springer-Verlag, New York, 2005.
[111] EN 10083-2: C45 steel plate high carbon steel.
[112] EN 10277-2: 2008 Bright steel products. Technical delivery conditions. Steels for general engineering purposes.
[113] A. Baldan, Adhesion phenomena in bonded joints, Int. J. Adhes. Adhes. 38 (2012) 95–116.
[114] J. Comyn, Diffusion and adhesion, in: J. Comyn (Ed.), Polymer Permeability, Chapman and Hall, London, 1985, pp. 177–215 (Chapter 5).
[115] J. Shang, M. Flury, J.B. Harsh, R.L. Zollars, Comparison of different methods to measure contact angles of soil colloids, J. Colloid Interface Sci. 328 (2008) 299–307.
[116] J. Comyn, Contact angels and adhesive bonding, Int. J. Adhes. Adhes. 12 (1992) 145–149.
[117] M.E. Diaz, J. Fuentez, R.L. Cerro, M.D. Savage, Hysteresis during contact angle measurement, J. Colloid Interface Sci. 343 (2010) 574–583.
[118] E. Chibowski, F. González-Caballero, Interpretation of contact angle hysteresis, J. Adhes. Sci. Technol. 7 (1993) 1195–1209.
[119] J.F. Joanny, P.G. de Gennes, A model for contact angle hysteresis, J. Chem. Phys. 81 (1984) 552–562.
[120] C.W. Extrand, Y. Kumagai, An experimental study of contact angle hysteresis, J. Colloid Interface Sci. 191 (1997) 378–383.
[121] P. Dalet, E. Papon, J.-J. Villenave, Surface free energy of polymeric materials: relevancy of conventional contact angle data analyses, J. Adhes. Sci. Technol. 13 (1999) 857–870.
[122] C.J. van Oss, Interfacial Forces in Aqueous Media, Dekker, New York, 1994.
[123] A. Rudawska, E. Jacniacka, Analysis of determining surface free energy uncertainty with the Owens-Wendt method, Int. J. Adhes. Adhes. 29 (2009) 451–457.
[124] A. Rudawska, Adhesive properties, in: V. Kazmiruk (Ed.), Scanning Electron Microscope, InTech Publisher, Rijeka, 2012, pp. 101–126 (Chapter 7).
[125] S. Ahadian, M. Mohseni, S. Morawian, Ranking proposed models for attaining surface free energy of powders using contact angle measurements, Int. J. Adhes. Adhes. 29 (2009) 458–469.
[126] V.S. Ajaev, T. Gambaryan-Roisman, P. Stephan, Static and dynamic contact angles of evaporating liquids on heated surfaces, J. Colloid Interface Sci. 342 (2010) 550–558.
[127] ASTM D 5946-04. Standard test method for corona-treated polymer films using water contact angle measurements.
[128] S.J. Hitchcock, N.T. Caroll, M.G. Nicholas, Some effects of substrate roughness on wettability, J. Mater. Sci. 16 (1981) 714–732.
[129] M.E.R. Shanahan, Sessile drops on heterogeneous surfaces: static and dynamic behavior, in: K.W. Allen (Ed.), Adhesion 15, 28th Ann Conf on Adhesion and Adhesives held at the City University, Elsevier, London, Amsterdam, 1990, pp. 116–130.
[130] C.W. Extrand, A thermodynamic model for contact angle hysteresis, J. Colloid Interface Sci. 207 (1998) 11–19.

Chemical treatment

6

6.1 Characteristics of chemical methods

Chemical treatment is implemented to obtain the suitable surface texture and the surface layer whose chemical composition will ensure high physicochemical activeness of the surface to the applied adhesive [1–7]. In addition, this constitutes an effective anticorrosion solution [8, 9].

There are two general purposes for applying chemical surface treatments in the preparation of surfaces of adherends for adhesive processes (including adhesive joining) [3, 5, 7]:

- to remove surface contaminants and impart specific properties to the surface layer, and
- to produce a film on the surface of adherends of specific characteristics—beneficial from the viewpoint of surface energy properties.

Chemical surface treatment by means of liquid chemicals is a solution applied on adherends with contaminants strongly bound to their surface.

Regarding the first group of chemical treatments, the procedure of subjecting the workpiece to the treatment typically involves the application of a bath of suitable composition and temperature, performed over a fixed time [3, 8]. Recommendations for etching different substrates may be found in relevant literature in the field of adhesive joining technology [8–16].

6.1.1 *Chemical etching*

The primary objective of chemical etching is to remove oxides off the surface of metals by means of acid solutions, acid salt or alkalis [3, 8, 17, 18] (Fig. 6.1), and different techniques [8, 19] (Fig. 6.2) can be used for chemical etching. During chemical etching, the etching agent reacts with the oxides deposited on the metal substrate surface. In the case of steel products, chemical etching consists in rust and scale removal and by means of mineral acids [5, 8, 9].

Etching with a paste or gel is suitable for etching limited areas, e.g. heat affected zones in a welded joint. It is the best to apply the paste with an acid-resistant brush. The paste should be rinsed with water before it dries. If, due to environmental protection or for other reasons, the paste is neutralized on a metal surface, it should be thoroughly rinsed with water. In addition, adequate time must be allowed for the paste to react with the substrate. At elevated temperature and where an extended digestion time is required after a certain time, reapplication of the etching agent may be required as it may dry out, making the digestion less effective.

The etching solution (or gel etching) in the spray form is suitable for etching large surfaces, e.g. when removal of iron inclusions is required. The etching gel in the form of a spray enables, among others, limiting the harmful emission of nitric acid vapours

Surface Treatment in Bonding Technology. https://doi.org/10.1016/B978-0-12-817010-6.00006-0

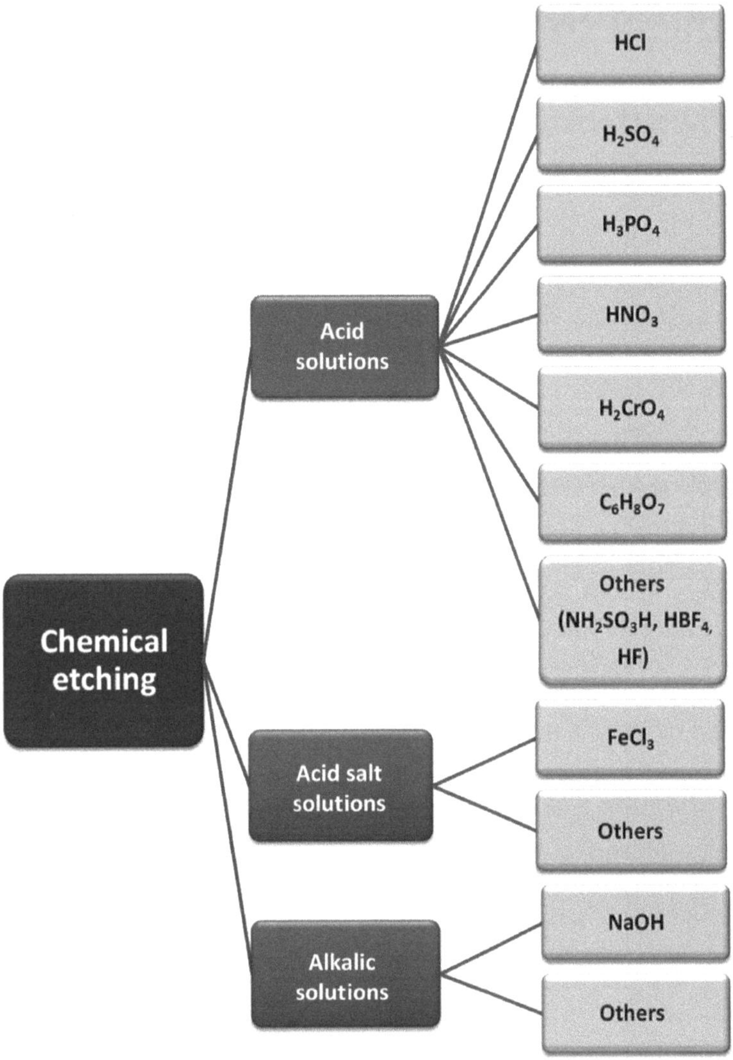

Fig. 6.1 Chemical etching methods.

present in the traditional submersion digestion process (as well as the paste etching). By means of pumps with acid-resistant elements, the etching solution is sprayed until a layer of even thickness is obtained over the entire surface, and the etching operation should not be carried out directly in the sun. Adequate response time should be provided, but it should not be allowed to dry out. After the etching process, thoroughly rinse the cleaned surface with a stream of clean water under pressure. To minimise acid spray, it is recommended to prerinse with lower pressure water. It should also be ensured that there are no traces of residues on the surface, and in the case of particularly sensitive surfaces, finally rinse with demineralised water.

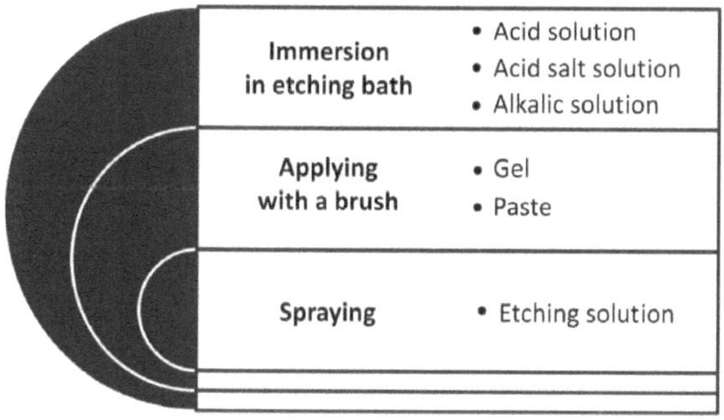

Fig. 6.2 Chemical etching techniques.

During immersion etching, the object is immersed in etching baths at the right time and temperature, which is described later in the chapter.

The type of substrate subjected to treatment (steel, copper, brass, aluminium, etc.) and the type of corrosion product on its surface decide of the composition of the bath and the process parameters [8, 20–22].

Steel substrates are typically treated in etch bath based on hydrochloric acid or sulphuric acid, nonferrous metals—in compositions of these acids with the addition of nitric acid. Other acids are considerably less frequently applied in etching, and their use is limited to specific substrates.

The recommended solution for etching aluminium is the alkaline bath composition of NaOH and other chemicals. With regards to etching aluminium it has been shown that the bath composition exerts a significant effect on the final joint strength in adhesive joining; for instance, the adhesive joint shear strength of aluminium alloy adherends pretreated in a 4% NaOH is by 40% higher than following etching in a 20% HNO_3 [8].

In practice, however, etching is implemented into the pretreatment process of steel substrates (iron alloys), whose corrosion products may differ depending on their formation conditions, taking the form of [20]:

- scale—oxides FeO, Fe_3O_4, Fe_2O_3, corrosion products formed at high temperatures in the production process; and
- rust—oxides, hydroxides, basic iron salts, atmospheric corrosion products formed in the presence of air and water media.

The effect of etching is the removal of corrosion products from the metal surface, which are [20–22]:

- dissolved in acid (e.g. atmospheric corrosion products in hydrochloric acid), and
- detached from the substrate surface by the dissolved hydrogen (e.g. scale in sulphuric acid).

The substrate surface having been exposed begins to dissolve in the process of dissolution of metal. The process occurs as a result of penetration of the metallic substrate by the acid solution through pores and cracks in the layer of corrosion products. The dissolution of metal is accompanied by the production of hydrogen bubbles.

The examples of reactions occurring during etching are as follows [20, 23]:

(a) $FeO + H_2SO_4 \rightarrow Fe\ SO_4 + H_2O$
(b) $Fe_3O_4 + 4H_2SO_4 \rightarrow Fe_2(SO_4)_3 + FeSO_4 + 4H_2O$
(c) $FeO + 2HCl \rightarrow FeCl_2 + H_2O$
(d) $Fe + 2HCl \rightarrow FeCl_2 + 2H$ (atomic in iron) or $\ldots \rightarrow H_2$
(e) $Fe + H_2SO_4 \rightarrow FeSO_4 + 2H$ (atomic in iron) or $\ldots \rightarrow H_2$

The dissolution of metal during etching is highly undesirable for the following reasons:

- deterioration of the end-product quality (loss of mass, hydrogen brittleness),
- increased acid consumption,
- deterioration of working conditions (the released hydrogen is flammable and explosive), and
- increased number of bath dumps, leading to rise in etch waste salinity.

The negative impact of the above-mentioned phenomena is counteracted by the introduction of etch inhibitors [20].

Prior to etching very oily substrates it is necessary to subject them to degreasing [3, 8]. The degreasing process is carried out in the surface treatment baths or by means of organic solvents. If the degree of grease/oil contamination is not significant the etch bath may be enriched by the addition of a composition of surfactants, which, being resistant to acid, enable conducting both etching and degreasing simultaneously; such a process is frequently carried out electrolytically.

Efficiency of etching (i.e. of corrosion products removal) is directly related to the concentration of acid in the bath, as well as bath temperature. During etching, the acid concentration level decreases while the concentration of iron salts rises—this subsequently leads to reducing the etching efficiency. In order to counter the process efficiency loss, the etch bath is refilled with concentrated acid and inhibitor; however, when the iron salts reach the concentration limit the bath ought to be replaced.

Etching acid efficiency increases with temperature, therefore the etch time and rate may be improved by increasing the ambient temperature. Nevertheless, there exist temperature limits not to be exceeded. It is predominantly during bath treatment that the risk of excessive etching as a result of excessive temperature is particularly high. In the case when etching is conducted with the application of a paste, gel, spray, or solution of elevated temperature, the etching effect may be impaired by the generated steam. Potential undesirable impact of the situation in question is, apart from the nonuniform etching, an additional problem in the rinsing stage. In order to avoid the said problems, the substrates should be etched at temperatures lower than 45°C, and away from direct sunshine [3, 8].

Some authors emphasises that the most effective of many pretreatments are complex, multistage processes involving a combination of degreasing, etching, and anodising [14]. In some cases further controlled etching is carried out after anodising.

6.1.2 Etching solutions

Etching is typically performed with the following acids [8, 13, 20–22, 24–26]:

- hydrochloric acid HCl,
- sulphuric acid H_2SO_4,

- chromic acid H_2CrO_4,
- phosphoric acid H_3PO_4,
- nitric acid HNO_3,
- citric acid $C_6H_8O_7$,
- chromic-sulphuric acid.

Hydrochloric acid etch

Hydrochloric acid HCl ($d = 1.18\,g/cm^3$, approximately 36% wt) is a light yellow liquid. Its vapour has a detrimental effect on mucous membranes and is highly corrodible for the exposed instrumentation and equipment. The aggressiveness of this acid is higher than that of the sulphuric acid; in addition, it shows higher efficiency and uniformity of the dissolution of corrosion products and etching. Substrates subjected to etching by means of hydrochloric acid exhibit superior surface state and lower extent of hydriding. Iron chlorides are easily rinsed off the substrate surface; nevertheless, inaccurate rinsing could lead to emergence of corrosion centres [20–22].

Operating concentrations are obtained by mixing concentrated acid and water. Considering the work safety, the water is added first and subsequently, with utmost care, the concentrated commercial acid. The bath is further enriched with inhibitors and composition of surfactants (Table 6.1).

Etching in hydrochloric acid should be performed in an open dip tank in the temperature range of 20–30°C. Throughout etching, the acid concentration levels are monitored and if necessary the acid is supplemented together with a suitable amount of the inhibitor. However, the bath ought to be dumped when the limit concentration level of iron is reached—$120\,g/dm^3$, which is when the bath loses its etching properties.

Sulphuric acid etch

The sulphuric acid H_2SO_4 ($d = 1.84\,g/cm^3$, approximately 98% wt) is an oily liquid; however, it may occur in the form of mist, which exhibits cancerogenic properties. The surface of workpieces etched by means of sulphuric acid shows nonuniform etching and higher hydriding. Sulphates prove harder to rinse off the substrate surface than chlorides. In bath, the sulphuric acid is a more economical solution than the hydrochloric acid. The sulphuric acid etch bath is prepared analogically to the previously described hydrochloric acid bath.

The etch process in sulphuric acid is conducted at ambient or elevated temperature (60–70°C); *nota bene*, elevated temperature accelerates the process. The process may

Table 6.1 Example of hydrochloric acid etch bath composition

Bath components	Concentration
HCl	10–20% wt
Inhibitor	0.2–2.0 g/dm^3
Nonionic surfactants	2.0–5.0 g/dm^3

Source: Own work based on Poradnik galwanotechnika (Handbook of Electroplating), Wydawnictwa Naukowo-techniczne, Wa rszawa, 2002 (in Polish).

be furthermore quickened by incorporating electrolytic, cathodic, or anodic etch. The recommended current density is in the range of 5–10A/dm^2 for steel, 3–7A/dm^2 for copper, and 2–4A/dm^2 for brass. Etching may be accelerated by adding 10–15g/dm^3 of NaCl to the bath [20]. While etching, the acid concentration levels are monitored in order to add acid in case of excessive loss, together with a suitable amount of the inhibitor. However, the bath should be dumped when the limit concentration level of iron is reached—90g/dm^3 [8, 20].

Chromic-sulphuric acid etch

One on the most commonly applied chemical treatments are based on a chromic-sulphuric acid etch [14, 15, 24, 25, 27, 28]. This treatment consists of immersion of the substrate in a solution of sulphuric acid and potassium dichromate. There are different chromic-sulphuric acid treatments and they differ in the compounds concentration and also the temperature and the used immersion time of etching process. Forest Products Laboratory (FPL) is known as the original etching [15, 24, 27, 28]. It involves low chromate concentrations and short time of treatment.

Nitric acid etch

Nitric acid HNO_3 (d=1.42g/cm^3, approximately 63% wt and 98% wt), is a highly corrosive mineral acid. Nitric acid after dilution is used for the etching of magnesium and magnesium alloys. Concentrated nitric acid is used as an ingredient in etching baths for etching copper and some grades of alloy steels. It is very rarely used to etch carbon steel [20].

Citric acid etch

Citric acid is a weak organic acid that has the chemical formula $C_6H_8O_7$. Etching process can also be carried out in a ferric-based etching solution with citric acid, which is a nontoxic additive. The addition of citric acid allows the sediment in the solution to be dissolved as it is formed, as there is no crystalline sedimentation to impede the biting process. This solution will also erode metal faster than pure ferric chloride and will stay active for a much longer period of time. These properties result from two following reasons. The citric acid will lower the pH of the bath (making it more acidic). The sediments released from the metal tend to form a complex with the citric acid as it is released from the surface of the metal, also increasing its solubility. This acid solution is used to dissolve rust from steel and passivate stainless steel [29].

Etch inhibitors

Etch inhibitors are added to baths with a task of inhibiting the reaction between the acid and metal substrate. Their implementation produces the following effects [20–22]:

- reducing acid use,
- improving product quality (reducing the metal mass loss and metal hydriding),
- reducing the number of bath dumps, and etch waste salinity, and
- improving working conditions,

There is a wide range of etch inhibitors available on the market. A properly selected etch inhibitor will impede the acid-metal reaction by 80%–95% [20].

6.1.3 Characteristics of etching various materials

Etching stainless and acid-resistant steel

Etching stainless and acid-resistant steel involves certain difficulties due to the presence of alloy components that are resistant to acid. Table 6.2 shows examples of solutions and etch parameters during treatment of stainless and acid-resistant steel.

Etching is performed by means of a mixture of acids, which typically involve the sulphuric and hydrochloric acids with the addition of oxidising acid, such as HNO_3 [8, 20].

Etching copper and alloys

Corrosion products on copper substrates are dissolved by a 10% sulphuric acid, however, this proves insufficient. The highest efficiency of treatment is obtained by etch with nitric acid (Table 6.3).

The process is relatively quick and stable, however, the surface following treatment is visually unaesthetic; an additional disadvantage of the process is high emission of hazardous nitrogen oxides. Therefore, the treatment is typically conducted in the solution of sulphuric acid, nitric acid, and other oxidising agents, such as H_2O_2, CrO_2, etc.

Etching zinc and cadmium

Zinc and cadmium are predominantly etched in a diluted hydrochloride acid (5%–10%) or sulphuric acid (2%–7%). The etch is carried out by a short-time immersion of the samples while constantly observing the surface of the substrate. Simultaneous etch with bright dip polishing treatment is performed in sulphuric and nitric acid mixtures (1:1) and lasts several seconds. Zinc alloys are etched at ambient temperatures in

Table 6.2 Etch solutions and parameters for stainless steel treatment

I	II	III	IV	V
Content in the mix (% by volume)				
H_2SO_4—15 HCl—5 H_2O—80	HCl—10 HNO_3—5 H_2SO_4—10 H_2O—75	HCl—30 HNO_3—8 H_2O—62	HCl—10 HNO_3—5 H_3PO_4—10 H_2O—75	HCl—10 HNO_3—22 H_3PO_4—43 H_2O—25
Temperature (°C)				
60	80	50	80	50

Source: own work based on Poradnik galwanotechnika (Handbook of electroplating), Wydawnictwa Naukowo-techniczne, Wa rszawa, 2002 (in Polish).

Table 6.3 Etch solutions for copper substrate treatment

Bath I	Bath II
H_2SO_4—10% by vol. HNO_3—3% by vol. HCl—1% by vol. Water—86% by vol.	H_2SO_4—1 dm^3 HNO_3—1 dm^3 HCl—10–20 mL Or NaCl—20–30 g

Source: Own work based on J. Godzimirski, J. Kozakiewicz, J. Łunarski, W. Zielecki, Konstrukcyjne połączenia klejowe elementów metalowych w budowie maszyn (Structural adhesive joint of metal elements in machine construction), Oficyna Wydawnicza Politechniki Rzeszowskiej, Rzeszów, 1997 (in Polish); Poradnik galwanotechnika (Handbook of Electroplating), Wydawnictwa Naukowo-techniczne, Wa rszawa, 2002 (in Polish).

H_2SO_4, in a process lasting 1–2 min. It is recommended that etch treatment should be followed by an additional rinse and dry of treated substrates [23].

Etching aluminium and aluminium alloys

Aluminium is etched in (4) 5%–10% solutions of NaOH at the temperature of 50–60°C for 2–3 min [8, 20]. This treatment may be conducted at ambient temperature, in which conditions, however, it does last longer. Table 6.4 shows typical etch baths for the treatment of aluminium and its alloys.

Occasionally etching aluminium substrates may cause discolouring to their surface, which becomes grey or black instead of silver-white. In such circumstances the surface may be brightened with a 30% solution of HNO_3 [20].

Digby and Packham [14] reported the results of a detailed examination of the effectiveness of nine different pretreatments for aluminium, including chemical treatments (FPL etching and ferric sulphate/sulphuric acid treatment in different combinations). The treatment using ferric sulphate/sulphuric acid has been shown to be broadly as effective as chromate-containing FPL etch.

Table 6.4 Etch solutions and parameters for treatment of aluminium and its alloys

Bath I	Bath II
H_2SO_4—10 by weight $Na_2Cr_2O_7{\cdot}2H_2O$—1 by weight H_2O—30 by weight	$Na_2Cr_2O_7$—160 g H_3PO_4 (85%)—320 g H_2O—do 1 dm^3
Temperature (°C)	
65–70	90–100
Time (min)	
10–12	3–5

Source: Own work based on J. Godzimirski, J. Kozakiewicz, J. Łunarski, W. Zielecki, Konstrukcyjne połączenia klejowe elementów metalowych w budowie maszyn (Structural adhesive joint of metal elements in machine construction), Oficyna Wydawnicza Politechniki Rzeszowskiej, Rzeszów, 1997 (in Polish).

Leena et al. [27] investigated three different methods of surface treatments such as solvent degreasing FLP etching and priming using an epoxy based primer of aluminium alloy. The FPL etching of substrate resulted in good wetting by the epoxy adhesive, highest surface energy, and surface roughness thereby highest lap shear strength properties than the only solvent degreasing used.

Prolongo and Urena [24] investigated the effect of the etching treatment (FPL and chromate-free treatment based on the sulphuric acid-ferric sulphate etch) on the some properties of two different aluminium alloys—A1050 and A2024, such as density, porosity, and composition of the formed oxide layer. The authors [24] proved that the free-chromium etching pretreatment, based on a sulphuric acid-ferric sulphate solution, provides higher strength values than the common dichromate-sulphuric acid etching treatment. This is justified because of the higher porosity of the formed oxide layer.

In work prepared by Lunder et al. [30] AA6060-T6 aluminium extrusions were subjected to various chrome-free pretreatments, including thin film AC anodising in hot sulphuric acid, alkaline etching, a phosphate-permanganate conversion coatings and as a reference—a chromic-sulphuric acid pretreatment (FPL etch).

Etching titanium alloys

Table 6.5 presents typical etch baths for the treatment of titanium alloys.

Some information of chemical treatment of titanium and titanium alloy was presented in Refs. [3, 8, 18, 31, 32]. Molitor et al. [18] mentioned both acid treatment (including nitric-hydrofluoric acid, hydrochloric-orthophosphoric acid, hot sulphuric acid and sodium dichromate-chromic acid) and alkaline treatment (alkaline peroxide etch) of titanium surface. Cotter and Mahoon [32] presented the results of surface pretreatments, based on alkaline hydrogen peroxide solutions, for adhesive bonding of titanium.

Both stressed and unstressed peel and lap shear joints after acid etching have shown to perform equal or better than phosphate fluoride or modified phosphate fluoride processes in the tests performed by Critchlow and Brewis [15].

Table 6.5 Etch solutions and parameters for treatment of titanium alloys

Bath I	Bath II
CrO_3—5 by weight H_2O—95 by weight	H_2SO_4—9.6 by vol. NaF—3.0 by vol. CrO_3—1.4 by vol.
Temperature (°C)	
60–70	20
Time (min)	
5	8

Source: Own work based on C.V. Cagle, Handbook of Adhesive Bonding, McGraw-Hill, New York, 1973.

6.1.4 Phosphatising

Phosphatising is a process of chemical or electrochemical deposition of matt grey phosphate coating on the surface of metal substrates, conducted in the hot solution of phosphates and phosphoric acid [20–22].

Molitor et al. [18] mentioned about the phosphate fluoride process [33] of titanium treatment. The phosphate-fluoride-base treatments are based on combination of trisodium phosphate, disodium tetraborate, potassium fluoride, and hydrofluoric acid.

The process of phosphatising was described in Chapter 7 (Section 7.1.4).

6.2 Experimental test

The information presented in this subchapter was prepared base on article [34].

6.2.1 Methods

Material and adhesive joint

The tests were conducted on samples made of EN AW-2024 aluminium alloy sheet. Selected mechanical properties were presented in Table 7.4.

The single-lap adhesive joint of EN AW-2024 aluminium alloy sheet were subjected to shear strength testing. The scheme of the tested adhesive joint with averaged dimensions was shown in Fig. 6.3.

The adhesive joint had the following average dimensions presented in Table 6.6.

Adhesive joints of aluminium alloys were formed with a Loctite Hysol 9484 two component epoxy adhesive, at weight ratio of 1:1. The selected properties of which

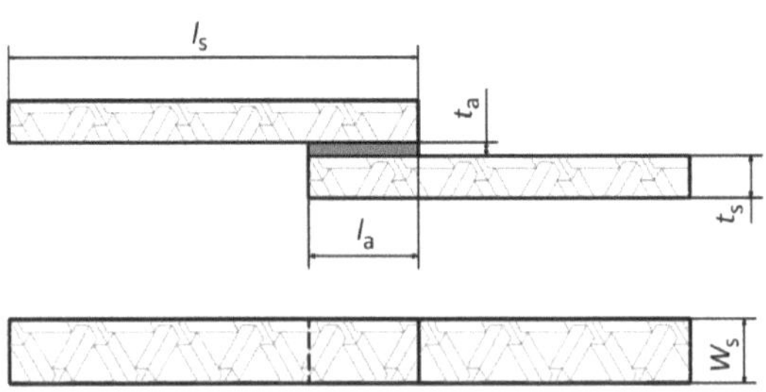

Fig. 6.3 Scheme of the tested single-lap adhesive joint.

Table 6.6 Theoretical dimensions of single lap joints

Theoretical dimensions of single-lap joints (mm)				
Specimen length (l_s)	**Specimen width (w_s)**	**Specimen thickness (t_s)**	**Adhesive layer thickness (t_a)**	**Lap length (l_a)**
100	25	1	0.15	14

Table 6.7 The selected properties of uncured Loctite Hysol 9484 adhesive [35]

Properties	Description/value
Chemical type (resin)	Epoxy
Specific gravity @ 25°C	1.30
Viscosity, Brookfield—RVT, 25°C, mPas (cP)	60,000÷175,000 (Spindle 7, speed 10 rpm)
Chemical type (hardener)	Polyamide
Mix ratio, by volume—Resin:hardener	1:1
Mix ratio, by weight—Resin:hardener	100:85
Working life, 22°C, 100 g	40

Table 6.8 The selected properties of cured Loctite Hysol 9484 adhesive [35]

Properties	Value
Tensile strength, ISO 527-3, N/mm^2	15
Tensile modulus, ISO 527-3, N/mm^2	161
Shore hardness, ISO 868, Durometer D	55
Elongation, ISO 527-3, %	50

were presented in Table 6.7 (for uncured material) and in Table 6.8 (for cured material). Data collated in Table 6.8 regard properties of adhesive cured for 7 days at 22°C. Loctite Hysol 9484 adhesive requires 24 h in room temperature in order to create high strength. The process, however, can be accelerated by applying higher temperature.

Adhesive used in the tests is delivered in dual packages (cartridges), facilitating dosing and preparation. Mixing and dosing was performed with special dispensing device—a double syringe and a static mixer. After mixing the adhesive was applied on one of the adhered surfaces. The samples of adhesive joints were subjected to pressure of 0.18 MPa. The single-stage cure process at ambient temperature (22 ± 1°C) for 48 h was used.

Surface treatment methods

Four surface treatment methods (chemical and electrochemical) of EN AW-2024 aluminium alloy sheet were applied, which were collated in Table 6.9.

The treatments parameters were presented in Table 6.10. The information was prepared on the basis of aviation company standards.

Table 6.9 The tested surface treatment methods of EN AW-2024 aluminium alloy sheet

Surface treatment variant	Designation
Degreasing	D
Etching	E
Anodizing	A
Chromate treatment	Ch

Table 6.10 Technological parameters of aluminium alloy surface treatment

Surface treatment variant	Technological parameters
Degreasing	Type of degrease agent: Loctite 7061 Chemical base: Acetone Technology: Three times degreasing by spraying Evaporating time: 3 min
Etching	Etching bath: NaOH and H_2O Bath composition: 40–60 g/L Bath temperature: 40–60°C Treatment time: 3–4 min Rinsing in warm water Drying time: 8–10 min
Anodizing	Type of electrolyte: H_2SO_4 Bath composition: 180–200 g/L Electrolyte temperature: 40°C Anodising time: 35–40 min Voltage: 50 V
Chromate treatment	Bath composition: (a) $Na_2Cr_2O_7$—5.5 g/L (b) Na_2SO_4—4.0 g/L (c) H_3BO_3—4.4–5.5 g/L (d) HNO_3—1.5 mL/L Bath temperature: 20–30°C Optimum pH 1.4–1.6 Chromate coating time: 0.5 min

Surface roughness parameters

Surface roughness parameters of adherends surfaces after tested surface treatments methods were measured using the Perthometer M2 profilometer. The surface roughness parameters were determined in accordance with ISO 4278 standard and ISO 25178 requirements. In each batch, 5 specimens at 3 different measuring positions were examined. The highest and lowest values were rejected and not taken into account in further analyses.

During the tests the following surface roughness parameters were examined:

R_a—arithmetic mean deviation,
R_z—ten-point height,
R_{max}—maximum roughness depth,
R_p—maximum peak height of the profile within the sampling length,
R_{pk}—reduced peak height,
R_k—core roughness depth, and
R_{vk}—reduced valley depth,

Strength tests

The tested adhesive joints of aluminium alloy sheet were subjected to shear strength tests which were conducted in compliance with the EN DIN 1465 standard using the Zwick/Roell Z150 Allround-Line testing machine.

6.2.2 Results

Surface roughness parameters

The results of surface roughness parameters of EN AW-2024 aluminium alloy following four different treatment methods (chemical and electrochemical) were collated in Table 6.11. The views of surface roughness profiles of tested adherends surfaces after four variants of surface treatment were presented in Table 6.12.

On the basis of obtained surface roughness parameters measurements (Table 6.11), it can be noticed that the highest roughness parameters were characterised by the surfaces of samples prepared by the use of etching operation. A group of treatments methods, using which much lower surface roughness parameters were obtained included degreasing, anodising, and chromate treatment.

In the case of etching treatment a highly developed surface structure with a large number of quite deep irregularities can be observed (Table 6.12). Close to etching treatment, the surface irregularities are characterised by surfaces after chromate treatment. Slightly different geometry is characterised by surfaces after anodising, while the surface of the examined materials after degreasing presents the actual shape of surface micro-perforations, from which various types of contaminants on the surface of the tested materials have been removed.

Shear strength of adhesive joints

The comparison of EN AW-2024 aluminium alloy sheet adhesive joints shear strength with regard to the four surface treatment methods (chemical and electrochemical) was presented in Fig. 6.4.

Comparing the results (Fig. 6.4) it should be noted that the highest strength of adhesive joints of aluminium alloy sheets was obtained after anodising, as a surface treatment of adherends surfaces, before boding. It was also observed that the smallest values were obtained in the case of etching and chromate treatment operations. Comparing the obtained values of shear strength of adhesive joints with surface roughness parameters, it can be assumed that not in every case these parameters can provide

Table 6.11 Surface roughness parameters of EN AW-2024 aluminium alloy following different treatment methods

Surface treatment variant	Designation	Surface roughness parameters (μm)						
		R_a	R_z	$R_{z,max}$	S_m	R_p	R_{pk}	R_{vk}
Degreasing	D	0.50	4.01	7.09	217	2.51	1.12	0.68
Etching	E	2.51	16.70	22.43	125	7.69	3.33	5.39
Anodizing	A	0.41	3.31	6.22	246	1.66	1.09	0.77
Chromate treatment	Ch	0.32	2.61	4.48	231	1.25	0.66	0.78

Based on A. Rudawska, M. Chruściel, Wpływ sposobu przygotowania powierzchni na wytrzymałość połączeń klejowych lotniczego stopu aluminium (The influence of the surface preparation method on the strength of glue joints of an aviation aluminum alloy) Technologia i Automatyzacja Montażu 2 (2011) 42–46 (in Polish).

Table 6.12 The views of surface roughness profiles of tested adherends surfaces after four variants of surface treatment

Surface treatment variant	Examples of surface roughness profile, $L_c = 8\,mm$
Degreasing	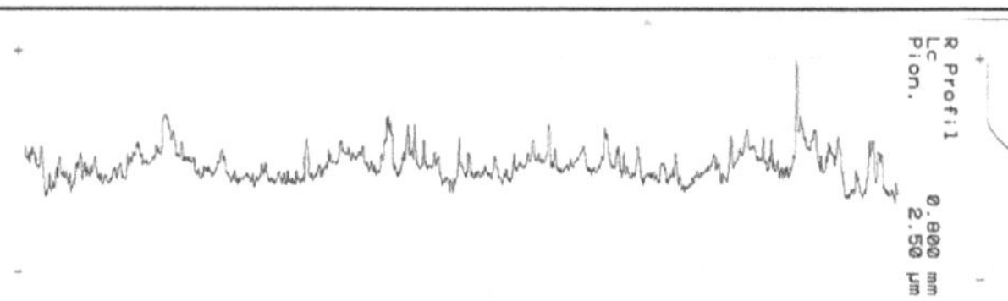
Etching	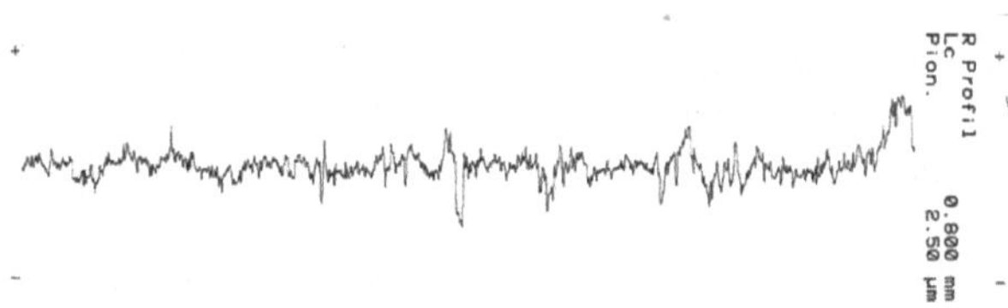
Anodizing	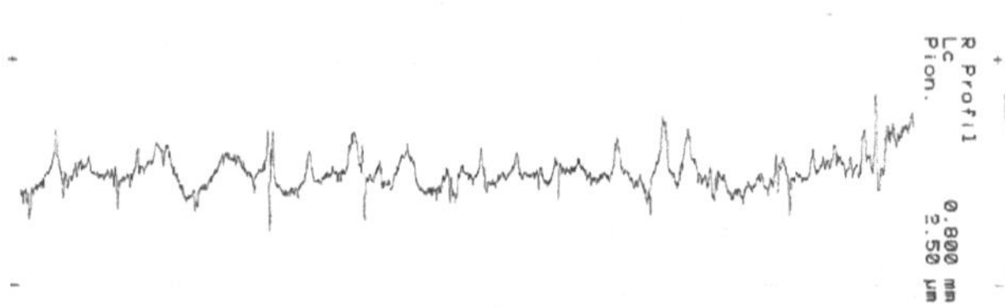
Chromate treatment	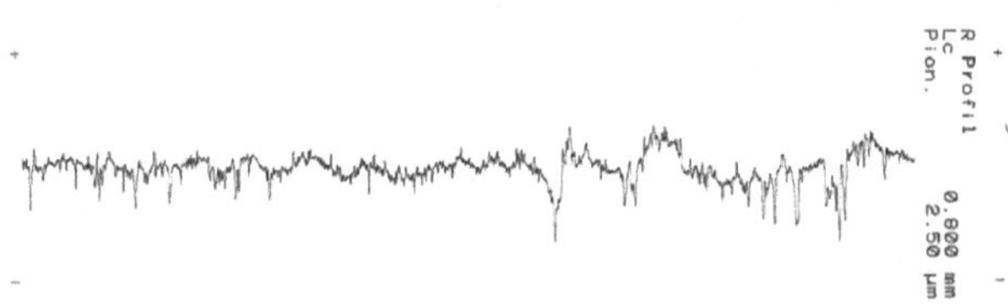

Based on A. Rudawska, M. Chruściel, Wpływ sposobu przygotowania powierzchni na wytrzymałość połączeń klejowych lotniczego stopu aluminium (The influence of the surface preparation method on the strength of glue joints of an aviation aluminum alloy) Technologia i Automatyzacja Montażu 2 (2011) 42–46 (in Polish).

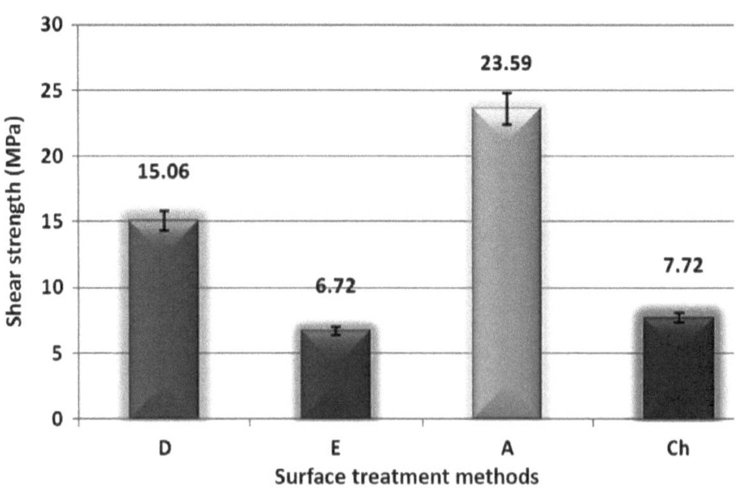

Fig. 6.4 Shear strength of EN AW-2024 aluminium alloy sheet adhesive joints after chemical and electrochemical surface treatment methods.

a guide with regard to the method of surface preparation for bonding. It turns out that in spite of not very high altitude parameters in the case of anodising, the highest value of strength of adhesive joints was obtained. It is also necessary to take into account the geometric structure of the surface, including the number and geometrical dimensions of micro-lines, which are extremely important due to the possibility of their penetration through an adhesive with a certain viscosity. Analysing the surface profiles, it can be seen that the surface structure after the anodising operation differs from the structure obtained after the other methods of surface preparation. Perhaps this is the structure that provides much better mechanical adhesion compared to other types of machining. But in addition to the geometrical structure, chemical activation of adherends surfaces is extremely important, which leads to high strength. This is, among others, the effect of chemical bonds in the adhesion between the surface of bonded material and adhesive.

6.2.3 Conclusions

The conducted research concerned the comparison of shear strength of EN AW-2024 aluminium alloy sheet, the surface of which has been subjected to chemical and electrochemical surface treatment methods used at one of the aviation plants. The surface treatment methods of aluminium alloy sheets surface include: degreasing, etching, anodising, and chromate treatment Surface roughness parameters measurements were also carried out after the four surface treatment methods.

On the basis of the research, it was noticed that the highest strength was characterised by adhesive joints of EN AW-2024 aluminium alloys, whose surfaces were subjected to the anodising process. The obtained results are in accordance with information published in the literature regarding the positive impact of the anodising operation as a method of treatment the surface of aluminium alloys for bonding [15, 36–42], although own tests have shown that strength values may depending on the type of anodising bath composition. Analysing the surface profiles, it can be seen that the surface structure after the anodising operation differs from the structure obtained after the other methods of surface preparation. Perhaps this is the structure that provides much better mechanical and appropriate adhesion compared to other types of surface treatment methods, and thus it is possible to achieve high strength, although the aspect of energetic properties of adherends surface is also very important, including adhesive properties that play an important role in the performance of, among others, adhesive joints.

The degreasing operation turned out to be advantageous in terms of the obtained resistance value, which seems to be particularly important in some cases due to technological and workshop possibilities [3, 5, 26, 43–46]. It does not require special equipment and devices. Not without significance is the possibility of applying this method to differently shaped surfaces. After the use of etching and chromate treatment (using specific factors and process parameters), the lowest strength results of the tested joints were obtained. These results were obtained using specific technological parameters of the process. Probably the use of other parameters will allow obtaining different strength of adhesive joints.

Summarising, when designing the surface preparation technology, it is necessary to analyse both the geometric structure of the surface (due to the phenomenon of mechanical adhesion) and energy properties (due to the participation of appropriate adhesion including chemical bonds).

References

[1] S. Ebnesajjad, C. Ebnesajjad, Surface Treatment of Materials for Adhesive Bonding, second ed., William Andrew Inc., Norwich, NY, 2013.
[2] S. Ebnesajjad, A.H. Landrock, Adhesives technology handbook, third ed., William Andrew, London, 2009.
[3] C.V. Cagle, Handbook of Adhesive Bonding, McGraw-Hill, New York, 1973.
[4] S. Ebnesajjad, Adhesives Technology Handbook, second ed., William Andrew Inc., Norwich, NY, 2008.
[5] R.D. Adams (Ed.), Adhesive Bonding. Science, Technology and Applications, Woodhead Publishing Limited, Cambridge, 2010.
[6] N. Brack, A.N. Rider, The influence of mechanical and chemical treatments on the environmental resistance of epoxy adhesive bonds to titanium, Int. J. Adhes. Adhes. 44 (2014) 20–27.
[7] W. Wu, M. Chen, S. Liang, X. Wang, J. Chen, F. Zhou, Superhydrophobic surface from Cu-Zn alloy by one step O2 concentration depending etching, J. Colloid Interface Sci. 326 (2008) 478–482.
[8] J. Godzimirski, J. Kozakiewicz, J. Łunarski, W. Zielecki, Konstrukcyjne połączenia klejowe elementów metalowych w budowie maszyn (Structural adhesive joint of metal elements in machine construction), Oficyna Wydawnicza Politechniki Rzeszowskiej, Rzeszów, 1997 (in Polish).
[9] S. Wirbilis, Chemiczne przygotowanie powierzchni metali przed obróbką galwaniczną (Chemical preparation of metal surfaces before galvanic treatment), Mechanik 9 (1985) 473–480. (in Polish).
[10] J.R.J. Wingfield, Treatment of composite surfaces for adhesive bonding, Int. J. Adhes. Adhes. 13 (1993) 151–156.
[11] P. Hass, O. Kläusler, S. Schlegel, P. Niemz, Effects of mechanical and chemical surface preparation on adhesively bonded wooden joints, Int. J. Adhes. Adhes. 51 (2014) 95–102.
[12] R.M. Podhajny, Comparing surface treatments, Converting 8 (1990) 46–52.
[13] D.M. Brewis, I. Mathieson, M. Wolfensberger, Treatment of low energy surfaces for adhesive bonding, Int. J. Adhes. Adhes. 15 (1995) 87–90.
[14] R.P. Digby, D.E. Packham, Pretreatment of aluminium: topography, surface chemistry and adhesive bond durability, Int. J. Adhes. Adhes. 15 (1995) 61–71.
[15] G.W. Critchlow, D.M. Brewis, Review of surface pretreatments for alluminium alloys, Int. J. Adhes. Adhes. 16 (1996) 255–275.
[16] F.T.M. Ariff, N. Yunus, M.R. Baig, The effect of mechanical roughening and chemical treatment on shear bond strength of urethane dimethacrylate denture base resin, Int. J. Adhes. Adhes. 31 (2011) 112–118.
[17] O. Lunder, F. Lapique, B. Johnsen, K. Nisancioglu, Effect of pre-treatment on the durability of epoxy-bonded AA6060 aluminium joints, Int. J. Adhes. Adhes. 24 (2004) 107–117.
[18] P. Molitor, V. Barron, T. Young, Surface treatment of titanium for adhesive bonding to polymer composites: a review, Int. J. Adhes. Adhes. 21 (2001) 129–136.

[19] Etching guide, https://www.eltrex.net.pl/img/products_files/poradnik-trawienia-bohler-avesta.pdf, 2018 (Accessed October 17, 2018).
[20] Poradnik galwanotechnika, (Handbook of Electroplating), Wydawnictwa Naukowo-techniczne, Wa rszawa, 2002 (in Polish).
[21] L.H. Durney (Ed.), Electroplating Engineering Handbook, fourth ed., Van Nostrand Reinhold, New York, 1984.
[22] A.K. Graham, H.L. Pinkerkton, Electroplating Engineering Handbook, Reinhold Publishing Corporation, New York, 1971.
[23] Electroplating, http://galwanizernie.pl/, 2018 (Accessed October 17, 2018).
[24] S.G. Prolongo, A. Ureña, Effect of surface pre-treatment on the adhesive strength of epoxy-aluminium joints, Int. J. Adhes. Adhes. 29 (2009) 23–31.
[25] A. Bishopp, Handbook of Adhesives and Sealants, vol. 1, Elsevier, Amsterdam, 2005.
[26] L. Kozma, I. Olefjord, Surface treatment of steel for structural adhesive bonding, Mater. Sci. Technol. 3 (1987) 954–962.
[27] K. Leena, K.K. Athira, S. Bhuvaneswari, S. Suraj, V. Lakshmana Rao, Effect of surface pre-treatment on surface characteristics and adhesive bond strength of aluminum alloy, Int. J. Adhes. Adhes. 70 (2016) 265–270.
[28] S. Tennyson, Surface treatment for aluminium bonding (STAB), J. Adhes. 8 (1977) 313–320.
[29] Metal salt etching, https://www.ganoksin.com/article/metal-salt-etching-resists-resist-removal/, 2018 (Accessed October 17, 2018).
[30] O. Lunder, B. Olsen, K. Nisancioglu, Pre-treatment of AA6060 aluminum alloy for adhesive bonding, Int. J. Adhes. Adhes. 22 (2002) 143–150.
[31] M. Davis, D. Bond, Principles and practices of adhesive bonded structural joints and repairs, Int. J. Adhes. Adhes. 19 (1999) 91–105.
[32] J.L. Cotter, A. Mahoon, Development of new surface pretreatments, based on alkaline hydrogen peroxide solutions, for adhesive bonding of titanium, Int. J. Adhes. Adhes. 2 (1982) 47–52.
[33] US Patent 2 864 732 Etching in phosphate fluoride (PF) of titanium to reduce bond degradation.
[34] A. Rudawska, M. Chruściel, Wpływ sposobu przygotowania powierzchni na wytrzymałość połączeń klejowych lotniczego stopu aluminium (The influence of the surface preparation method on the strength of glue joints of an aviation aluminum alloy), Technologia i Automatyzacja Montażu 2 (2011) 42–46 (in Polish).
[35] Loctite Hysol 9484 adhesive, http://www.brammer.nl/Downloads/Epoxy_1202.pdf, 2018 (Accessed October 18, 2018).
[36] L. Anicai, C. Trifu, L. Dima, Anodic oxidation and coloring of aluminum powders, Met. Finish. 98 (2000) 20–25.
[37] A. Rudawska, I. Danczak, Adhesive properties and adhesive joint strength of aluminium alloys following surface pre-treatment, Recent Pat. Mech. Eng. 6 (2013) 152–165.
[38] S.-M. Moon, S.-I. Pyun, The mechanism of stress generation during the growth of anodic oxide films on pure aluminium in acidic solutions, Electrochim. Acta (21-22) (1998) 3117–3126.
[39] L. Domingues, J.C.S. Fernandes, M. Da Cucha Belo, M.G.S. Ferreira, L. Guerra-Rosa, Anodising of Al 2024-T3 in a modified sulphuric acid/boric acid bath for aeronautical applications, Corros. Sci. 45 (2003) 149–160.
[40] A. Bjørgum, F. Lapique, J. Walmsley, K. Redfird, Anodising as pre-treatment for structural bonding, Int. J. Adhes. Adhes. 23 (2003) 401–412.
[41] G.W. Critchlow, K.A. Yendall, D. Bahrani, A. Quinn, F. Andrews, Strategies for the replacement of chromic acid anodizing for the structural bonding of aluminium alloys, Int. J. Adhes. Adhes. 26 (2006) 419–453.

[42] B.B. Johnsen, F. Lapique, A. Bjørgum, The durability of bonded aluminium joints: a comparison of AC and DC anodizing pretreatments, Int. J. Adhes. Adhes. 24 (2004) 153–161.
[43] A. Rudawska, J. Nalepa, M. Müller, The effect of degreasing on adhesive joint strength, Adv. Sci. Technol. Res. J. 11 (2017) 75–81.
[44] W. Brockmann, P.L. Geiß, J. Klingen, B. Schröder, Adhesive Bonding. Materials, Applications and Technology, Wiley-VCH Press, Weinheim, 2009.
[45] L.C. Jackson, Surface cleanliness measurements, contaminant characterization and control, in: Society of Manufacturing Engineers' Substrate Selection and Conditioning Seminar, 3–4 October, Chicago, Illinois, 1978.
[46] M.J. Troughton, Handbook of Plastics Joining: A Practical Guide, second ed., William Andrew, New York, 2008.

Electrochemical treatment

7

7.1 Characteristics of electrochemical methods

Electrochemical treatment methods are generally classified into (Fig. 7.1): electrochemical etching and electrochemical conversion coating methods (e.g. anodising, chromatising, and phosphating) [1–9].

7.1.1 *Electrochemical etching*

There are two types of etching processes: chemical etching and electrolytic etching. The chemical etching uses chemical-resist patterns drawn on metal substrates where the exposed metal is removed by the etchant. The electrolytic etching electrochemically removes materials by using electrodes formed into intended shapes as cathodes and the object itself as an anode. Electrolyte fills the space between the two and electrolysis using a direct current removes the intended material [10–12].

The process of electrochemical etching of metals is a complex electrochemical phenomenon. As electric current flows through an electrolyte, strong polarisation, and proper distribution of current density trigger electrochemical phenomena which cause the surface of a metal, i.e. the anode, to dissolve. Cathodes play here the role of an element which enables closing the current circuit by the electrolyte and are responsible for proper distribution of the current density. Cathodes used in electrolytic etching do not wear even after a longer period of time [11].

One of the most popular theories explaining the phenomenon of electrolytic etching (and also electrolytic polishing) is the theory developed by Jacquet (called the 'viscous film theory') [11, 13, 14], according to which etching/polishing is caused by the anodal action of electrolyte on, first of all, the peaks of surface roughness of a metal subjected to etching/polishing and to a lesser degree, on cavities. Such diversified dissolving can be explained by the formation of an anode film known also as a viscose film. This film covers the metal surface, with projecting surface roughness peaks being covered by a thinner film layer than the cavities (Fig. 7.2). The resistance of the thinner film layer (on the peaks) is lower than that of the thicker film layer (in the cavities).

For this reason the peaks of surface roughness dissolve to a greater extent than the cavities. In this manner, the surface irregularities are gradually made even until polishing or etching. This theory clarifies the basic course of electrolytic etching or electrolytic polishing, yet it does not explain individual cases. The Jacquet theory is right only for a certain group of phenomena; it properly explains the processes of electrolytic etching/electrolytic polishing of copper and steel in an electrolyte composed of phosphorus acid and organic additions as well as electrolytic etching/electrolytic polishing of steel in a chloric-acetic electrolyte. In these cases one can observe the formation of a dense colloidal or liquid film from the reaction products. Many theories (Elmore theory, Hoar theory, and Edwards' theory) have been developed in order

Surface Treatment in Bonding Technology. https://doi.org/10.1016/B978-0-12-817010-6.00007-2

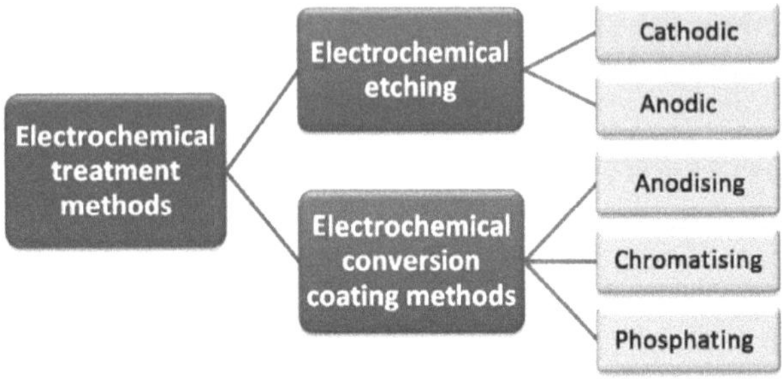

Fig. 7.1 Electrochemical treatment methods.

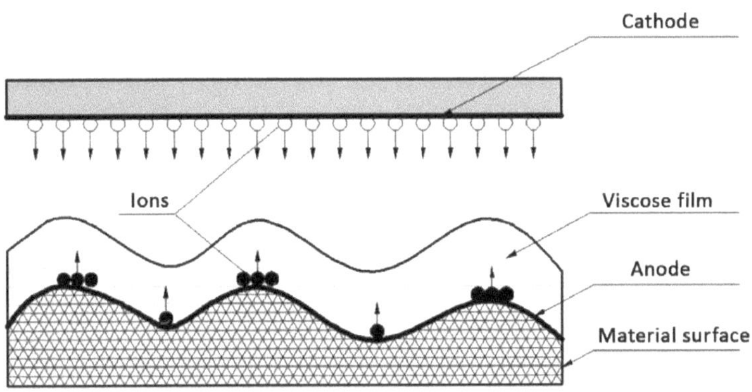

Fig. 7.2 Formation of anode film.

to explain this and other phenomena [13]. Yet they all share the same disadvantage, i.e. they explain phenomena in a temporary manner and are usable only in individual cases [13].

The most important factors affecting etching conditions include the potential difference (PD) and the density difference of current, electrolyte temperature, electrolyte stirring, surface pretreatment, heat treatment of the specimens, treatment time, dimensions of the electrolyser used for etching, and electrolyte consumption degree [11].

The process of electrochemical etching, which may be conducted cathodically or anodically, utilises the electrical current action. Cathodic etching involves reduction of metal oxides by the hydrogen released in the process. In addition, the method exhibits mechanic properties consisting in removing oxides off the surface of metals, which further facilitates etching. In anodic etching, the metal substrate becomes electrolytically melted at its surface while oxides become detached from the surface as a result of oxygen release. As a rule this is sulphuric acid at various concentrations that is the main component of the etching electrolyte [15].

The electrochemical methods are also employed to develop conversion coating on the surface of adherends in preparation for adhesive processes. The beneficial properties of these coatings have a positive effect on, *interalia*, the conditions for joining by adhesive or other bonding methods.

7.1.2 Anodising

Anodising (electrolysis process), also referred to as eloxation, or electrolytic oxidation, is one of the most widely applied electrochemical treatment methods [16–22] and during this process oxide coatings are produced (one of the types of conversion coatings). The anodising process is predominantly employed in treatment of aluminium and its alloys, as well as certain types of steel, titanium, and magnesium alloys [9, 16, 18, 23–25].

The electrolysis process is carried out under specific current conditions (current and voltage) and at a constant temperature. Sample parameters of industrial processes carried out using direct current are presented in Table 7.1 (prepared on the basis on Refs. [6, 7, 19, 22]).

The anodising process is twofold and utilises two simultaneous electrochemical reactions; one is the build-up of the oxide layer, which melts as a result of the other involved reaction. The process is realised in a bath of specific chemical composition [3, 15, 16, 20, 26, 27].

The metal subjected to anodising is immersed in a properly selected electrolyte solution. During the current flow on the anode, the formation of the oxide film takes place (7.1) [25]

$$\mathrm{Me} + n\mathrm{H_2O} \rightarrow \mathrm{MeO}_n + 2n\mathrm{H}^+ + 2ne^- \tag{7.1}$$

where Me—anodised metal.

At the same time hydrogen is formed at the cathode (Eq. 7.2)

$$2n\mathrm{H_2O} + 2ne^- \rightarrow n\mathrm{H_2}\uparrow + 2n\mathrm{OH}^- \tag{7.2}$$

The properties and the final shape of the conversion coating depend on the type of treated material (type of aluminium), the coating formation method, and the preliminary surface treatment employed in the process. Surface porosity is regulated by modifying the basic conditions of anodising, i.e. voltage, temperature, electrolyte concentration, and anodising time [6, 7, 25]. Furthermore, porous surface enables applying various colouring techniques on anodised aluminium objects.

The thickness of the oxide surface developed on the surface of metals in the anodising process is slightly bigger; hence it provides superior corrosion protection to the natural passive layer. The schematic representation of the surface structure of anodised aluminium alloy and stages of oxide layer development in aluminium anodising process is presented in Fig. 7.3 (prepared based on Refs. [6, 7]).

The thickness of the coating depends on the type of metal, the applied voltage, the temperature of the electrolyte, and the time of anodising [25].

During anodising, the aluminium at the surface becomes converted into the aluminium oxide as a result of solution of OH– electrolyte ions action, which is represented by the following Eq. (7.3) [6, 7]:

$$2\mathrm{Al} + 6\mathrm{OH}^- \rightarrow \mathrm{Al_2O_3} + 3\mathrm{H_2O} + 6\mathrm{e}^- \tag{7.3}$$

Table 7.1 Conventional anodising process (examples)

Type of process	Parameters/properties				
	Electrolyte	**Current conditions**	**Temperature (°C)**	**Time (min)**	**Oxide coating properties/ coating thickness**
Alumilite	10%–20% H_2SO_4	1–2 A/dm^2 10–20 V	20–30	10–30	Transparent 5–30 μm
Hard anodising	7%–15% H_2SO_4	2–5 A/dm^2 20–120 V	−10 to +10 (usually 0)	15–90	Transparent, gray, black 3–150 μm
Eloxal GXh	3%–5% $(COOH)_2$	1–2 A/dm^2 40–60 V	18–20	40–60	Yellowish 10–60 μm
Bengough-Stuarta process	2.5%–3% CrO_3	0.1–0.5 A/dm^2 40 V	40	40	Gray, opaque 2.5–15 μm
Boeinga process	10% H_3PO_4	10–12 V	23–25	20–30	Colourless 1–2 μm
Condenser process	9%–15% H_3BO_3	50–500 V	90–95	–	Colourless

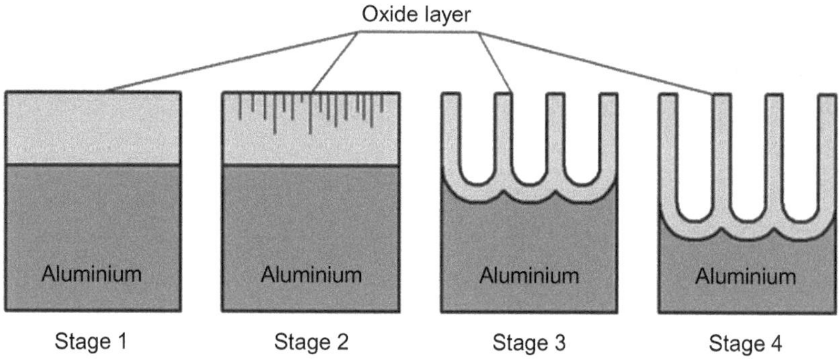

Fig. 7.3 Stages of oxide layer development in aluminium anodising process: Stage 1, development of barrier layer; Stage 2, crack propagation; Stage 3, creation of pores; Stage 4, growth of pores.

The thickness of the oxide layer increases with anodising time [18, 28], and ranges from several tens of millimetres, in the case of protective and decorative layers, to over 100 mm, in robust technical layers. The structure of produced coatings is determined by the power of electrolyte solution to dissolve the oxide layer. In the initial stage of the oxide coating conversion process, a thin layer of Al_2O_3, of 0.01–0.1 μm, is produced. This layer, referred to as the barrier layer [6, 7, 22], is then transformed into a porous layer when the barrier layer converts at the oxide-electrolyte interphase The barrier layer is formed when Al^{3+} ions migrate in the electric field and become involved in the reaction with O^{2-}or OH^- ions to produce an anhydrous Al_2O_3. Finally, the porous layer is subject to further growth in thickness (from several to 100 μm) [6, 15, 22] (Fig. 7.4, prepared based on Refs. [6, 7]).

Oxide coatings converted by means of solutions exhibiting medium dissolving capacity are characterised by a specific porous structure: the cylindrical pores are structures normal to the penetrated surface, arranged in a regular array, whose diameter ranges from several to several dozen nanometres, depending on the conditions of anodising. The cylindrical whiskers rest on the thin nonporous barrier layer developed on the metal surface.

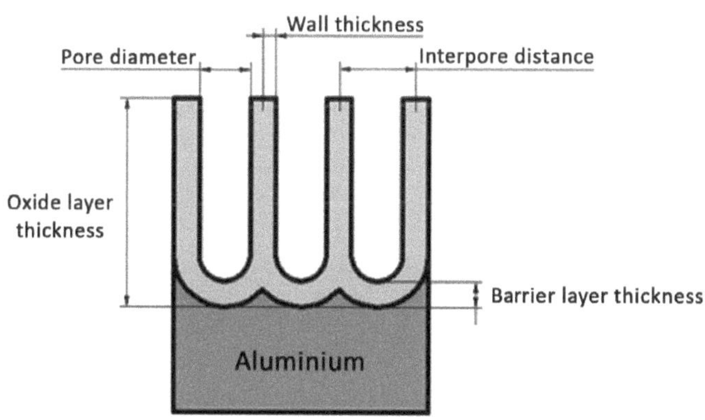

Fig. 7.4 Schematic parameters of porous anodic oxide layer of aluminium.

Dense concentration of a large number of small-diameter pores causes that the surface of thus prepared alumina is of suitable texture and exhibit good properties in terms of adsorption, which are favourable regarding preparation for dyeing [7, 8, 22, 26, 29].

Protective and decorative anodising processes are often combined with aluminum dyeing process.

There are three methods to obtain coloured oxide coatings (Fig. 7.5):

(1) inseparable dyeing (self-coloured),
(2) dyeing with organic and inorganic dyes, and
(3) electrolytic dyeing.

Inseparable dyeing (self-coloured) is one technological operation with anodising the surface. Bath components, in addition to sulphuric acid, are usually organic acids. At high voltages and current densities, the resulting oxide coating incorporates decomposition products of organic acids, aluminum alloy components as well as non-oxidised aluminum grains (Fig. 7.5A). In this method, light-black to black-coloured coatings are obtained that are hard and abrasion resistant.

Dyeing with dyes is carried out as a separate operation after completion of the anodising. The dyes contained in the dyeing baths penetrate the pores in the oxide coating and adsorb on the surface of aluminum oxide there. Organic dyes enable the surface to be given a wide range of colours [25].

Dyeing operations are commonly performed with the use of organic dyes, which are easily adsorbed by the alumina surface [15, 22, 29]. The dyeing process is generally carried out in a diluted solution of a dyeing agent, at elevated temperature and maintaining the pH of the solution within the set range. One disadvantage, which significantly limits the scope of dyes used in architectural aluminium applications, is that the layer of colorant applied on the surface shows little resistance to light exposure, particularly sunlight. Therefore, these are the nonorganic salt solutions, providing superior dye fastness that may be applied. Golden colouring is obtained by the use of ferric ammonium oxalate.

If the dyeing bath is a solution of inorganic compounds, then during the dyeing the specific ions are approved in the pores of the coating (Fig. 7.5B). There, compounds of the desired colour are lost as a result of chemical reactions [25], e.g. golden colouring is obtained by the use of ferric ammonium oxalate.

Electrolytic dyeing is applied on previously anodised surfaces. During electrolysis, a specific metal (tin, nickel, cobalt, and copper) is released in the pores of the anodic oxide coatings. When the colouring metal is tin, the objects after anodising are immersed in a tub containing bathing; stannous sulphate solution with stabilising additives. The bath is supplied with 50 Hz alternating voltage of 25 V. One pole is electrodes made of, e.g. stainless steel, located in the bathtub. The second pole consists of stained objects. During dyeing, metal particles deposit on the oxide coating at the base of the pores (Fig. 7.5C). The colour of the coating is the result of light scattering through the metal pores. Thus, the larger the number of metallic particles found in the pores, the darker is the colour of the coating.

Test results assessing the addition of dyes during anodising are more extensively described in work [30].

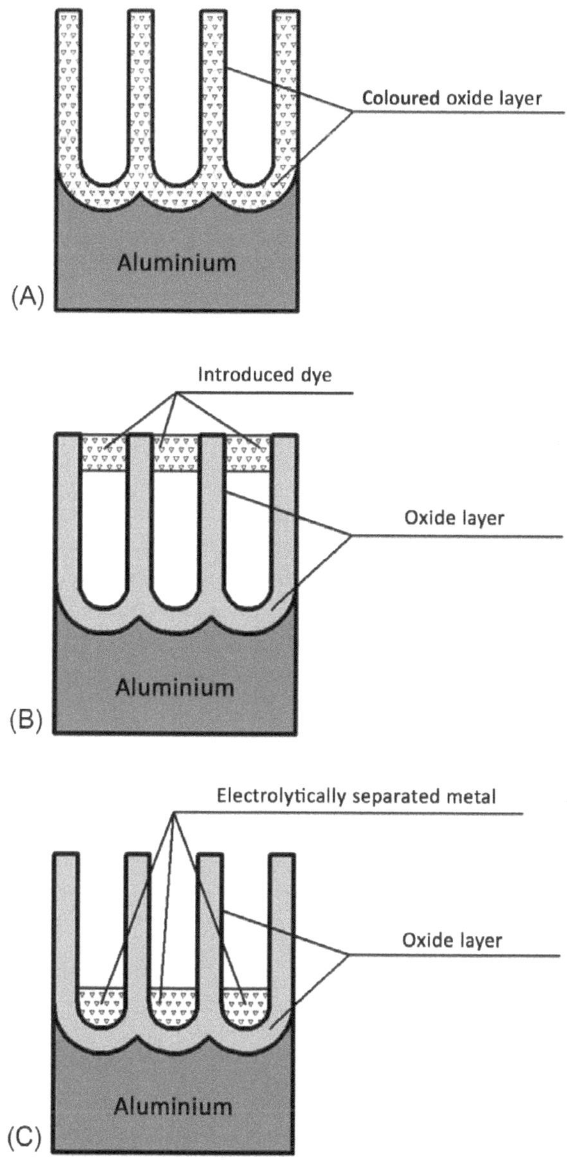

Fig. 7.5 Anodic oxide coatings on aluminum: (A) inseparable dyeing (self-coloured), (B) dyeing with dyes, (C) electrolytic dyeing.

After dyeing, the anodic oxide coatings are subjected to a sealing process, e.g. by immersing the anodised elements in hot distilled water [19, 25]. During sealing, aluminium oxide attaches the water molecules, the pores on oxide film are filled with aluminium hydroxide as volume expansive transformation from aluminium oxide to hydroxide takes place [19]. The best results are obtained when the temperature is near the boiling point of water, because only slightly above 80°C is formed a sparingly soluble and corrosion resistant $Al_2O_3H_2O$. It is also possible to apply sealing in solutions of nickel acetate or sodium dichromate [25].

Protective and decorative layers are predominantly applied in sulphuric acid baths. The coating thickness is specified in norms to fit a particular application and an expected service life of a bath-treated object. Anodising is commonly followed by a layer sealing operation (mentioned above), which is aimed at closing the capillary pores of the coat [22].

Anodising methods are classified into three major groups (Fig. 7.6), which additionally include subgroups [15, 22, 31, 32]:

- Type I anodising,
- Type II anodising,
- Type III anodising.

Type I anodising

The oxide coating is produced in the chromic acid solution, which acts as electrolyte. This anodising type is commonly found in applications on the objects of complex geometry, hence it is frequently used to anodise casts. The oxide layer, which is thin and dark but relatively stronger than the identical one obtained in anodising type II, penetrates the element in 50%, whereas the other half builds up on its surface.

We can distinguish the following subtypes of type I anodising [22, 31, 32]:

- Type I A—chromic acid anodising,
- Type I B—low voltage chromic acid anodising, and
- Type I C—nonchromic acid anodising, an alternative to I and I B processes—e.g. boric acid anodising method developed by Boeing [7, 8, 22].

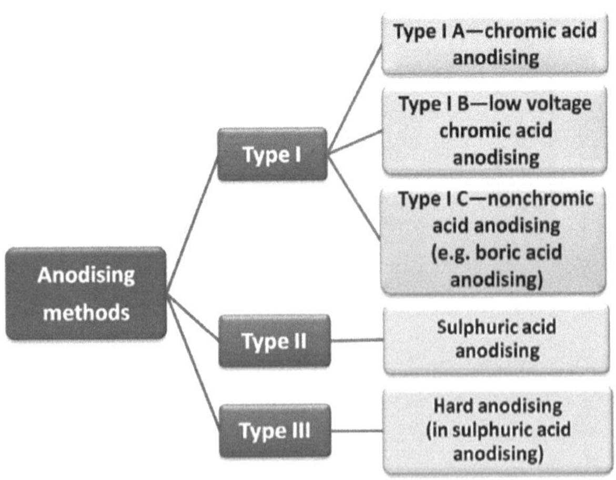

Fig. 7.6 The examples of anodising methods.

Type II anodising

Type II anodising is by far the most widespread anodising method in use. In this procedure the oxide coating is generated in the sulphuric acid bath, which acts as an electrolyte. The oxide layer penetrates the substrate in 67%, whereas the remaining 33% grows on the surface. The characteristic features of the type II oxide coating is its porous surface and light colouring, both of which facilitate applying dyes on anodised elements, in addition the layer exhibits considerable thickness, which translates to good abrasion resistance [22, 32].

Type III anodising

Type III anodising, which tends to be referred to as *hard anodising* [22, 32, 33] on account of the significant hardness of the conversion coating, is similarly carried out in the sulphuric acid bath. However, the acid concentration level is higher than in the case of anodising type II. The temperature of the process is maintained in the range of −3°C to 0°C. The initial voltage of 12V is gradually increased until it reaches 60V. A similar procedure is applied to the current density, which is gradually raised from the initial 2.5 A/dm^2 to approximately 3.5 A/dm^2. The sulphuric acid bath is typically supplemented with the addition of the oxalic acid. The oxide coating penetration is 50%, and 50% of the layer is built up on the surface of the substrate. The surface of objects subjected to anodising acquires hardness equivalent to the upper limit for the Rockwell C hardness scale = 65 HRC, which is considerably harder than the coating hardness developed during anodising type II. The hardness of coating obtained in anodising type III corresponds to hardened carbon steel.

The characteristics of the anodic oxide coating on the surface of aluminium have been presented in, e.g. Z. Wiącek's works [34, 35]. G.W. Critchlow et al. [9] have investigated the impact of surface treatment of aluminium adherends on the strength of joints. The latter study implemented, *interalia*, the boric and the sulphuric acid anodising methods.

7.1.3 Chromatising

Chromatising is a process during which chromium oxides are generated on the surface of metals [31, 36], in one of the two basic methods of chromatising:

- the electrochemical method—the objects are immersed in the chromatising solution in the presence of electrical current and
- the chemical method—the objects are immersed in the chromatising solution (which may contain, e.g. dichromians).

Chromate coatings are representatives of conversion coatings, i.e. layers obtained in the process of chemical or electrochemical conversion of the subsurface layer of metal in the presence of necessary chemical agents, which involves formation of salt that is practically insoluble in the medium where the reaction occurs.

Chromate treatment of electrolytic coatings and metals is applied, e.g.

- to increase the corrosion resistance of metals or protective metal coatings,
- to apply the antifingerprint coating,

- to apply the ground coat on certain metal substrates in preparation for painting or bonding with other organic coatings, and
- for decorative or colour effect.

The chromate conversion coating bath is typically composed of chromates or chromic acid. Reactions that constitute the essence of the chromatising process occur in the presence of activating anions: fluoride or sulphate. Even in face of local damage, chromate coatings retain their protective properties, which is the consequence of the fact that the coat contains chromium compounds that are anodic inhibitors of corrosion.

In addition, chromatising produces coating that may be [28]:

- yellow—the amorphous yellow coat of specific weight between 0.1 and 1.0 g/m^2, containing hexavalent chromium,
- green—the amorphous chromium phosphate of specific weight between 0.1 and 2.0 g/m^2, the colour of the coating from colourless to rich green. Unlike the yellow ones, these coats do not contain hexavalent chromium.

Immediately upon deposition, chromate conversion coatings acquire gel-like characteristics: they are most soft, easily abradable, and exhibit absorptive properties. Thus prepared coating develops its complete properties on drying.

7.1.4 Phosphating

Phosphating is the most widely used metal pretreatment process for the surface treatment and finishing of ferrous and nonferrous metals. Due to its economy, speed of operation, and ability to afford excellent corrosion resistance, wear resistance, adhesion, and lubricative properties, it plays a significant role in the automobile, and process and appliance industries [37].

Phosphation (phosphatisation) is a process of chemical or electrochemical production of a protective matte-gray phosphate coating on the surface of metals, carried out in hot solutions of phosphates and phosphoric acid [31, 37, 38]. This is mostly used as a pretreatment method in conjunction with another method of corrosion protection. Phosphating is the process of producing conversion coatings of zinc phosphate and/or iron phosphate with certain properties, including those that improve the adhesion of adhesive materials to the substrate being prepared. As a result of this process, the anaerobic surface (e.g. steel) is transformed into an amorphous or crystalline phosphate coating, this is characterised by a highly developed surface.

The phosphate layer contributes to the advantageous properties of the surfaces on which it is produced, such as [31, 38, 39]:

- ensures surface roughness, thanks to which it increases the adhesion of adhesive materials (such as adhesives, paints, and varnishes) to the substrate,
- acts as a barrier layer separating metal from oxygen and ambient water, and
- contributes to the slowing down of galvanic corrosion due to the formation of a strong bond between the conversion layer and the substrate.

The phosphatising process consists in immersing the phosphated metal in an aqueous solution of monosubstituted $Me(H_2PO_4)_2$ phosphate containing free phosphoric acid (Me^-, $Fe^{2+,}$ Zn^{2+}, and Mn^{2+} lub Ca^{2+}). The phosphatising process consists in the reaction of a metal dihydrogen orthophosphate. At the interface between the metal and solution phases, there is a shift in the balance of soluble salts towards the formation of sparingly soluble mono- or neutral salts. This can be represented by reactions (7.4, 7.5) [40]:

$$Me(H_2PO_4)_2 \rightarrow MeHPO_4 + H_3PO_4 \tag{7.4}$$

$$3Me(H_2PO_4)_2 2 \rightarrow Me_3(PO_4)_2 + 4H_3PO_4 \tag{7.5}$$

The mechanism of formation of phosphate coatings is reduced to the formation of a thin layer of iron phosphates in the first phase as a result of the reaction between the base metal and the phosphoric acid, which proceeds until a uniform film of these phosphates is formed. In the second stage, crystallisation takes place on the surface of the metal from inert solutions and monoacid phosphates, which in the presence of phosphoric acid tend to switch to a water-soluble, two-sourced form.

In the phosphatising bath, the concentration of Me^{+2}, H^+, $H_2PO_4^-$, and $H_2PO_4^{-2}$ should be selected so that for the reaction temperature a small change in the concentration of hydrogen ions leads to the formation of insoluble phosphates. The process can be carried out in two types of bath. In the first case, alkaline phosphates are formed on steel iron (V) phosphate, in the second environment containing heavy metals such as Zn, mainly a coating composed of zinc phosphate (V) is formed [31].

Phosphate coatings produced on steel, zinc or zinc coated objects are divided into thick, medium, thin, and very thin phosphate coatings. The thickness of the manganese phosphate coating is approximately 1/3 of the unit weight of the coating. Thick, medium, and thin coatings have a crystalline structure, while very thin coatings can be amorphous, crystalline or biphasic, involving both crystalline and amorphous phases. All phosphate coatings are crystalline coatings but with very small grain size. In general, the thickness of the phosphate coating varies from 1 to 20 μm, with individual crystals reaching the dimensions of 100 μm and more in the plane of the sample [25, 39, 40].

Depending on the technological requirements, phosphate coatings are applied by immersion or spraying. Depending on the purpose, different coating thicknesses are given in the Table 7.2 [25].

Typically, the phosphate coating is amorphous and consists mainly of various ferric phosphates, which depends to a large extent on the basic components of the phosphatising bath, the type of substrate, and the method of surface preparation. In baths containing

Table 7.2 Phosphate coating thicknesses

Type of coating	**Unit weight (g/m^2)**
Very thin coating	>2.0
Thin coating	2.0–4.5
Medium coating	4.5–7.5
Thick coating	<7.5

nitrite accelerators in the form of nitrites, chlorates or other oxidants, very fine crystals are formed. In baths containing mono-substituted alkaline phosphates, a coating with a content of up to 60% $FePO_4$ is deposited, and the remaining 40% is Fe_2O_3 ferric oxide.

Phosphate coatings have a relatively high porosity of 0.5%–2% of the total surface depending on the type of coating and to some extent on the thickness of the coating. It can be assumed that with a properly conducted phosphating process, the porosity decreases with the increase of the treatment time and the increase of the coating thickness. Above all, however, the porosity is influenced by the chemical composition and structure of the coating obtained from various baths. The porosity of the coating lowers its protective value and therefore, for anticorrosive purposes, the phosphate coatings are treated as a primer, for example under paint coatings. In addition, the anticorrosive properties of the phosphate coating-paint coatings are influenced by the quality of the phosphate coating itself, as an additional protective barrier and a factor that increases the adhesion of the paint coat to the substrate. The affinity of the organic lacquer film with the nonmetallic conversion coating is much higher than with the clean metal surface, and not only due to the highly developed surface of the coating but also due to the chemical bonding of phosphate crystals to organic resins. In addition, the phosphate coating prevents the spread of rust as a result of subcoating corrosion, the so-called filiform corrosion [25, 31].

The typical technological processes of phosphating involve the following operations [37, 39]:

- cleaning the surface,
- rinsing,
- surface activation,
- phosphating,
- rinsing,
- neutralising rinse (optional),
- drying, and
- application of supplemental coatings: lubricants, sealers, oil, etc.

After rinsing, the other operation can be used like pickling the coating in alkaline phosphate baths or zinc phosphates, subsequent rinsing, and passivation in chromic acid (VI). The quality of the coating is assessed on the basis of a visual inspection of 5 corrosion or rust deposits, a sample immersed in 1% NaCl solution for 5 min for thin coatings or 3% NaCl for 15 min for coarse coatings (PN-81/H-97016).

The main uses of phosphating are [31, 36, 37, 39, 40]:

- corrosion protection in conjunction with organic coatings, such as paints and polymer films,
- facilitation of cold-forming processes, such as wire drawing and tube drawing, or deep drawing,
- corrosion protection in conjunction with oils and waxes,
- corrosion protection with no subsequent treatment,
- improving antifriction properties, such as break-in, wear resistance, antigalling and coefficient of friction, and
- providing strong adhesion bonding for subsequent painting or other organic coating.

Lots of information of phosphating process issues and test is presented in the works prepared by Hagans and Haas [36], Sankara Narayanan [37], and Hassib Abbs [38].

7.2 Experimental test—Anodising and chromatising

7.2.1 Methods

Adherend and adhesive joints

Tests were conducted on shear loaded single-lap adhesive joints of 2024 aluminium alloy sheet. The scheme of the tested adhesive joint was shown in Fig. 7.7, and the adhesive joint real dimensions were presented in Table 7.3.

The selected properties of 2024 aluminium alloy were presented in Table 7.4.

Test specimens were joined with a Loctite 3430 two component epoxy adhesive, at weight ratio of 1:1. Loctite 3430 adhesive develops the high strength at the room temperature within 2–4 h. The properties of uncured components of Loctite 3430 adhesive were shown in Table 7.5.

The resin and the curing agent were applied directly from the dual cartridges by dispending through the static mixer head supplied. The mixing time was approximately 15 s after uniform colour (ultraclear) was obtained. The adhesive was applied immediately after mixing to one of the surfaces (after suitable surface treatment process) to be joined. Adherends were assembled immediately after mixed adhesive has been applied. Working life of the mixed adhesive is 4–7 min at 22°C. The samples of adhesive joints were subjected to pressure of 0.2 MPa. The single-stage cure process at ambient temperature (22 ± 1°C) for 48 h was used.

Surface treatment methods

Four surface treatment methods of AW 2024 aluminium alloy sheet were applied, viz. self-colour anodising (SCA), yellow and black anodising in sulphuric acid as well as chromate conversion coating, which were collated in Table 7.6.

SCA process consists of the following, subsequent operations of set parameters [30]:

- pretreatment in nitric acid,
- rinsing with cold running water,

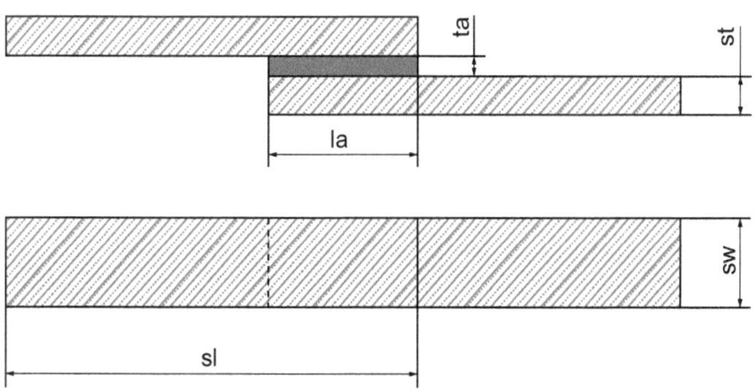

Fig. 7.7 Tested adhesive joints of aluminium alloy sheet.

Table 7.3 Dimensions of adhesive joint of aluminium alloy sheet

Sheet length s_l (mm)	**Sheet width s_w (mm)**	**Sheet thickness s_t (mm)**	**Adhesive joint lap length l_a (mm)**	**Adhesive thickness t_a (mm)**
100 ± 1.10	20 ± 0.10	0.7 ± 0.02	14 ± 0.75	0.1 ± 0.02

Table 7.4 The selected properties of 2024 aluminium alloy sheet [41]

Properties	**Value**
Young modulus	73.1 GPa
Poisson's ratio	0.33
Elongation at break	19%
Tensile yield strength R_m	324 MPa
Ultimate tensile strength	469 MPa

Table 7.5 The selected properties of uncured components of Loctite 3430 adhesive [42]

	Description/value	
Properties	**Resin**	**Curing agent**
Chemical type	Epoxy	Epoxy
Appearance	Ultraclear	Ultraclear
Viscosity, 25°C	20,000–30,000	20,000–30,000

Table 7.6 The tested surface treatment methods of AW 2024 aluminium alloy sheet

Surface treatment variant	**Designation**
Self-colour anodising	SCA
Yellow anodising	YA
Black anodising	BA
Chromatising	CH

- etching in sodium hydroxide,
- rinsing with warm running water,
- rinsing with cold running water,
- bleaching in nitric acid,
- rinsing with cold running water,
- anodising in sulphuric acid,
- rinsing with cold running water,
- rinsing with hot running water,
- compacting layer in hot water,
- rinsing with hot running water, and
- drying.

The technology process of colour anodising in sulphuric acid comprises the following operations [30]:

- etching in sodium hydroxide,
- rinsing with warm running water,
- rinsing with cold running water,
- bleaching in nitric acid,
- rinsing with cold running water,
- anodising in sulphuric acid,
- rinsing with cold running water,
- dyeing black or yellow,
- rinsing with cold running water, and
- drying.

In the case of anodising in sulphuric acid, anodic coating is generated at the aluminium-anodic layer interface. Anodic layer thickness depends on the ratio of electrochemical anodic film formation rate to the rate of its chemical dissolution in electrolyte. The prerequisite for film development is that pores are generated in its weakest spots. The pores volume amounts to approximately 30% of the anodic film, according to the Polish Norm PN-80/H-97023.

Chromate conversion coating technology of the analysed aluminium alloy samples consists of the following operations [30]:

- solvent vapour degreasing,
- alkaline degreasing,
- rinsing with raw water,
- rinsing with demineralised water,
- etching in sulphochromic bath,
- rinsing with raw water,
- rinsing with demineralised water,
- chromate coating,
- rinsing with demineralised water, and
- drying.

The chromate coating bath method applied in tests generated chromate conversion layer approximately 0.25–1 μm thick.

The basic parameters of main operations of particular technological processes, which are to form the layers on the tested aluminium alloy surface, were presented in Table 7.7. The information was prepared on the basis of engineering company standards.

Surface roughness

Surface roughness parameters of tested adherends surfaces were measured using the T8000 RC120-400 Hommel—Etamic-3D profilometer. The surface roughness parameters were determined in accordance with ISO 4278 standard and ISO 25178 requirements. In each batch, 5 specimens at 3 different positions were examined. On the whole, 15 measurement results were obtained per each batch. The highest and lowest values were rejected and not taken into account in further analyses.

Table 7.7 Technological parameters of aluminium alloy surface treatment [30]

Surface treatment operation	Technological parameters
Anodising in sulphuric acid	Type of electrolyte: H_2SO_4 Bath composition: 180–200 g/L Electrolyte temperature: 10–15°C Anodising time: 35 min
Coating compaction with hot water	Medium—demineralised water Water temperature: 90–95°C Treatment time: 25 min Optimum pH 4.5–6.5
Dyeing yellow	Dyeing agent—$K_2Cr_2O_7$ Bath composition: 40–55 g/L Bath temperature: 90–95°C Dyeing time: 30 min Optimum pH 4.5–6.2
Dyeing black	Dyeing agent—Acid Black 2ATT 100% Bath composition: 1–5 g/L Bath temperature: 62–67°C Dyeing time: 20–30 min Optimum pH 4.5–5.0
Chromate conversion coating	Factor type: ALODINE 1200 Bath composition: 11–23 g/L Demineralised water Bath temperature: 20–30°C Chromate coating time: 0.5–5.0 min

The following surface roughness parameters were examined:

R_a—arithmetic mean deviation,
R_z—ten-point height,
R_{max}—maximum roughness depth,
R_p—maximum peak height of the profile within the sampling length,
R_{pk}—reduced peak height,
R_k—core roughness depth,
R_{vk}—reduced valley depth,
R_t—maximum height of profile,
S_m—mean spacing of profile peaks, and
S—mean spacing of local peaks of the profile.

Determination method of surface free energy

The determination of the surface free energy (SFE) of aluminium alloy sheets was performed with the application of the Owens–Wendt method [43–46]. The Owens–Wendt method consists in determining the polar and dispersive components of the SFE. The polar component is the sum of polar, hydrogen, induction, and acid-base interactions. The equation for calculation of the surface free energy takes the following form (7.6) [43]:

$$\gamma_S = \gamma_S^d + \gamma_S^p \tag{7.6}$$

where: γ_S is the SFE; γ_S^d is the dispersion component of SFE; γ_S^p is the polar component of SFE.

Determining the SFE was conducted with two measuring liquids (polar and apolar) of known surface tension values as well as the polar and dispersive component values of the SFE. In order to determine the SFE, distilled water and diiodomethane were used as measuring liquids (O–W method). The values of necessary SFE γ_S and its components of the measuring liquids are presented in Table 7.8.

Direct measurement of contact angle of a liquid drop on the analysed surface of adherends after considered surface treatments was applied. The contact angle was measured at a temperature of 20±1°C and an air humidity of 30±2%. The volume of drops of the measuring liquids ranged from 1.4 to 1.7 μL. To every sample surface (there were five samples per each batch) five drops of the measuring liquids were applied. The measurements were made using a PGX goniometer manufactured by Fibro System (Sweden) and a PG programme for computer image analysis.

The SFE components are derived from relations (Eqs. 7.7, 7.8) based on [47]:

$$\left(\gamma_s^d\right)^{1/2} = \frac{\gamma_d\left(\cos\theta_d + 1\right) - \sqrt{\frac{\gamma_d^p}{\gamma_w^p}}\gamma_w\left(\cos\theta_w + 1\right)}{2\left(\sqrt{\gamma_d^d} - \sqrt{\gamma_d^p\frac{\gamma_w^d}{\gamma_w^p}}\right)} \tag{7.7}$$

$$\left(\gamma_s^p\right)^{1/2} = \frac{\gamma_w\left(\cos\theta_w + 1\right) - 2\sqrt{\gamma_s^d\gamma_w^d}}{2\sqrt{\gamma_w^p}} \tag{7.8}$$

where γ_S^d is the dispersion component of the SFE of the solid material; γ_S^p is the polar component of the SFE of the solid material; γ_d is the SFE of diiodomethane; γ_d^d is the dispersion component of diiodomethane SFE; γ_d^p is the polar component of diiodomethane SFE; γ_w is the SFE of water; γ_w^d is the dispersion component of water SFE; γ_w^p is the polar component of water SFE; θ_d is the contact angle of diiodomethane; and θ_w is the contact angle of water.

Strength tests

The tested adhesive joints of aluminium alloy sheet were subjected to failure testing, which allowed determining values of the strength of tested adhesive joints. The tests of shear strength of adhesive joints were conducted in compliance with the EN DIN 1465 standard using the Zwick/Roell Z150 Allround-Line testing machine.

Table 7.8 The surface free energy γ_L and its components of the measuring liquids [44]

Measuring liquids	The surface free energy and its components		
	γ_L (mJ/m^2)	γ_L^d (mJ/m^2)	γ_L^p (mJ/m^2)
Distilled water	72.8	21.8	51.0
Diiodomethane	50.8	48.5	2.3

7.2.2 Results

Surface roughness

The evaluation of ad surface treatment was derived from the test results of surface irregularities structure state. The height of surface irregularities was analysed based on measurements of selected parameters of surface roughness profile, while the irregularities structure on 3D visualisations. The number of analysed parameters was limited to eight height parameters and two parameters referring to spacing of peaks (Tables 7.9 and 7.10). Height parameters comprised two parameters frequently applied in general analysis of surface roughness and six parameters for in-depth surface roughness profile analysis.

Results collated in Table 7.9 and Table 7.10 represent mean values of nine profiles measurements. Examples of observed profiles as well as visualisations of surface roughness structure of test samples are presented in Figs 7.8–7.15.

Figs 7.8 and 7.9 show examples of surface roughness profiles of 2024 aluminium alloy surface sample after SCA in sulphuric acid.

Figs 7.10 and 7.11 present surface roughness profiles of 2024 aluminium alloy surface sample after yellow anodising in sulphuric acid (YA).

The examples of surface roughness profiles of 2024 aluminium alloy surface sample after black anodising in sulphuric acid (BA) were shown in Figs 7.12 and 7.13.

Examples of surface roughness profiles of 2024 aluminium alloy surface sample, subjected to chromate conversion coating (CH) were presented in Figs 7.14 and 7.15.

The results of surface roughness parameters of samples subjected to anodising indicate that anodising has an impact on the height of surface irregularities. Black anodising (BA) produced the highest decrease in roughness parameters values. The difference between SCA and mean values ranged from +7% to −10% approximately.

However, none of the anodising methods had a considerable influence on the surface irregularities structure. This observation is, furthermore, confirmed by profiles and 3D visualisations of test samples surface structure presented in Figs 7.8–7.15. The analysis of correlation between anodising technology and surface roughness profile proves that the application of technologies in question resulted in decreasing the R_{vk} and the R_k parameters from +(9.5–6.2)% to −(16.5–8.8)% respectively.

Among analysed samples surface treatment technologies, it was chromate conversion coating (CH) which proved to be of the greatest impact on both the height and the structure of surface irregularities, Figs 7.14 and 7.15. In the case of this method, the decrease in irregularities height parameters value ranged between −6.2% for the R_k and −42.6% for the R_{vk}, while there was practically no difference in the values of two parameters, R_p and R_{pk}. As a result, noticeable smoothing of the test sample surface was achieved. Reduced mean spacing of profile peaks S_m by 18% and the local S by 12.5% entails increased number of irregularities, as confirmed by representation of the roughness profile in Fig. 7.14 and the structure in Fig. 7.15.

The comparison of topographic features of 2024 aluminium alloy sheet samples subjected to surface treatment operations, which aimed at developing different films and are presented in Figs 7.10, 7.11, 7.13, and 7.15, concludes that the chromate conversion coating has a considerably more oriented structure (Fig. 7.15) as compared to

Table 7.9 Surface roughness parameters of 2024 aluminium alloy following anodising [30]

Surface treatment variant	Designation	Surface roughness parameters (μm)									
		R_a	R_z	R_z max	R_p	R_{pk}	R_k	R_{vk}	R_t	S_m	S
Self-colour anodising	SCA	0.45	4.12	6.30	1.40	0.47	1.21	1.26	6.57	127	36
Yellow anodising	YA	0.38	3.49	5.53	1.27	0.47	1.04	0.96	5.77	122	32
Black anodising	BA	0.44	3.99	5.87	1.46	0.57	1.16	1.24	6.37	129	35
Mean values for test samples		0.423	3.87	5.9	1.38	0.5	1.14	1.15	6.24	126	34.3

Table 7.10 Surface roughness parameters of 2024 aluminium alloy following chromatising [30]

Surface treatment variant	Designation	Surface roughness parameters (μm)									
		R_a	R_z	R_z max	R_p	R_{pk}	R_k	R_{vk}	R_t	S_m	S
Chromatising	CH	0.34	3.14	4.44	1.39	0.52	1.07	0.66	5.00	103	30

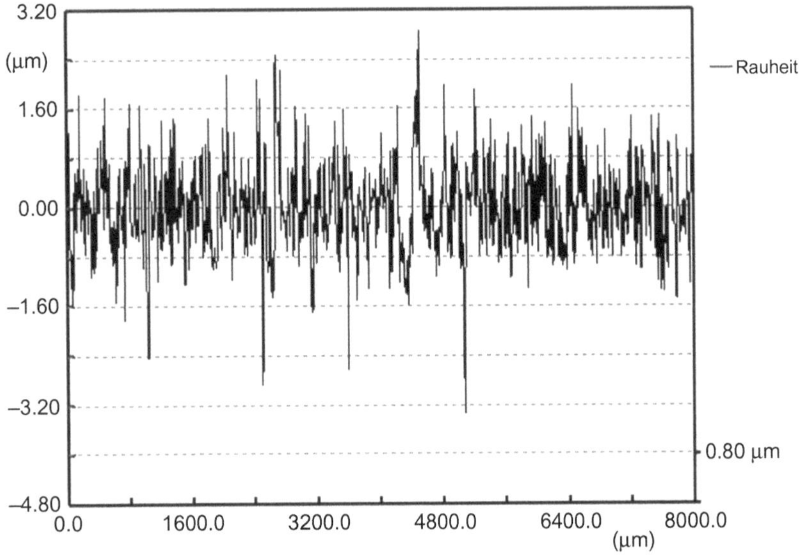

Fig. 7.8 Surface roughness profile of 2024 aluminium alloy sheet sample, self-colour anode (SCA) [30].

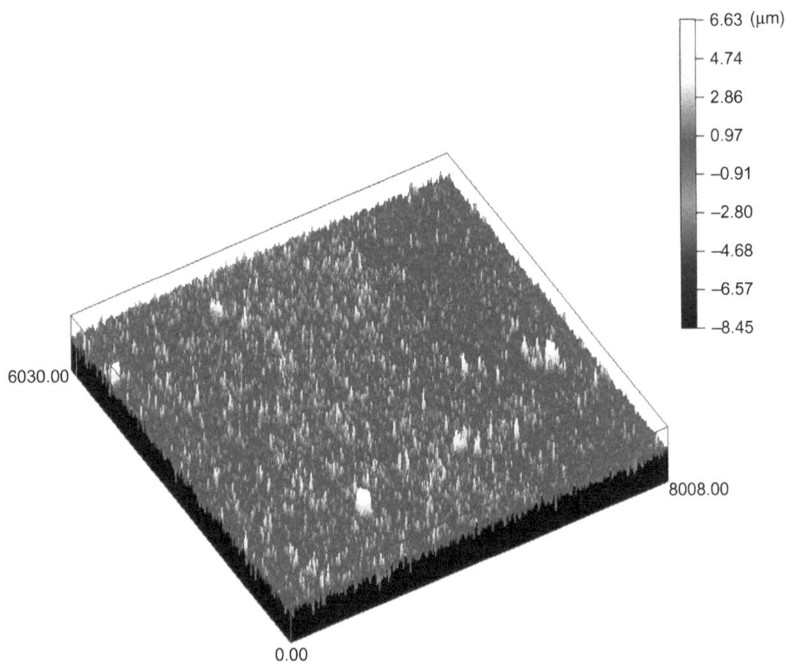

Fig. 7.9 Visualisation of surface roughness profile of 2024 aluminium alloy sheet sample, self-colour anode (SCA).

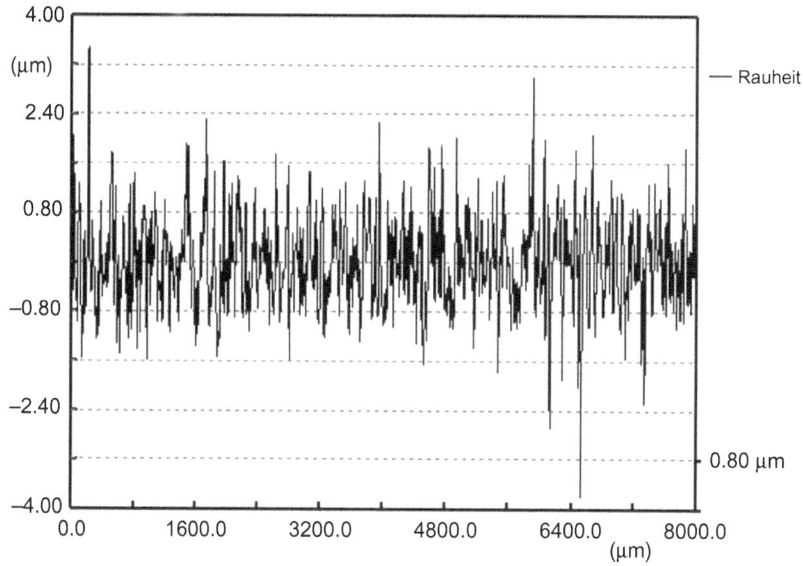

Fig. 7.10 Surface roughness profile of 2024 aluminium alloy sheet sample, yellow anode (YA) [30].

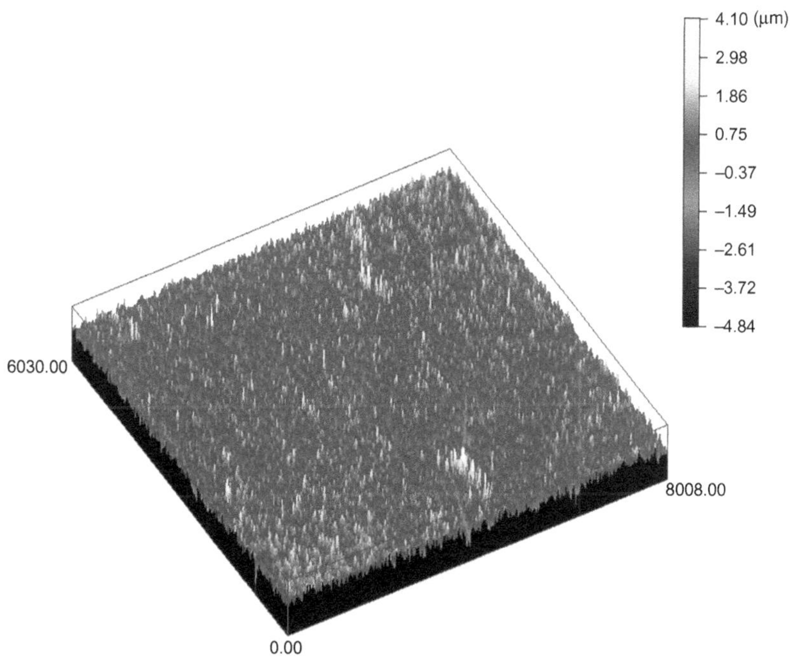

Fig. 7.11 Visualisation of surface roughness profile of 2024 aluminium alloy sheet sample, yellow anode (YA).

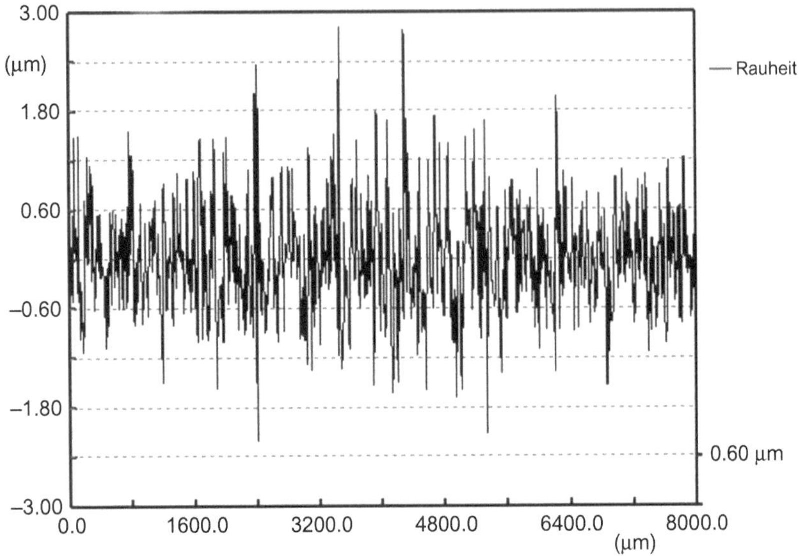

Fig. 7.12 Surface roughness profile of 2024 aluminium alloy sheet sample, black anode (BA) [30].

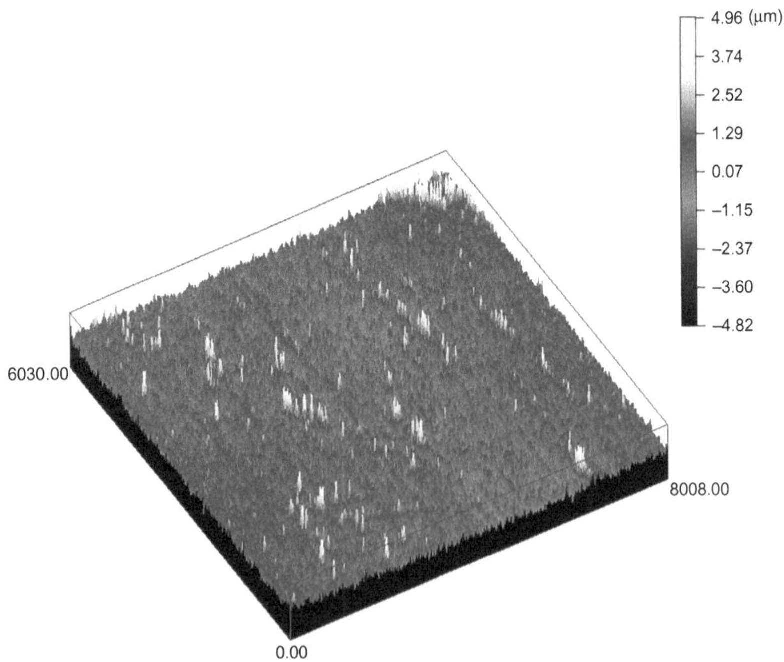

Fig. 7.13 Visualisation of surface roughness profile of 2024 aluminium alloy sheet sample, black anode (BA).

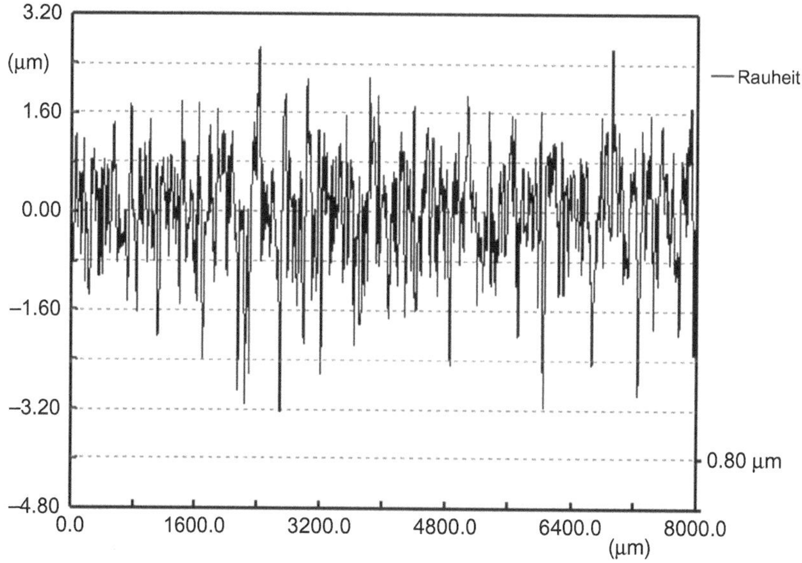

Fig. 7.14 Surface roughness profile of 2024 aluminium alloy sheet sample, chromate conversion coating (CH) [30].

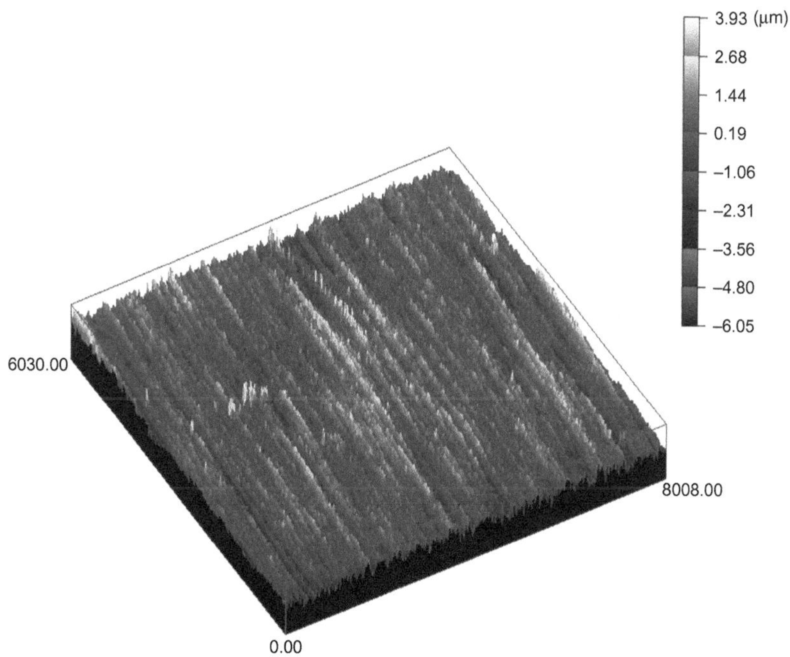

Fig. 7.15 Visualisation of surface roughness profile of 2024 aluminium alloy sheet sample, chromate conversion coating (CH).

anodic film. It might be the reflection of the process of sheet production from aluminium alloy. Chromate coating is thinner in comparison to anodic film; therefore, surface roughness mapping is greater than for anodic layers.

Surface free energy

The SFE and its components values, obtained from the contact angle measurement with distilled water and diiodomethane (the Owens–Wendt method), of specimens subjected to surface treatment operations are presented in Fig. 7.16.

The results of the SFE and its components calculations presented in Fig. 7.16 indicate that the highest SFE value was obtained in the case of chromate coating (69.8 mJ/m^2) and SCA in sulphuric acid (68.5 mJ/m^2). At the same time, the lowest value of the SFE was observed for black anodising.

Furthermore, considerable differences in the polar component values of the SFE after surface treatment operations were noted. The highest value of the polar component was observed in SCA, while the lowest in the case of black anodising, 42.7 and 11.5 mJ/m^2, respectively. The indices for particular surface treatment operations according to the highest polar component are as follows: yellow anodising—37.7%, black anodising—73%, and chromate coating—14.3%.

In terms of the dispersive component, it was noted that in the majority of cases (3 out of 4) comparable values were measured, however, they constitute a different percentage of the SFE. Following SCA the dispersive component of the SFE amounts to 37.7%, after yellow anodising—58.9%, 76.6% after black anodising and finally 49% after chromate conversion coating.

Another parameter under analysis was the contact angle value. In the case of SCA considerable standard deviation was observed for wetting with both water—4.11 degrees

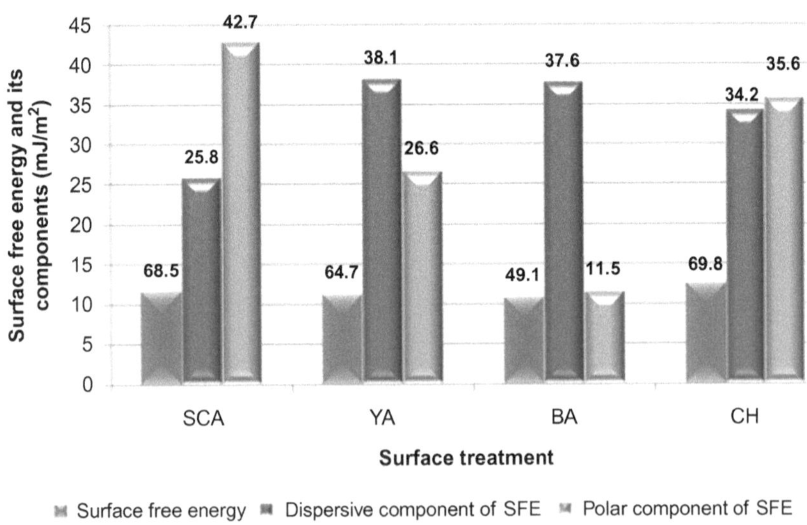

Fig. 7.16 SFE and its components of 2024 aluminium alloy surface following different surface treatment.

(for the angle of 20.9 degrees) and diiodomethane 1.61 degrees (for the angle of 38.4 degrees). Consequently, the scatter of results of the surface free energy measurements equalled 4.3 mJ/m^2.

With regard to anodising with dyeing the anodic film yellow, standard deviation of the contact angle in the case of distilled water was equal to 3.23 degrees for average contact angle of 27.9 degrees; for diiodomethane—0.77 degrees for the angle of 38.5degrees. The scatter of the SFE results was 4.2 mJ/m^2.

Subjecting surfaces to black anodising produced the lowest value of the surface free energy among the surface treatment operations analysed in the tests. It was observed that the standard deviation of the wetting angle of CH_2J_2 was small with diiodomethane as the measuring liquid, viz., 1.12 degrees for the average contact angle value of 27.9 degrees. Whereas the scatter of results was appreciable in the case of distilled water, as it ranged between 53 and 71.7 degrees. The measurements of the contact angle with distilled water as the measuring liquid showed high instability of the angle value. This probably has impact on the value of the polar component of the SFE, which for this surface treatment method was the lowest measured in the tests. The scatter of results of the SFE for this variant of surface treatment was 8.2 mJ/m^2.

Finally, in terms of chromate coating, the standard deviation of water contact angle equalled 2.68 (for the angle of 22.9 degrees), while in the case of diiodomethane—1.18 degrees (for the angle of 16.4 degrees). The difference in the SFE values, resulting from variable results of contact angle measurements, amounted to 3.2 mJ/m^2.

Adhesive joint strength

Comparative analysis of single-lap shear loaded adhesive joints of 2024 aluminium alloy sheet the surface of which was subjected to different surface treatment operations is presented in Fig. 7.17.

The highest shear loaded joint strength of 2024 aluminium alloy sheet was observed in the case of chromate conversion coating (11.96 MPa), and the lowest for the process of anodising accompanied by dyeing the anodic film black (10.59 MPa). Nevertheless, the values of adhesive joint strength of the analysed materials after certain surface treatment operations were similar, therefore, the test results were analysed statistically, based on algorithms presented in literature [48, 49].

Statistical analysis comprised tests of significance, which allow comparison of a mean value of a given feature in two statistical populations. The tests accounted for a type I error, occurring when a tested hypothesis, null hypothesis, is rejected when true. The alpha level accepted was $\alpha = 0.05$. The conducted tests were treated as small samples due to their number—below 30. The statistical verification was carried out with: the Student's t-test, the F-test and relevant statistical models. In the case of equality of standard deviations, Student's t-test was applied; otherwise F-test was used. The results of tested hypothesis of equality of variances based on proper models for the analysed adhesive joint variants are presented in Table 7.11, where S symbol means variance and m—mean.

The statistical analysis of test results reveals that the majority of mean values of shear loaded adhesive joints strength do not show statistically significant discrepancies, with an alpha level of significance of $\alpha = 0.05$. A notable difference was recorded

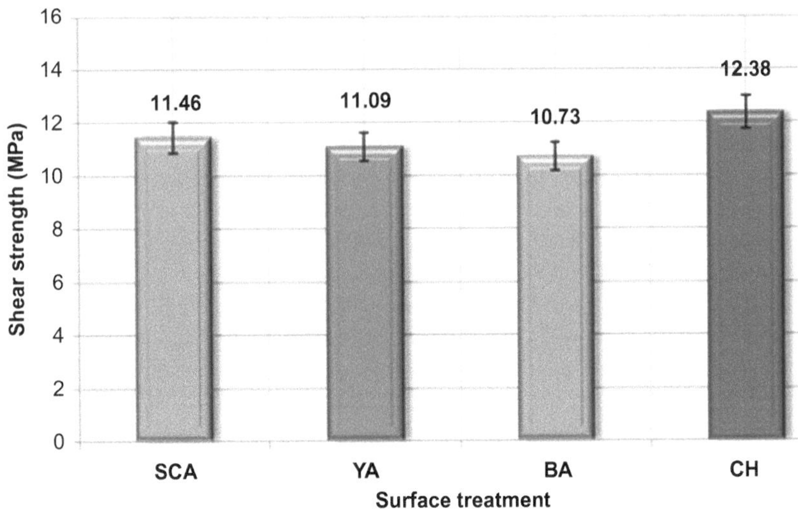

Fig. 7.17 2024 Aluminium alloy sheet adhesive joint strength following surface treatment: SCA, self-colour anodising; YA, yellow anodising; BA, black anodising; CH, chromate conversion coating.

only in the case of adhesive joint strength of 2024 aluminium alloy sheet the surface of which was subjected to black anodising and chromate conversion coating.

The conclusion that can be therefore drawn from the analysis is that the type of anodising is of no influence on adhesive joint strength, as no statistically significant differences at the accepted risk level were observed. Similarly in the case of anodising and chromate conversion coating comparison in most instances there were no notable differences in joint strength.

7.2.3 Discussion

The surface roughness test results demonstrated the correlation between the technology of anodising and the height of surface irregularities. Nevertheless, the examined anodising variants had practically no significant impact on the surface irregularities structure, as verified by observed profiles representations and 3D visualisations of the structure of samples surface, presented in Figs 7.6–7.13. Finally, it was chromate conversion coating which proved to be the most effective in changing the height and the structure of irregularities of the surface among the analysed surface treatment methods.

The SFE calculations results showed in some cases minor differences in the SFE values. However, when considering anodising in sulphuric acid it was observed that the modification of the process with dyeing applied is reflected in both total value of the SFE and the percentage distribution of its dispersive and polar components.

Test results in Fig. 7.14 regarding the polar component demonstrate considerable differences in its values. Simultaneously, no significant changes were noted as for

Table 7.11 Hypothesis of equality of two means tested with the student's ***t***-test with an alpha level of significance of $\boldsymbol{\alpha}=0.05$

Surface treatment Fig. 7.11	*F* statistic value	Critical value $F_{(\alpha;n_1-1;n_2-1)}$	Conclusion	*T* statistic value	Critical value $t_{(\alpha;n_1+n_2-2)}$	Conclusion
SCA-YA	1.866	3.388	$S_1^2=S_2^2$	−1.490	2.447	$m_1=m_2$
YA-BA	1.036	3.388	$S_1^2=S_2^2$	1.345	2.447	$m_1=m_2$
BA-CH	5.739	6.041	$S_1^2=S_2^2$	3.469	2.228	$m_1\neq m_2$
YA-SCH	5.945	6.041	$S_1^2=S_2^2$	2.188	2.228	$m_1=m_2$

the dispersive component. This seems to indicate that particular surface treatment methods affect the polar component, producing varying weight percentage of different interactions of the polar component (determined according to the Owens–Wendt method). The dispersive interactions, however, are practically not influenced by the chosen surface treatment methods. In addition, notable differences of the polar component of the SFE following pretreatment operations were observed.

Chromate conversion coating produced almost identical values of the dispersive and the polar component of the SFE. It may indicate similar contribution of dispersive and polar interactions. This may result in strong interactions between the material surface, with chromate coating, and adhesive medium (e.g. glue, paint, or varnish), provided that such a medium contains polar groups. Then strong polar bonds, which influence the adhesive strength of such a bond, can be formed. Comparing the values of the SFE and its components following chromate conversion coating with the surface roughness parameters it can be observed that the effect of decrease in roughness depth (at approximately 36% of decrease in height of irregularities of the profile) has beneficial influence on the value of the surface free energy as well as on the distribution of the SFE components.

The highest scatter of results of the SFE values was observed in the case of black anodising, and amounted to $8.2\,mJ/m^2$, which constitute nearly twice the value noted in other variants. Supposedly, after this surface treatment operation the surface of anodic film dyed black may be characterised by certain inhomogeneity of adhesive properties, which may contribute to its lower strength as compared to other analysed surface treatment variants.

The highest strength was manifested by sheet joints formed on the surface subjected to chromate conversion coating. In the case in question the distribution of components of the SFE was nearly equal; this might be of beneficial effect on the strength of adhesive joints. Furthermore, it appears that it was the surface structure of samples subjected to chromate conversion coating that additionally increased the joint strength. The increase in the number of irregularities was coupled with the decrease in the height parameter (the smoothing of the test sample surface following chromate conversion coating as compared to anodic film). In the case of black anodising, the distribution of the SFE components was 76.6% for the dispersive SFE component and 23.4% for the polar component.

7.2.4 Conclusion

The application of surface treatment operations on aluminium alloy sheets leads to certain changes in the SFE values. However, different variants of anodising in sulphuric acid resulted in changing percentage distribution of the dispersive and the polar component of the SFE.

The comparison of the SFE and its components values with the surface roughness parameters indicates that increasing the number of irregularities together with decreasing their height parameters has a positive effect on both the SFE value and the proportions of the dispersive γ_S^d and the polar γ_S^p components.

In terms of the strength of adhesive joints of aluminium alloy sheets the surface of which was subjected to anodising or chromate conversion coating operations it was observed that in a number of cases there were no statistically significant differences in the mean values recorded. It emerged, however, that anodising with dyeing the anodic layer black, for which the greatest scatter of the SFE calculation results was noted, produced the lowest joint strength (statistically significant as compared with chromate conversion coating). It appears that the selection of the dye may determine adhesive properties. In addition, the greatest decrease in the roughness parameters values was observed for this anodising method. Most likely the inhomogeneity of adhesive properties contributes to lower joint strength.

Another observation concerns the increase in the number of irregularities with simultaneous decrease in their height parameters, induced by chromate conversion coating operation, which improved the strength of adhesive joints.

To conclude, the choice of surface treatment affects the surface roughness, adhesive properties as well as adhesive joints strength. It appears that the analysis of all factors ensuring strong adhesion between the adhered surface and the adhesive, together with high joint strength, is required.

References

[1] C.V. Cagle, Handbook of Adhesive Bonding, McGraw-Hill, New York, 1973.

[2] S. Ebnesajjad, Adhesives Technology Handbook, second ed., William Andrew Inc., Norwich, NY, 2008.

[3] S. Ebnesajjad, C. Ebnesajjad, Surface Treatment of Materials for Adhesive Bonding, second ed., William Andrew Inc., Norwich, NY, 2013.

[4] C.A. Grubbs, Anodizing of aluminum, Met. Finish. 99 (2001) 478–493.

[5] F.W. Eppensteiner, M.R. Jenkins, Chromate conversion coatings, Met. Finish. 98 (2000) 497–509.

[6] G.E. Totten, D.S. MacKenzie, Handbook of Aluminum, Marcel Dekker Inc, New York, 2003.

[7] A. Eftekhari, Nanostructured Materials in Electrochemistry, Wiley-VCH, Weinheim, 2008.

[8] J.W.W. Willstrop, H. Sutton, Anodic films on aluminium and its alloys for protection against corrosion, Trans. IMF 15 (1939) 53–68, https://doi.org/10.1080/00202967.1939.11869374.

[9] G.W. Critchlow, D.M. Brewis, Review of surface pretreatments for alluminium alloys, Int. J. Adhes. Adhes. 16 (1996) 255–275.

[10] C. Nanev, K. Dichera, Electrochemical etching, Acta Phys. Hung. 57 (1985) 271–283.

[11] M.S. Węglowski, Electrolytic etching in welding metallography, Bull. Inst. Weld. (Biuletyn Intsytutu Spawalnictwa—polish journal) 2 (2013) 5–10.

[12] V. Mardirosevici, The Application of Electro-Etching in the Manufacturing Process for Cardiovascular Stents, Dissertation, Galway Mayo Institute of Technology, Ireland, 2014.

[13] R. Rokicki, T. Hryniewicz, Enhanced oxidation–dissolution theory of electropolishing, Trans. Inst. Met. Finish. 90 (9) (2012) 188–196.

[14] J. Dobrowolski, Electropolishing, Państwowe Wydawnictwa Techniczne, Warszawa, 1952 (in Polish).

[15] A.K. Graham, H.L. Pinkerkton, Electroplating Engineering Handbook, Reinhold Publishing Corporation, New York, 1971.

[16] L. Domingues, J.C.S. Fernandes, M. Da Cucha Belo, M.G.S. Ferreira, L. Guerra-Rosa, Anodising of Al 2024-T3 in a modified sulphuric acid/boric acid bath for aeronautical applications, Corros. Sci. 45 (2003) 149–160.
[17] A. Bjørgum, F. Lapique, J. Walmsley, K. Redfird, Anodising as pre-treatment for structural bonding, Int. J. Adhes. Adhes. 23 (2003) 401–412.
[18] G.W. Critchlow, K.A. Yendall, D. Bahrani, A. Quinn, F. Andrews, Strategies for the replacement of chromic acid anodizing for the structural bonding of aluminium alloys, Int. J. Adhes. Adhes. 26 (2006) 419–453.
[19] B.B. Johnsen, F. Lapique, A. Bjørgum, The durability of bonded aluminium joints: a comparison of AC and DC anodizing pretreatments, Int. J. Adhes. Adhes. 24 (2004) 153–161.
[20] A.C. Kennedy, R. Kohler, P. Poole, A sodium hydroxide anodize surface pretreatment for the adhesive bonding of titanium alloys, Int. J. Adhes. Adhes. 3 (1983) 133–139.
[21] J.-s. Zhang, X.-h. Zhao, Y. Zuo, J.-p. Xiong, The bonding strength and corrosion resistance of aluminium alloy by anodizing treatment in a phosphoric acid modified boric acid/sulfuric acid bath, Surf. Coat. Technol. 202 (2008) 3149–3156.
[22] M.F. Stevenson Jr., Anodizing. in: C.M. Cotell, J.A. Sprague, F.A. Smidt Jr. (Eds.), ASM Handbook, Surface Engineering, vol. 5, 1944, pp. 405–411, https://doi.org/10.1361/asmhba0001275. (Rev.).
[23] A. Acar, Ö. İnan, S. Halkaci, Effects of airborne-particle abrasion, sodium hydroxide anodization, and electrical discharge machining on porcelain adherence to cast commercially pure titanium, J. Biomed. Mater. Res. B Appl. Biomater. 82B (2007) 267–274.
[24] C. Ingram, K. Ramani, The effect of sodium hydroxide anodization on the durability of poly(etherketonetherketoneketone) adhesive bonding of titanium, Int. J. Adhes. Adhes. 17 (1997) 39–45.
[25] B. Tendaj, Powłoki ochronne I ochronno-dekoracyjne (Protective and Protective-Decorative Coatings), Wydawnictwa Uczelniane, Lublin, 1988 (in Polish).
[26] J. Edwards, Coating and surface treatment systems for metals, A comprehensive guide to selection, ASM International, Stevenage, Hertfordshire, Materials Park, OH, 1997.
[27] J.L. Stauffer, Finishing Systems Design and Implementation, A Guide for Product Parameters, Coatings, Process, and Equipment, Society of Manufacturing Engineers, Dearborn, MI, 1993.
[28] S.-M. Moon, S.-I. Pyun, The mechanism of stress generation during the growth of anodic oxide films on pure aluminium in acidic solutions, Electrochim. Acta 21–22 (1998) 3117–3126.
[29] L. Anicai, C. Trifu, L. Dima, Anodic oxidation and coloring of aluminum powders, Met. Finish. 98 (2000) 20–25.
[30] A. Rudawska, I. Danczak, Adhesive properties and adhesive joint strength of aluminium alloys following surface pre-treatment, Recent Pat. Mech. Eng. 6 (2013) 152–165.
[31] L.H. Durney (Ed.), Electroplating Engineering Handbook, fourth ed., Van Nostrand Reinhold, New York, 1984.
[32] Anodising, https://pl.wikipedia.org/, 2018 (Accessed August 23, 2018).
[33] K. Mądry, Anodising of aluminium, http://www.galwanizernie.pl, 2018 (Accessed August 20, 2018).
[34] Z. Wiącek, Struktura warstewek tlenkowych na Al i jej wpływ na wytrzymałość połączeń klejowych. Cz. I. Wpływ warunków formowania warstewek tlenkowych na ich właściwości adhezyjne (The structure of oxide coatings on Al and its effects on strength of adhesive joints. Part I. Influence of conditions of forming oxide coatings on their adhesion properties), Polimery 2 (1977) 61–65 (in Polish).

[35] Z. Wiącek, Struktura warstewek tlenkowych na Al i jej wpływ na wytrzymałość połączeń klejowych. Cz. II. Wpływ uszczelniania anodowych warstewek tlenkowych na ich właściwości adhezyjne (The structure of oxide coatings on Al and its effects on strength of adhesive joints. Part II. The impact of anode oxide coating sealing on their adhesion properties), Polimery 3 (1977) 97–98 (in Polish).
[36] P.L. Hagans, C.M. Haas, Chromate conversion coatings. in: C.M. Cotell, J.A. Sprague, F.A. Smidt Jr. (Eds.), ASM Handbook, Surface Engineering, vol. 5, 1944, pp. 405–411, https://doi.org/10.1361/asmhba0001275.
[37] T.S.N. Sankara Narayanan, Surface pretreatment by phosphate conversion coatings—a review, Rev. Adv. Mater. Sci. 9 (2005) 130–177.
[38] M.H. Abbas, Kinetics of aluminum chromatizing, Met. Finish. 12 (1995) 8–13.
[39] Phosphating, https://www.corrosionpedia.com/definition/876/phosphating, 2018 (Accessed August 29, 2018).
[40] Formation of phosphate coatings, http://www.galwano.pl/slabid-92, 2018 (Accessed August 21, 2018).
[41] Aluminium 2024-T4, http://asm.matweb.com/search/SpecificMaterial.asp?bassnum=ma2024t4, 2018 (Accessed June 21, 2018).
[42] Loctite 3430, Technical Data Sheet, http://www.industrial-gp.ro/media/22344/loctite_3430_-_en.pdf, 2018 (Accessed July 19, 2018).
[43] C.D. Volpe, S. Siboni, Some reflections on acid-base solid surface free energy theories, J. Colloid Interface Sci. 195 (1997) 121–136.
[44] P. Dalet, E. Papon, J.-J. Villenave, Surface free energy of polymeric materials: relevancy of conventional contact angle data analyses, J. Adhes. Sci. Technol. 13 (1999) 857–870.
[45] C.W. Extrand, A thermodynamic model for contacts angle hysteresis, J. Colloid Interface Sci. 207 (1998) 11–19.
[46] J. Shang, M. Flury, J.B. Harsh, R.L. Zollars, Comparison of different methods to measure contact angles of soil colloids, J. Colloid Interface Sci. 328 (2008) 299–307.
[47] A. Rudawska, Surface free energy and 7075 aluminium bonded joint strength following degreasing only and without any prior treatment, J. Adhes. Sci. Technol. 26 (2012) 1233–1247.
[48] W. Krysicki, J. Bartos, K. Królikowska, M. Wasilewski, Rachunek Prawdopobieństwa I Statystyka matematyczna w zadaniach. Cz. II Statystyka matematyczna (Probability and Mathematical Statistics Tasks. Vol. II. Mathematical Statistics), PWN, Warszawa, 1997 (in Polish).
[49] E.L. Lehmann, J.P. Romano, Testing Statistical Hypotheses, Springer-Verlag, New York, 2005.

Further reading

[50] R.D. Adams, J. Comyn, W.C. Wake, Structural Adhesive Joints in Engineering, second ed., Chapman & Hall, London, 1997.

Other treatment

8

8.1 Characteristics of special treatment methods

8.1.1 Primer application

Characteristics of primers and the method of primer application

The scientific literature employs various synonyms of the term referring to the (pro) adhesion substances, such as, adhesion promoters [1–5], primers [2, 6–8], activators and priming paints [1], and coupling agents [9–14].

Primers intermediate between the adhesive and the adherend and offer an alternative surface preparation solution of polymer materials in particular [6, 15]. The term has been used to define substances that increase the adhesion of the surface of substrates or which accelerate the cure of, e.g. adhesive. These chemicals increase the share of chemical, mainly covalent bonds. The energy of these bonds is far greater than that of the basic bonds occurring in adhesive joints (i.e. the Van der Waals forces). In adhesion technology the most commonly used terms are: adhesion promoter and primer [15–24].

The surface free energy of adhesion promoters in the solid state (i.e. following solvent evaporation) is higher than that of the adherend surface; therefore, it promotes adhesive processes such as adhesive joining, printing, laminating, etc. The joints exhibit higher adhesive strength when bonded in the presence of primers [12, 15, 25–59]. The structure of a good adhesion promoter should ensure interaction with both the adherend and the intermediate substance (e.g. adhesive) [1, 30–35].

Rider and Chalkley [36] proved that the addition of primer was an efficient method to improve toughness and durability of bonded aluminum joints.

Brack and Rider [35] reported that the addition of thin primer film to the titanium slowed degradation rates and led to higher fracture toughness at longer humid-exposure times.

He et al. [24] investigated the effect of primer and annealing treatments on the shear strength between anodised titanium alloy (Ti6Al4V) and epoxy. The shear strength for the specimen using anodised titanium alloy with primer was improved by 51.6% as compared with that of untreated specimen.

In general, two types of adhesion promoters can be distinguished: one of them is applied to the surfaces of adherends, while the other is introduced directly into bonding materials [37–40] (Fig. 8.1).

Walker [34] investigated the effect of organosilanes as adhesion promoters (for surface coatings, adhesives, and syntactic foams), when used as pretreatment primers and additives, on the bond strength of two pack epoxide and polyurethane paints applied to aluminium and mild steel. Walker [34] stated that silanes when used as additives to structural epoxide and polyurethane adhesives were less effective than

Surface Treatment in Bonding Technology. https://doi.org/10.1016/B978-0-12-817010-6.00008-4

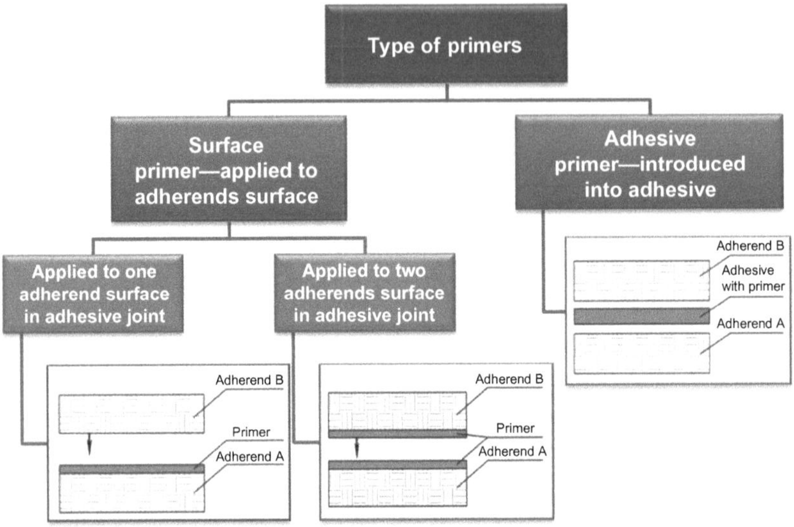

Fig. 8.1 Type of primer considering the criterion—the method of using in the elements of adhesive joint (adhesive and adherend) and the method of application.

used as pretreatment primers on metals but are highly effective on glass substrates. The compressive properties of glass microballon/epoxide syntactic foams were shown to be improved by the addition of silanes.

To give an example, if the task of an adhesion promoter is to increase the strength of adhesive joint, the adhesion promoter is applied to both surfaces of adherends [3, 37, 41]. During the bonding of the adhesion promoter with the surfaces of the joined materials, the promoter diffuses deeply into the surface it was applied to, which leads to chemical bonds with molecules of this substrate [7, 17, 42, 43].

The application of adhesion promoters should account for the type substrate, particularly in terms of the adherend-adhesive material system, bond formation conditions, the form of adhesion promoter and its properties (e.g. bonding capability, sensitivity to temperature rise, volatility, etc.) [1, 37, 40, 44].

Primers are predominantly selected to simultaneously chemisorb on the surface of adherends and chemically bond with the adhesive. Prior to the application of primers, it is necessary to choose a primer suitable for a given adherend-adhesive system, and resolve certain technological parameters of bonding, particularly, the time between the application of the primer and the deposition of the adhesive onto the joined surfaces and the form in which the proadhesive is applied [1, 37].

Primers are typically applied with a brush on one or both adherend surfaces (Fig. 8.1) to produce the potentially thinnest layer. Upon drying (after 10–60 s) a layer of adhesive is applied and further joint assembly activities follow. Application of primers may also be done by spraying (primers in the form of an aerosol with various nozzles) or in an automated manner—by means of special dispensers (Fig. 8.2).

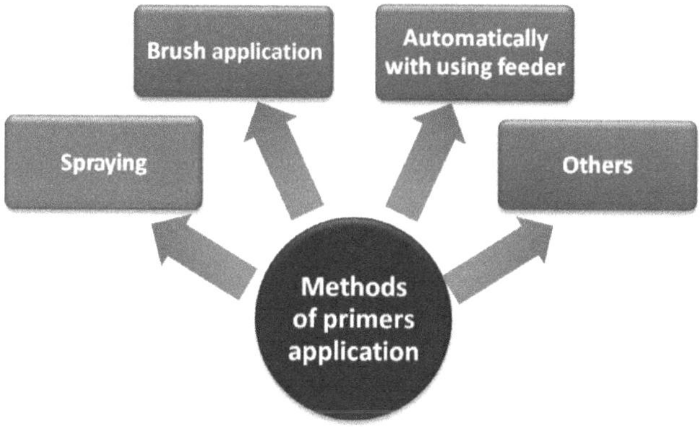

Fig. 8.2 Methods of primer application.

Chemical type of primers

The list of available primers is certainly quite extensive. Commercially available adhesion promoters include silanes (most widely applied, e.g. in bonding metals and glass), titanates, and metal zirconals [1, 2, 15, 25, 45–50].

Other notable adhesion promoters include: phenolic-formaldehyde resins, chlorinated rubbers (for polypropylene bonding applications), heterocycles containing nitrogen (as a copper bonding solution), polyurethane compounds, acrylic compounds, nitrocellulose compounds, boranes compounds, and natural resins [11, 21, 40, 49–53] (Fig. 8.3).

Silanes

Silanes are organosilicon compounds. Organosilicon compounds are organometallic compounds and is one important type in the organic compound of elements, refers to the organic compound containing carbon—silicon bond inside the molecule structure. It has similarity in its structure with the organic carbon compounds but also has difference with organic carbon compounds. For example, the silicone molecule silicon—only contains single bond between silicon atoms without containing double and triple bonds [45, 55].

Their increased use compared to other types of primers is primarily due to the chemical structure that allows chemical binding of a variety of inorganic and organic structures. The silane groups present on the silica surface favourably affect their interaction with polymers. A large variety of reactions on the surface of silica allows the surface properties to be changed by the formation of new atomic groups [8, 56].

Many studied showed that the use of silane primers for adhesives could improve the durability of adhesively bonded metal joints [9, 11, 12, 31, 40, 48]. Additionally, silane surface treatment could be an alternative for toxic and carcinogenic chromate-base treatments [46]. Silane primers were successfully used as primers on substrates to increase hydrolytic stability of metal/adhesive interface [11]. Proper formulation of silane adhesion promotores may provide very-resistant bonds between epoxies and metals or glass [48]. Silanes were used also to improve the adhesion between resin (as matrix) and additives in composites [13].

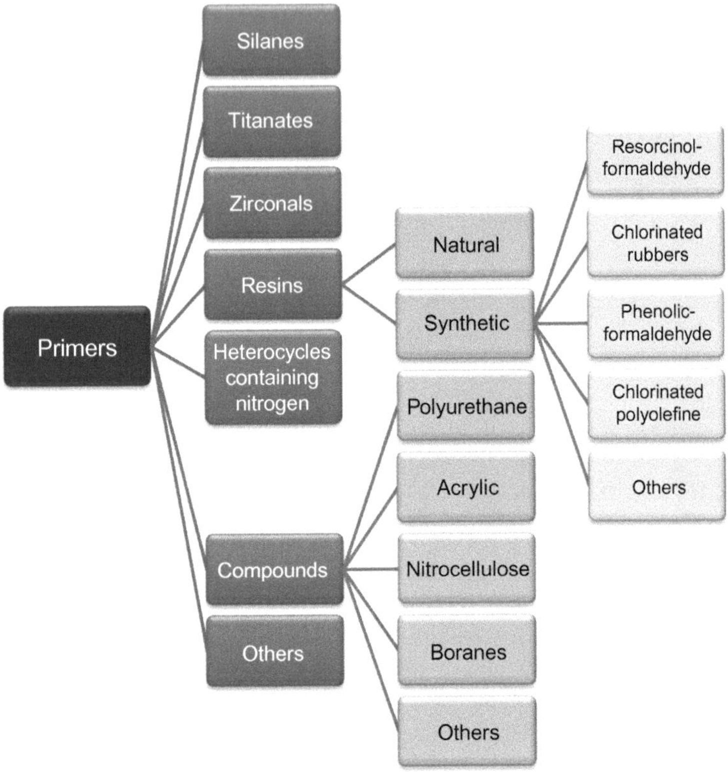

Fig. 8.3 Examples of chemical type of primers.
Based on works prepared by A. Krysztafkiewicz, B. Ragner, M. Maik, Wpływ modyfikacji powierzchni na właściwości fizykochemiczne krzemionki strąconej (The effect of surface modification on the physicochemical properties of precipitated silica), Fizykochemiczne Problemy Mineralurgii 28 (1994) 177–186 (in Polish); V. Ducháček, A. Kuta, J. Šimek, Roles of alkoxy titanates and zirconates as multifunctional agents for polymers, Polimery 49 (2004) 159–162; W. Brockmann, P.L. Geiß, J. Klingen, B. Schröder, Adhesive Bonding. Materials, Applications and Technology, Wiley-VCH Press, Weinheim, 2009; W.M. Rzymski, M. Włóka, T. Grzeżółkowska-Rosiak, Łączenie elastomerów z metalami (Joining elastomers with metals) Polimery 50 (2005) 530–536 (in Polish).

Characteristics (properties, chemical reaction with other materials, applications) of a lot of type of silanes are presented by Witucki [22], Xie et al. [14], V. Ya Lee [45], M.-L. Abel et al. [32], Fotea and D'Silva [6], Palanivel et al. [46], Walker [34], Owen and Williams [49], Gledhill et al. [12], Mohseni et al. [31], Cohen [5], Jang and Garton [47], and others.

Krysztafkiewicz [3] presented the characteristics of numerous adhesion promoters, pointing to the positive effect of titanium, silane and borate pro-adhesive substances on the adhesion phenomenon. And for instance, if the task of a given adhesion promoter is to increase the adhesive strength of the adhesive joint, the adhesion promoter is applied to both adherend surfaces [1].

The tests of organofunctional silanes, as adhesion promotores, which were employed to improve the organic coatings to the AA1050 aluminium surface, were presented by Mohseni et al. [31]. The increase in adhesion strength and surface energy values confirms the effectiveness of silane promoters treatment in improving the wettability characteristics and adhesion properties.

Lin [38] presented the characteristic of organofunctional silicone modifiers and three types of silicone release-coated film substrates.

Other chemical base of primers

Titanates are of limited use because they can only be used in some inorganic-organic systems ($CaCO_3$, metal oxides-PVC, polyester and epoxy resins) [10, 54, 58].

Zircoaluminate adhesion promoters as metal organic adhesion promoters were presented by Cohen [5].

Resorcinol-formaldehyde resins are one of the most important groups of adhesion promoters. They are used for joining, for example, brass or galvanized steel with rubber [40, 44]. The others resin and resin compounds are used as the primers to joining of various materials [23, 58]. Rider and Chalkley [36] stated that application of an epoxy-modified phenolic primer containing strontium chromate pigments improves the durability of bonded aluminium joints.

Another group is photoprimers, like variety of hydroxyl functionalised aryl azides (investigated by [6], which on photoactivation could covalently bond to the surface and introduce hydroxyl groups to facilitate covalent bond formation with the adhesives.

Leal et al. [59] studied the self-etch primers.

Several commercially available primers have sulphur-based monomers and their compositions [19]. Piva et al. [19] stated that these monomers are capable of increasing the bond strength to precious metals, mainly silver-palladium alloys, with higher bond strength durability in comparison with other gold alloys.

Type of materials for which the use of primers is recommended

Considering the properties of the adhesive and the type of adherend, there is no one universal proadhesion agent. Manufacturers typically offer primers dedicated to specific types of adhesives and surfaces. Table 8.1 shows an exemplary datasheet for the chemical base of primers.

Table 8.1 Chemical base characteristics and purpose of primers

Chemical base	Application
Heptane	Nonpolar polymers, such as: polyethylene, polypropylene, silicone, polytetrafluoroethylene, thermoplastic elastomers
Water/glycol	Nonpolar polymers, such as: polyethylene, polypropylene, silicone, polytetrafluoroethylene, thermoplastic elastomers
Propanol	Glass, metals, certain polymers
Methyl ethyl ketone	Glass, metals, polymers, wood

Source: Own work based on Primers. www.loctite-kleje.pl, 2018 (Accessed 10 September 2018).

The primes are applied during the adhesive (e.g. bonding) process of the following examples materials:

- metals and metals alloys:
  - aluminium alloys [27–29, 31, 43, 58],
  - hot-dipped galvanised steel [11, 53],
  - titanium and titanium alloy [24],
  - nickel-chromium, silver-palladium, and silver-gold [19],
  - electroplated-phosphated steel [11], and
  - copper [2],
- polymers [37, 60, 61]:
  - polypropylene (PP) [47],
  - low-density polyethylene (LDPE) [47],
  - polyimide (PI) [2],
  - polyethylene (PE) [62],
- composites (e.g. glass fibre/polyester composites [41], and
- leather [6, 7].

8.1.2 Plasma treatment

Low-temperature plasma characteristics

Currently, the following types of plasma are distinguished [15, 63]:

- low-temperature plasma ('cold')—covers the plasma range below tens of thousands of Kelvin,
- high-temperature plasma ('hot')—covers the range from several to several hundred million Kelvin,
- ring discharge—is a low-pressure plasma variety created in a variable electromagnetic field with a frequency of about 100 MHz and a pressure of about 100 Pa.

Plasma treatment consists in subjecting the surface layer of the modified material to a low-temperature plasma, which is generated by high-frequency partial discharges (up to 2.45 GHz), occurring in the vacuum chamber the time of 15 s to several minutes [39, 64–67].

The low-temperature plasma is generated at atmospheric pressure or pressure typically reduced to 0.1–1.0 hPa [68]. In low-temperature plasma, the temperature of light T_e particles, or electrons is much higher than that of heavy T_h particles, i.e. neutrons, protons, metastable forms of the excited atoms of this gas, and atoms and fragments of electrically neutral molecules. The energy of electrons in low-temperature plasma, employed in the surface treatment of polymeric materials, is usually between several to a dozen or so eV (1 eV corresponds to 11,600 K) [68].

During plasma treatment several factors are of significance [15, 39, 64–87]: the duration of plasma operation, the energy and unit power of partial discharges, the pressure and temperature in the discharge chamber, gas type, and flow intensity, partial discharges frequency, electrical voltage, as well as temperature of the modified plastic, etc.

Atmosphere of plasma treatment

Plasma is a partially ionised gas (or gas mixture) consisting of approximately equal number of electrons and ions as well as atoms, neutral molecules, and electromagnetic radiation [37, 63, 68, 84, 86–88].

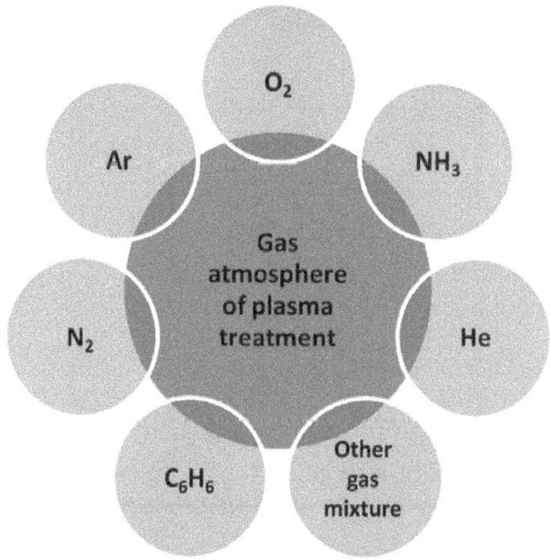

Fig. 8.4 Gas atmosphere of plasma treatments.

Incomplete discharges are generated by a fast-changing electromagnetic field in the atmosphere of various gases (oxygen, nitrogen, helium, argon, xenon, chlorine, and air) [15, 68, 89–99] and also gas mixture (e.g. Ar/N_2, SF_6 SO_2, CO_2, NH_3, Ar/NH_3) [64, 76, 91, 100–108], which were presented in Fig. 8.4.

The type of created functional groups depends on the mixture of gases composing the plasma. The oxygen plasma generates groups containing oxygen, the nitrogen plasmas generate groups containing nitrogen (e.g. amino –NH, $-NH_2$, and –CN). Halogen plasma leads to the emergence of chlorine atoms on the surface.

Akram et al. [100] investigated the plasma ion implantation under atmospheric pressure plasma as surface treatment of titanium in order to increase the adhesive bond strength. A substantial improvement in the surface energy of titanium was observed after the atmospheric pressure plasma treatment and the surface energy increases with increasing exposure time of this treatment.

Akram et al. [76] highlighted the rationale of adhesive bonding of atmospheric pressure plasma treated high-temperature-resistant polymeric sheet such as polyimide sheet, with titanium sheets.

In work presented by Sheu and Shyu [89] Kevlar 149 fibres were surface-modified by NH_3, O_2, and H_2O plasmas to improve the adhesion to epoxy resin and poly(*p*-phenylene terephthalamide) (PPTA) film prepare from Kevlar 149 fibres was also modified to estimate the changes in surface energy by the plasma treatments.

Rozovskis et al. [65] tested the oxygen plasma treatment of polyimide (PI) films on the adhesion of electroless copper coatings.

Kusano et al. [64] investigated the plasma treatment of glassy carbon plates for adhesion improvement using various gases and gas mixture. He gas, gas mixtures of He and reactive gases such as O_2, CO_2, and NH_3, Ar gas and Ar/NH_3 mixture were used as treatment gases.

LDPE and high density polyethylene (HDPE) were plasma-treated with N_2 and O_2 plasma and the effects of polar functional groups formed by plasma treatment on adhesion were studied by Choi et al. [98].

Type of low-pressure (cold) plasma

Plasma treatment (as electrical treatment which is one of the dry physical process) could be produced by (1) corona discharge [109–111] and (2) glow discharge [80, 85, 109, 110, 112, 113]. Three types of glow discharge are used for the surface modification of polymers, depending on the mode of power supply, such as (Fig. 8.5):

- direct current (DC) glow discharge [90, 109, 114–116],
- radio frequency (RF) glow discharge [67, 117, 118], and
- microwave glow discharge [91, 118, 119].

Among these three types of glow discharge the DC glow discharge method has been claimed to be the most effective for the surface modification of polymers to improve the adhesion properties [109]. Bhowmik et al. [109] investigated the surface modification of HDPE and PP by exposure to a DC glow discharge in air at different power levels to improve surface energy of these polymers.

Various authors emphasised that glow discharge treatment is the most effective method with respect to uniformity of surface modification, its precise control, and the absence of chemical hazards [109]. Moreover the polymer surfaces can be modified efficiently and effectively by plasmas which are composed of various active species including free radicals and ions, without inducting any environmental pollution [107, 113, 114].

Radio-frequency (rf) O_2 or Ar plasma treatment is provided to be a rapid, reproducible and contaminant-free method for the introduction and enhancement of oxidic functional groups on the glassy carbon surfaces, while the glassy carbon surface can be hydrophobic by rf SF_6 plasma treatment [64].

Effects of plasma treatment

Several processes occur on the surface of substrates during plasma activity: etching, cleaning, cross-linking of particles in the surface layer (causing increase in cohesion), and in particular the formation of new functional groups (Fig. 8.6) [77, 79, 102, 120–128].

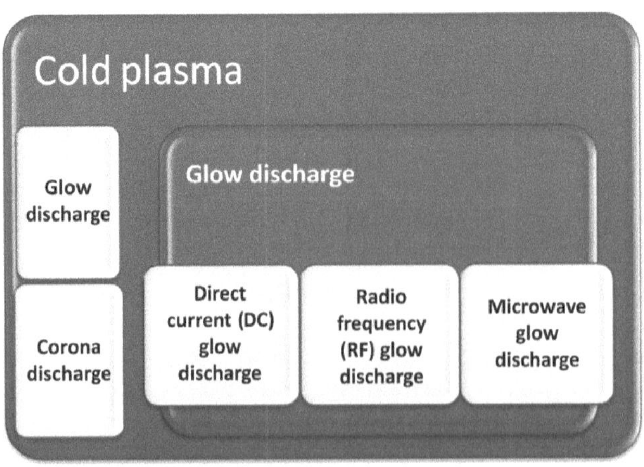

Fig. 8.5 Type of cold plasma treatment.

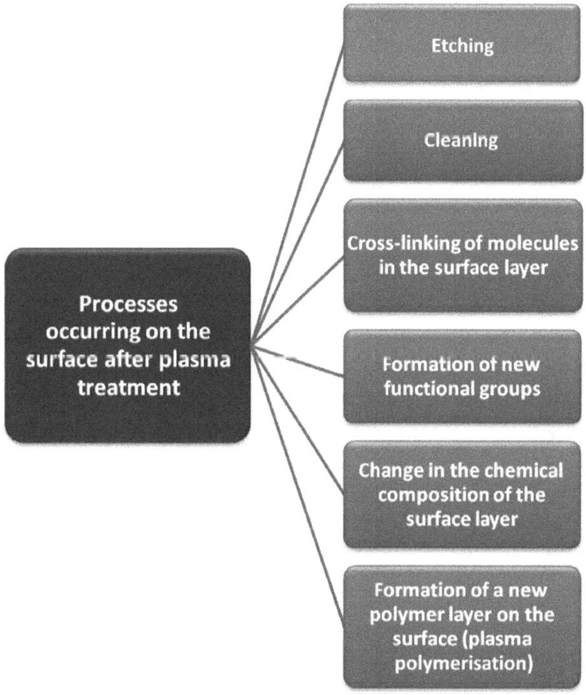

Fig. 8.6 Processes occurring on the surface after plasma treatment.

Model for plasma polymer surface treatment was presented by Dorai and Kushner [129].

Substrates are subjected to *etching* whenever the surface layer exhibits high degree of contamination. During etching, the weak boundary layer is removed along with certain fragments of the amorphous phase, while the properties of fillers and crystalline structures are modified. Further results of etching are: the increase in the surface roughness of the substrate, which results in the increase of the real contact surface with the adhesive, contributing to the increase in mechanical adhesion—an important factor in the adhesive strength of joints [15, 68, 79, 121, 126, 128].

Cleaning the surface with *partial discharges* leads to a deep removal of organic contaminants from the surface, which hence increases adhesion [121]. Plasma treatment is applied to both polymeric and nonorganic materials, and the time required to clean the surface of the materials is about several tens of seconds [15]. Due to the fact that during the treatment, much less material is removed from the surface unit than in etching, this method is used to remove weaker contaminants [15, 121–123, 128].

Cross-linking (curing) occurs during partial discharges in a noble gas atmosphere (e.g. argon or helium) and in oxygen. When subjected to UV radiation, the polymers are cured to form a thicker layer (of approximately 1–5 μm) than the one in which the oxidation occurs, because the photons are strongly absorbed by the polymer. Cross-linking leads to the emergence in the surface layer a structure strongly bound with the substrate, which exhibits good resistance to thermal energy and is characterised by relatively high mechanical strength. Therefore, during the process of adhesive joints failure the cracking processes are inhibited [15, 128, 129].

The emergence of new chemical structures, spurred by partial discharges, contributes to chemical changes in the surface layer of the polymer. In addition to the formation of new functional groups (such as, hydroxyl, carboxyl, aldehyde, and ketone groups), plasma polymerisation may also take place in the surface layer. As a result of this process, a thin layer of polymer is deposited, whose properties depend on the type of gas in use. A characteristic feature of this layer is its strong cross-linking, insolubility, and good barrier properties for gases and liquids [15, 68, 126].

Liston et al. [124] provided a critical review of the development and trends in the field of materials surface modification by 'cold', low-pressure plasma treatment. The interaction mechanisms between a plasma and a polymer surface are examined; these include physical bombardment by energetic particles and by ultraviolet (UV) photons, and chemical reactions at or near the surface. The resulting four main effects, namely cleaning, ablation, cross-linking, and surface chemical modification, occur together in a complex synergy, which depends on many parameters controlled by the operator. The authors foresee that the use of plasma processes will continue to expand, because they have unique capabilities, are economically attractive, and are 'friendly' towards the environment.

Considering the use of the plasma method as a method of surface/surface layer preparation for adhesive bonding, the following effects of plasma modification can be distinguished [64, 65, 78, 82, 102, 105, 112, 127, 130–132]:

- increase in the wettability,
- increase the surface free energy,
- increase adhesive strength, and
- enhanced adhesion.

Mandolfino et al. [133] investigated effect of cold plasma treatment on surface roughness and bonding strength of polymeric substrates. The experimental results showed that the optimised plasma process may remarkably change the surface morphology, increasing wettability properties of the surfaces and shear strength of the bonded joints.

Williams et al. [102] used atmospheric pressure helium and oxygen plasma for the surface treatment of 410 stainless steel and carbon-fibre laminates prior to bonding. The results showed that carbonaceous contamination was removed by plasma treatment, and specific functional groups (e.g. carboxylic acids) were formed on the surface of adherends. These functional groups promoted strong chemical bonding to epoxy film adhesive.

Kruse et al. [84] proved that the low-temperature plasma process resulted in excellent adhesion strength of the following polymers: PP, polybutylene terephthalate (PBT), PBT blends, and a high-temperature thermoplastic, polyetheretherketone (PEEK).

Vishnuvarthanan and Rajeswari [97] investigated the oxygen plasma treatment of PP films with different treatment times and with various plasma powers. The oxygen plasma treatment increased the film surface wettability and enriched the hydrophilicity of the PP film and it caused a decrease of contact angle values and the increase of surface free energy.

Encinas et al. [101] reported that atmospheric pressure air plasma torch (APPT) treatment activates the surface of the polyolefins HDPE, LDPE, and PP by means of surface free energy and wettability enhancement.

Akram et al. [76] proved that surface free energy was found to increase with increase in exposure time of atmospheric pressure plasma treatment of polyimide sheets surface.

Takagi et al. [119] presented that peel strength tests of Cu layers deposited on plasma-treated polyimide films showed that the plasma treatment significantly enhanced Cu layer adhesion.

The plasma surface treatment could be combining with other treatments methods like priming [30, 39, 77] or other treatment [39]. The main purpose of tests carried out by Pantoja et al. [30] was to improve the adhesion behaviour of PP by combining atmospheric pressure plasma surface activation and silane adhesion promoter. Experimental evidences presented by Encinas et al. [101] supported superiority of the APPT treatment to increase wettability and adhesion of polyolefinic surfaces, especially when combined with the use of a primer.

Type of activation materials

Cold plasma surface treatment is known to be powerful process to modify surface characteristics of both metallic and polymeric materials [39, 69–73]. Therefore, it represents an efficient, nonpolluting and economical alternative to clean, activate and improve the adhesion of substrate surfaces (e.g. aluminium [39, 74]. The cold plasma may also be used to improve the adhesion between different materials, such as aluminium and polybutylene terephthalate [75].

The scientific literature in the field contains a breadth of detailed information on plasma modification, especially in relation to polymeric materials (polymers [64–68, 81, 82, 112, 115, 120, 134–140] for example:

- polymers:
  - poly(ethylene terephthalate) (PET) [112, 139–141],
  - polyimide (PI) [65, 79, 91, 113, 131, 142],
  - polyethylene (PE) [94, 98, 101, 109, 139], and
  - PP [30, 77, 78, 88, 109, 112, 139],
  - others,
- polymer composites [64, 66, 67, 78, 89, 128, 143–145]:
  - Kevlar 149,
  - thermoplastic composite,
  - thermoset composite, and
  - **others.**

Activation in low-pressure plasma is effective and durable, especially for thermoplastics and hardening plastics. It is conducted at a temperature of 60–100°C in reactors with low-pressure oxygen, nitrogen, fluorine, or noble gases. The presence of extremely high voltage causes atoms to knock out which leads to activation of the surface by means of charged and accelerated anions, electrons, and atoms of gas in plasma, which are simultaneously subjected to the UV radiation. The monomers polymerise under such conditions to form a thin outer polymer coating [15, 68, 126].

Sheu and Shyu [89] proved that the surface free energy and the work of adhesion of water on the PPTA surface were markedly improved by the plasma treatment.

In the tests carried out by Wade et al. [67] surface modification by oxygen plasma treatment resulted in significantly improved adhesion between the unidirectional GFRP Nylon-6,6 thermoplastic composite, and an applied toughened epoxy adhesive.

Nakamura et al. [79] investigated and characterised the adhesion between polyimide thin films and copper after oxygen plasma etching, focusing on the effect of plasma treatment on the adhesion strength at ambient temperature.

The effect of oxygen plasma treatment on the surface properties of thin-doped indium oxide (ITO) substrates for polymer LEDs and the changes in the surface properties of treated ITO substrates with ageing time were investigated by You and Dong [83]. Experimental results showed that oxygen plasma treatment increases the oxygen concentration, decrease carbon concentration, and enhances the surface free energy and polarity, and thereby improves the surface properties of ITO substrates.

Meddeb et al. [142] investigated the effect of oxygen plasma treatment on the electrical and surface properties of polyimide. This treatment led to an increase in the oxygen presence on the polyimide surface and a marked s surface hydrophilicity and also an increase in dielectric breakdown at room temperature.

Ku et al. [78] considered the effect of atmospheric pressure plasma treatment of PP on the peel strength and shear strength between PP/aluminum composites.

Navaneetha Pandiyaraj et al. [112] investigated the adhesive properties of the plasma modified PP and polyethylene terephthalate (PET) film surface and the results showed a considerable improvement in surface wettability even for short exposure times.

Rezaei et al. [140] presented a comprehensive study of surface modification of PET film substrates to improve its adhesion properties using a large-area atmospheric plasma.

Van Deynse et al. [94] investigated the surface modification of polyethylene (LDPE) by an argon atmospheric pressure plasma jet.

Plasma can be also used to treatment the surface of metals and metals alloys, and different system materials (metal/polymer) [70, 75, 78, 145, 146], such as:

- titanium and titanium alloy [76, 100, 108, 117, 147],
- copper [79, 91],
- aluminium and aluminium alloy [69, 73, 74, 132],
- different types of steel [148–156].

Sorrentino and Carrino [81] published that they had employed the plasma treatment for the cleansing and activation of aluminium alloy surfaces as an alternative method to traditional electrochemical and chemical surface preparation methods. In their study, Nakamura et al. [79] focused on the effects of oxygen plasma etching as a surface treatment on the adhesion between a thin polyamide film and metal, and the adhesive strength of this assembly of substrates. Their investigation showed the correspondence between the adhesive strength of the polyamide film and metal assembly and the chemical state of the surface of the polyamide film. The hydrophilic bonds formed during oxygen plasma pickling, e.g. C-O, affect the adhesion between a thin layer of polyamide and metal.

Sperandio et al. [82] investigated the problem of aluminium surface preparation by means of the plasma method in the atmosphere of a gas mixture consisting of nitrogen and oxygen of various concentrations. It was noted that the use of plasma treatment increases the adsorption of polymer chains due to the increase in the surface polarity; as a result there is a marked increase in the Van der Waals forces between the adhesive substance and the adherend.

Rey et al. [52] investigated the effects of different surface treatments (primer, plasma and combination of both) on the mechanical resistance of interface between wires of NiTi-shape memory alloy and silicone rubber and proved that NiTi/rubber interface improved only by means of plasma offers a new way to obtain biocompatible interfaces in such composite material.

Polini and Sorentino [39] used the various surface treatment of 2024 aluminium alloy including cold plasma treatment.

In the study prepared by Usami et al. [91], a rapid enhancement of the adhesion force between the Cu film and the polyimide (PI) substrate was demonstrated using the microwave plasma.

Panousis et al. [108] presented the experimental test of the surface treatment of TiA6V4 titanium alloy using an atmospheric pressure plasma jet generated by a DBD discharge in nitrogen flow.

8.1.3 Corona treatment

Corona discharge

The corona discharge method (referred to as activation) is one of the types of surface activation, i.e. increasing its surface energy [15, 68, 157, 158]. The increase in adhesion is imparted by the cleansing effect of oxidation occurring on the substrate surface. Ozone is a side-effect generated during corona discharge [15].

The corona discharge treatment is a long established method for treating polymers prior to adhesive process (adhesive bonding, printing with inks, lamination to other films, vacuum metallisation, etc.) [157]. The two main advantages of corona discharge treatments are that the equipment investment is much lower than for ordinary low-temperature plasmas installations since no vacuum system is needed, and it is possible to treat large surface areas [159–161].

Corona discharge is a sequence of short high-frequency electric discharges carried out at a short distance from the surface of the material, particularly of polymer [15, 68, 162–166]. During corona discharge, a blue glow appears as a result of the current flowing in air medium or other neutral liquid from a high-voltage discharge electrode: ionisation of the medium enables the movement of the charge and generation of plasma around the electrode. Moreover, the generated ions lift the electric charge to the lower-potential area surrounding the collecting electrode, alternatively they recombine creating neutral atoms. The process parameters are as follows: voltage 15–30 kV, distance between electrodes and treated substrate surface 1–2 mm [68]. Increasing the electrode voltage frequency, which produces the best effects in the range of 15–40 kHz [68], positively affects the uniformity of corona discharges and improves the quality of surface treatment of polymers. This type of activation is often used for polymer films because it is a relatively simple method that allows maintaining the imparted properties to over an extended period of several months [15, 166]. An example of corona discharge effects is presented in Fig. 8.7.

Polymer materials are composed of long, homogeneous molecular chains. The chains are usually connected to each other and form even longer chains, leaving few free chain ends that constitute potential connection points. During crowning, electrons are accelerated on the surface of polymers and by impacting at the polymer chains in the surface layer they cause their chemical bonds to break. The kinetic energy of electrons in the interelectrode space (amounting to anywhere from a few to a dozen or so electronvolts) is generally greater than the energy of chemical bonds in polymer chains (e.g. the energy of C–C, C–H, C–N bonds is less than 5 eV), therefore, the electrons impacting on the macromolecules of the polymer cause dissociation of a certain portion of these chemical bonds [68, 167].

This contributes to the formation of a large number of radicals that participate in chemical reactions (mainly oxidation) in the surface layer of the substrate material and which react mainly with oxygen, ozone (generated during electrical discharges), water molecules, OH groups, and with each other. As a result, polar compounds are formed, containing *interalia* ketone, aldehyde, hydroxyl as well as carboxyl groups. Polar groups change the surface properties of the material, and these results in an increase in the surface free energy (particularly of its polar component [15]). A model scheme of corona discharge is shown in Fig. 8.8.

During this method, incomplete corona discharges are generated in the activation devices, filled with air at atmospheric pressure. In a corona treater, known as

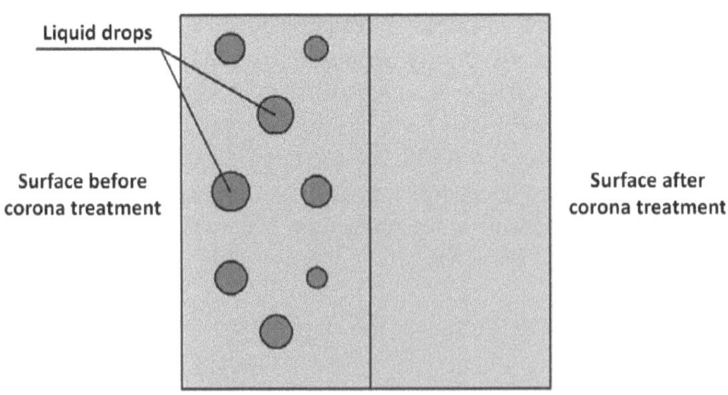

Fig. 8.7 Surface prior to corona treatment (on the left) and following the treatment (on the right). Source: Own work based on Koronowanie (Corona treatment) (in Polish). https://www.technograph.pl, 2018 (Accessed 20 August 2018).

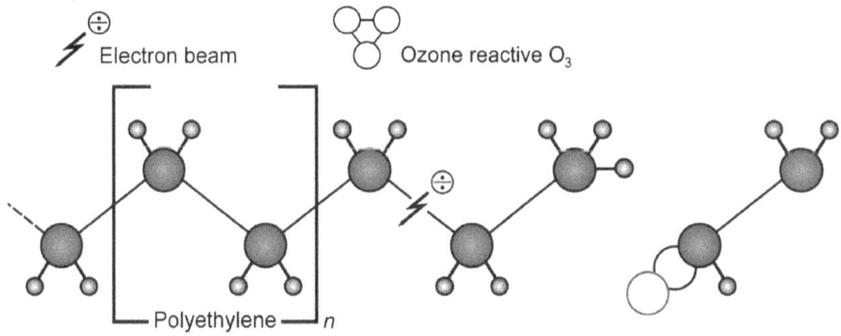

Fig. 8.8 Molecular chain of polyethylene (PE) broken by corona discharge.
Source: Own work based on Koronowanie (Corona treatment) (in Polish). https://www.technograph.pl, 2018 (Accessed 20 August 2018).

a *skototron*, the activated surface travels near high-voltage electrodes. Between the electrodes, the corona discharge forms plasma, i.e. a mixture of gas with active molecules: ions, electrons [68, 162, 165].

The electrical discharge between the electrodes causes ionisation of the surrounding atmosphere. The area of the ionised atmosphere, the corona, may take the form of a blue glow, occurring around high-voltage electrical equipment. The corona causes the emergence of ozone, which is an extremely aggressive form of oxygen. For this reason, large installations for corona activation should be provided with filters to capture the emitted ozone [168, 169].

The interaction of ionised particles on the surface of the treated substrate causes the oxidation of this surface by replacement of the hydrogen contained in the molecular bonds with oxygen. This increases the surface energy of the material, hence increasing its wettability. The measure of the degree of surface preparation is the amount of energy transmitted by the ionised particles. This energy is proportional to the output power of the generator, and depends on the gap between the electrodes and the interaction time [15, 68, 168].

To recapitulate, during corona discharge the following phenomena occur on the surface of the treated substrate [68, 159, 162, 165, 167]:

- partial cleaning and alteration of the polymer surface geometry,
- the breaking of chemical bonds, which is accompanied by the formation of radicals that constitute the basis for the formation of polar groups, e.g. ketone, carboxylic, and aldehyde,
- degradation and destruction of the surface layer,
- formation of, inter alia, oligomers, which are beneficial to the wetting processes; however, excessive share of these chemicals decreases the adhesive strength, even though the surface energy and wetting show improvement, and
- cross-linking (cure).

The efficiency of the corona discharge depends on several factors, including [15, 68, 160, 161]:

- process conditions (including individual activation energy, workpiece feed),
- properties, type, share of additional components, and structure (the share of the crystalline and amorphous phase) of the activated polymeric material,

- environmental conditions (humidity, temperature, share of ozone), and
- design features of the activator (thickness and type of electrode insulation, shape, dimensions, and the number of the high-voltage electrodes).

The corona discharge technology does not require the use of unloading chambers, as it is with other plasma methods. It is a continuous process characterised by remarkable efficiency, which can be directly incorporated into existing production lines, high reliability and simple construction of activators, as well as low investment and operating costs.

The corona discharge is an effective method for increasing the surface tension of polymeric films and metal-plated substrates, as well as of nonabsorbent paper and cardboard surfaces [170].

Gerenser et al. [166] and Strobel et al. [165] studied the aging of corona-treated polyethylene films. Strobel et al. [165] investigated the surface properties of air-corona-treated PP films on aging a variety of storage conditions.

Green et al. [164] examined the effects of 13 pretreatments of the homopolymer PP and one of the treatments was corona discharge.

Moghadamzadeh et al. [163] presented the different surface treatments techniques for adhesive bonding of Wood Polymer Composite (WPC) material and one of the tested surface treatments was corona discharge treatment.

Corona treatment was applied to modify surface of general purpose polystyrene (GPPS) film to improve its molecular interaction with lacquers, paint, and adhesives in work prepared by Gołębiewski and Żenkiewicz [162]. GPPS film treating by corona discharge caused increase of wettability of surface layer what can be confirmed by changes in the contact angle of this layer by different test liquids. Changes in wettability of treated GPPS film depend on specific treatment energy.

8.1.4 Flame treatment

In the flame method, the modified material (mainly polymer) is subjected to a short-time exposition (from a fraction of a second to a few seconds) to a gas flame formed as a result of combustion of gas in air [15, 68, 125, 157, 169]. The process is carried out with the use of special devices and a gas mixture of light saturated hydrocarbons with a predominant proportion of methane or propane.

A high-temperature gas flame causes the oxygen molecules to disintegrate into free oxygen atoms. The exhaust gas contains active carbon and nitrogen molecules, free electrons, and positively charged ions. Compared to the corona discharge, the generated gas plasma is of lower density and energy of electrons; however, its mass flow is considerably higher [68].

The application of the flame method requires considerable care when selecting the following process parameters [15, 168, 171]:

- composition of the mixture (air to gas ratio)—it is recommended that the mixture shows imbalance for the advantage of air, in the amount of 3%–5%,
- thermal power of the burner,

- distance between the treated surface and the active part of the burner—it is recommended to maintain the optimal distance (considering the criterion of the strongest interaction of the flame), i.e. 10–12 mm from the light blue flame cone, and for polymer materials, the distance is 5–15 mm [168], and
- time of substrate surface exposure to flame should be as short as possible so as not to cause damage to the surface of flame-treated elements.

Proper activation of the substrate surface is possible only by accounting for the above-mentioned parameters. It is recommended that during the modification of the surface layer of polymers, laminar combustion should be maintained; laminar burning conditions require the nonturbulent motion of the burned mixture [15].

Flame treatment induces chemical reactions on the surface of treated polymers, which produce the following effects presented in Fig. 8.9 [68, 86, 125, 157, 163].

The flame treatment process can be illustrated by the treatment of polymer material in the oxidising flame of a combusted gas-air mixture with a relative air-transfer coefficient (0.96–0.97), at a temperature of 1100–2800°C, generated at a distance of 5–15 mm between the flame cone and the substrate surface, over the time of 0.5–3.0 s [68]. The fuel is a natural gas, propane, butane, and the oxygen in the air is the oxidant. A distinction is made between laminar and turbulent combustion, *nota bene* the latter is the recommended type for flame treatment applications.

Flame treatment effects favourable changes in the molecular structure of the surface of treated polymeric materials, which result in an increase in the surface energy and wettability—important factors determining the quality of adhesive joints [68]. The polar component of the surface free energy increases with the rise in the number of surface treatments of polymeric materials, whereas the strength of the adhesive bond is improved already during the first treatment.

Flame treatment of polymer surfaces is applied to various substrates, including polyamide, PET, yet it was first introduced to the processing of polyolefin films [68, 171, 172].

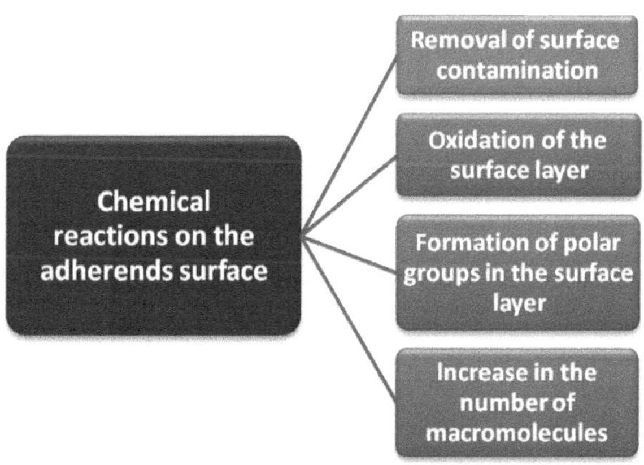

Fig. 8.9 Chemical reaction on the adherends surface after flame treatment.

In the case of steel surfaces, flame heating causes scale cracking, burning, or loosening of organic substances or old coats of paint [15]. The process of flame cleaning facilitates the removal of surface contaminants by means of various mechanical methods. The potential risk of thermal deformation, however, limits the scope of applications of this method as it is not recommended for treating thin sheets—less than approximately 4 mm.

The condition of the surface following thermal modification does not show satisfactory effects of treatment, which are typically evaluated with the following quality assessment factors for cleaned surfaces [15, 68]:

- general appearance of the surface,
- scale residue,
- corrosion marks, and
- other contamination.

Żenkiewicz [171] presented the general objectives of surface modification of plastics articles with particular reference to the flame modification technique. Flame modification effects exercised on various, mostly polyolefinic, products were analysed. The flame technique was advantageous and complementary to the corona-discharge treatment method.

Sutherland et al. [172] studied the changes inducted on the surface of PP homopolymer following flame treatment. The effect of air-to-gas ratio, total flow arte, contact time with the flame and the distance between the inner cone tip of the flame and the polymer were investigated.

Custódio et al. [157] reported the effect of two surface pretreatments: corona discharge and flame ionisation on three timbers (maritime pine, iroko, and European oak) prior to improving bond durability of adhesive joints of these materials.

Moghadamzadeh et al. [163] investigated the different surface treatments techniques for adhesive bonding of WPC material and one of tested surface treatment was flame treatment.

Flame treatment was one of surface treatment of the homopolymer PP presented by Green et al. [164].

8.1.5 Surface treatment with electromagnetic radiation

Several methods employing the effect of electromagnetic radiation on the surface of materials are distinguished. The crucial difference between them is in the range of electromagnetic wave length and the types of wave source [15, 68, 121, 126, 173–182], and are therefore classified into (Fig. 8.10):

- UV radiation,
- laser radiation,
- ionising radiation,
- electron radiation, and
- γ-radiation.

The presented methods of electromagnetic radiation interaction are used primarily for surface treatment of polymers and polymer composites [113, 175, 177, 183–185], but these methods are also used for surface treatment of metals and metal alloys [128, 174, 186–200].

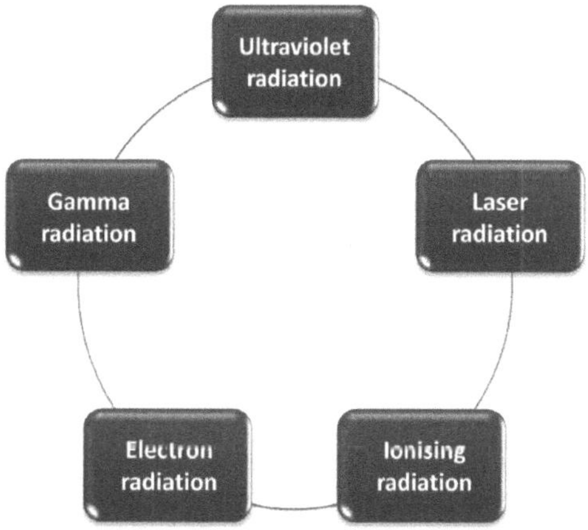

Fig. 8.10 Type of electromagnetic radiation.

UV radiation

Surface treatment with UV radiation (the wavelength region $\lambda = 150$–$250\,\text{nm}$) is carried out in an atmosphere of oxygen or ozone, and causes oxidation of the surface, reduction of the contact angle and a significant increase of adhesive strength [15]. UV radiation is emitted by discharge lamps, microwave discharges in special-purpose generators, or lasers [15,59]. Surface treatment with UV radiation is typically carried out in vacuum chambers due to the absorption of this radiation by air molecules.

Urbaniak-Domagala [185] presented pretreatment of PP films for the creation of thin polymer layers, among other things UV irradiation.

Laser radiation

Polymeric materials

Surface treatment with laser radiation is used in the preparation of polymeric materials [68, 173] and employs the generation of thermal effects, which causes the melting of several-micrometre-thick microscopic areas of the surface, the increase of the surface roughness and the removal of its fragments.

If carried out in the presence of oxygen, laser action leads to the oxidation of the surface [174, 201]. Laser radiation is used for fine machining of small surfaces of complex geometry. The modified surface exhibits a range of improvements: higher wettability and the surface free energy of substrates through implementation of polar functional groups mainly during oxidation, improved degree of polymer cure and the surface geometry, without changing the core of the substrate [15]. Physical changes in the surface layer produced by laser radiation predominantly çonsist in increasing the roughness of the polymer material. The quantitative effects of physical modifications are approximately proportional to the number of laser pulses [68]. Scheme of the laser cleaning process was presented by Madolfino et al. [121].

Baburaj et al. [173] used the laser ablation process to modify the surface of titanium, thus obtaining the beneficial properties of its surface; the effects included the increase in the real area of the contact surface, whose chemical composition was furthermore changed consequently improving the wettability and adhesion forces.

In work prepared by Belcher et al. [198] titanium alloy substrates were laser-etched using a neodymium-doped yttrium aluminium garnet (Nd:YAG) laser in order to clean the surface prior to adhesive bonding.

Köckritz et al. [202] examined the development of laser pretreatment method for glass fibre reinforced PP surfaces. The aim of the tests was to create a reproductible surface for bonding low-energy PP which adheres very poorly of most adhesives and foams to the matrix materials for plastic composites.

Mandolfino et al. [121] investigated the effect of laser and low-pressure plasma treatment on the mechanical properties of adhesive bonded joints.

Müller et al. [194] described the tests of the application of Mini-TEA-CO_2 laser for modification of titanium surfaces to improve the bonding stability to a PMMA polymer.

Bénard et al. [178] used an excimer laser treatment, as a way to control the adhesion performances of glass/epoxy and carbon/epoxy composites.

A high-power excimer laser was used to treat the adherend surfaces of a glass-fibre-reinforced poly(butyl terephthalate) composite for the purpose of improving adhesion quality in the work prepared by Man et al. [177].

Metals and metals alloy

Laser treatments are also powerful tools for modifying the surface composition of metals in order to improve mechanical, chemical, and optical properties [180, 181, 186, 191, 193, 199, 203]. Light elements (oxygen, nitrogen, and carbon) of the reactive atmosphere can be inserted in a surface layer whose composition and microstructure depend on the irradiation conditions and on the thermal properties of the target.

Different issues of laser treatments of titanium substrates were reported in the works prepared by [193, 204]. The laser technology with respect to titanium substrates have a lot of advantages, include a considerable shortening of the time required to manufacture coatings in the case when the surface area is small, significant cost reduction, and most importantly the possibility of having a selective, pointwise influence on the material.

Antończak et al. [204] presented analysis of the complex chemical structure of the layers made on titanium in the process of the laser heating of its surfaces with a nanosecond laser in the atmospheric environment. Laser irradiation in air allows to obtain surface layers containing different oxides and nitrides, in spite of their lower thermodynamic stability [180].

Pou et al. [191] considered the potential use of millisecond pulse laser for texturing commercially pure titanium (ASTM 1). In this work laser treatments were performed using two different laser sources in normal air under atmospheric pressure. Results indicated that the processing assisted by the Nd:YAG laser is more suitable to modify the surface topography of titanium samples.

Haubrich et al. [186] presented the study on the effects of the ambient gases of a laser pretreatment of titanium alloy TiAl3V2.5 on the interface and bond strength to a thermoplastic polymer, PEEK.

The information and test results of laser treatments of aluminium substrates were presented in the works prepared by [174, 188, 190].

Dodiuk et al. [190] investigated the excimer laser treatment of aluminium alloys as alternative to the use of ecologically and unfriendly chemicals involved in conventional anodising pretreatment. Experimental results indicated that preadhesion laser surface treatment significantly improved the shear strength of modified-epoxy bonded aluminium specimens compared with untreated and anodised substrates.

Critchlow et al. [188] reported the CO_2-laser ablation as treatment of aluminium alloy coupons and shown that following laser treatment a 22% increase in initial lap shear joint strength was observed compared with degreased-only controls.

Rechner et al. [174] investigated the surface pretreatment using laser to examine the influence of selected pretreatment parameters on the strength and aging resistance of the bonded joints of wrought aluminium alloy AW 6016.

Ionising radiation

One of the reasons for the use of ionising radiation is to shape the surface properties of polymers. Ionising radiation is obtained from different types of accelerators, including low-energy equipment with electron energy lower than 100 keV [68].

Electron radiation

In surface treatment with electron radiation [15, 68] the electron radiation is obtained from electron accelerators, whereas the ionising radiation—from isotopic γ-radiation sources. The basic parameters of accelerators are the power and kinetic energy of electrons, which determine the increase in the depth of penetration of electrons into the treated material and the efficiency of the process. During the irradiation of the polymer material in air, the surface layer becomes oxidised, which increases its wettability and the surface free energy. The latter also occurs at the workpiece boundary layer, thus improving the adhesive properties of the treated material.

In addition, electron radiation may be conducted in the atmosphere of reactive gases, including such as O_2 or NH_3, to produce polar functional groups in the surface layer of the treated substrate [15].

Żenkiewicz [176] investigated the effect of electron radiation dose on oxidation of surface layers of polyethylene, PP, and PET films. The results presented that the extent of oxidation of the surface layers of the polymer films increased monotonically with the absorbed electron radiation dose.

γ-radiation

Surface treatment with γ-radiation (in the wavelength region of 0.01 nm) is typically carried out in air [15, 68, 184, 205], whereas the actually treated material is significantly deeper than the reach of UV radiation (which results from much higher radiation energy than the latter). The source of radiation in this method is the excited nucleus of selected elements, mainly cobalt or less frequently caesium. During the exposure of the polymer material, oxidation, degradation, and cross-linking occur.

The treatment process using γ radiation is in general carried out in ambient conditions, due to the lack of its absorption by air molecules [15].

The application of the radiation grafting method has been broadly addressed in the literature [183, 205–210] to improve polymeric surface quality for various applications such as biocompatibility and wettability. Shojaei et al. [206] demonstrated the performance of the radiation grafting technique as a chemically based surface pretreatment method to adhesion modification of polyethylene surfaces towards the metallic layer. γ-Irradiation over a dose range 4–10 kGy is used for grafting.

Oxygen consumption and yield of oxidation products during γ-irradiation were studied on five types of polyethylene (PE), ethylene–butene copolymer (EB), and ethylene–propylene copolymer (EPR) using gas chromatography, mass spectrography, and high-resolution NMR in the work prepared by Arakawa et al. [183].

Imai et al. [205] investigated the surface treatment of PP film with γ-gamma radiation. PP films were irradiated with ^{60}Co γ-rays in vacuum or in air and stored in air. To clarify the difference of mechanical properties between the samples irradiated in vacuum and in air, the effect of cross-link was considered, together with the oxidation. The test presented that (among other things) the increasing rate of the concentration of carbonyl group during irradiation in air was higher than those during storage in air and the tensile strength and the elongation of the sample irradiated in air decreased with storage time.

Shen et al. [184] used the γ-rays at room temperature in air atmosphere to an average dose of 3.4 Mrad to treatment ultrahigh molecular weight polyethylene.

8.2 Experimental test

The information in this subchapter was prepared based on article [20].The purpose of this study is determination of the effect of primers on adhesive properties and strength of adhesive joints of aluminium sheets, using polyurethane adhesives.

8.2.1 Methods

Adherends

The experiments were performed on EN AW-5754-H18 (AlMg3) aluminum alloy sheet. The mechanical properties of EN AW-5754-H18 aluminum alloy sheet are presented in Table 8.2.

Table 8.2 Mechanical properties of EN AW-5754 aluminum alloy [211]

Property	Value
Young modulus E	68 GPa
Poisson's ratio ν	0.33
Shear strength	180 MPa
Tensile strength (ultimate)	320 MPa
Brinell hardness	88

The samples of EN AW-5754-H18 aluminum alloy sheet before tested surface treatment were subjected to cataphoretic painting (this process was described in Ref. [212]).

Adhesive joints—Shape and dimensions

The scheme of the single-lap adhesive joints made of EN AW-5754-H18 aluminum alloy sheet with averaged dimensions was shown in Fig. 8.11.

The single-lap adhesive joints had the following real dimensions presented in Table 8.3.

The thickness of the adhesive layer was set by means of a 3-mm diameter wire made of AlMg5 alloy, which was placed in the middle of the width of the adhesive layer (Fig. 8.11).

Surface treatment method

The surface treatment methods were presented in Table 8.4. The chemical properties of the degreasing agent and primer were presented in work [20]. Teroson FL is a liquid universal cleaner and thinner based on petroleum spirit (aromatic free and low *n*-hexane content) and contains no chlorinated hydrocarbons. Terostat 8519 P is a black liquid activating agent based on polyurethane. The product is characterised by very good coating properties, high stability, and resistance to UV radiation. Terostat 8519 P is also used to increase adhesion of adhesives and polyurethane such as polyurethane adhesive Terostat 8596. It is recommended that lacquered surfaces be cleansed using Teroson FL.

The first variant of surface treatment (Variant A) consisted in only degreasing. The degreasing process involved performing three steps. First and second step consisted

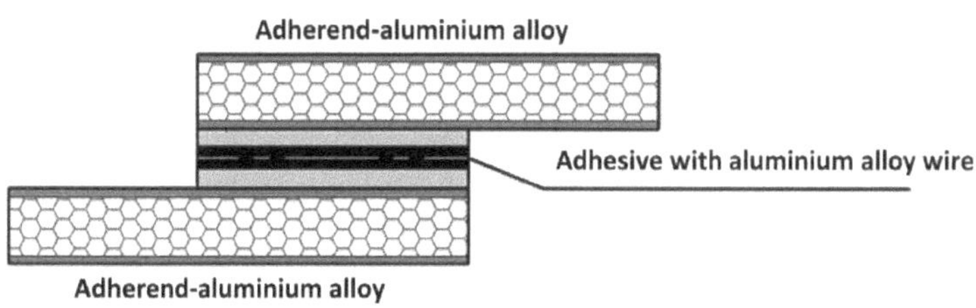

Fig. 8.11 Shape of the tested adhesive joint.

Table 8.3 Real dimensions of single-lap joints

Real dimensions of single-lap joints (mm)					
Specimen length	**Specimen width**	**Specimen thickness**	**Adhesive layer thickness**	**Lap length**	**Wire diameter**
100±1.19	20±0.95	3	3	51.9–59.3	3

Table 8.4 Variants of surface treatments

Variants of surface treatment	Description/agent	
	Degreasing	**Priming**
Variant A	Teroson FL degreasing agent	–
Variant B	Teroson FL degreasing agent	Terostat 8519 P primer

in applying the Teroson FL degreasing agent via wiping with a dust-free fabric twice on the surfaces of the sheets and then left for 6 min to evaporate in compliance with the product safety data sheet (third step).

The second variant of surface treatment (Variant B) involved degreasing (according the method presented above) and then the application of thin layer of primer Terostat 8519 P using a brush to both adherends surfaces. The time from the primer application on the adherends surface and adhesive application was 12–16 min.

Adhesive

Terostat 8596 one-component polyurethane adhesive/sealant was used to prepare the adhesive joints of EN AW-5754-H18 aluminum alloy sheet. This adhesive cures by a reaction with moisture. Terostat 8596 exhibits high shear strength, a very high cure rate, and good adhesion to painted surfaces with or without primer. The properties of the Terostat 8596 adhesive are presented in Table 8.5.

The adhesive joints after two variants of surface preparation (Variant A and Variant B) were made at a temperature of 22 ± 2°C and air humidity of $33 \pm 2\%$. The samples were cured under a pressure of 0.018 MPa Terostat 8596 polyurethane adhesive was applied and a spacing wire with a diameter of 3 mm was placed inside it. Two series of tests were performed: 15 h curing (two batches of 6 adhesive joints samples) and 64 h curing (two batches of 6 adhesive joints samples).

Table 8.5 Properties of Terostat 8596 [213]

Properties	**Description/value**
Chemical base	Polyurethane
Curing mechanism	Moisture curing
Cure rate	Approximately 5 mm/24 h
Consistency	Smooth, sag-resistant paste
Density	Approximately 1.2 g/cm^3
Viscosity, 25°C	5000–6000 Pa s
Tensile strength (DIN 53504)	Approximately 8.5 MPa
Elongation at break (DIN 53504)	Approximately 300%

Experimental tests

During the experiment part the following tests are made: strength test, surface roughness parameters measurement, determination of contact angle and surface free energy calculation, DSC test, and microscopic test.

Strength test

The single-lap adhesive joints made of EN AW-5754-H18 aluminum alloy sheet were subjected to destructive testing—shear strength testing, according to the DIN EN 1465 standard [106] and using the ZWICK/ROELL Z150 materials testing machine.

Surface roughness parameters measurement

Surface roughness parameters were measured using the T8000 RC120-400 Hommel—Etamic-3D profilometer. The surface roughness parameters were determined in accordance with standard ISO 4278 and ISO 25178 requirements. In each batch, five specimens at three different positions were examined (15 measurement results were obtained per each batch). The highest and lowest values were rejected in further analyses.

Contact angle and surface free energy

In order to estimate adhesive properties of aluminium alloys sheet sample surfaces after two variants of surface treatment, the surface free energy was calculated. To calculate surface free energy, the contact angle θ of adherends surfaces was measured. The measurements were made using the method for direct measurement of the angle formed by a measuring liquid drop and the surface being examined. The measurements of the wetting angle were performed at a temperature of $20 \pm 1°C$ and air humidity of $30 \pm 2\%$. The volume of drops of the measuring liquids ranged from 0.6 to 1.4 µL. To each sample surface (there were five samples per batch) five drops of the measuring liquids were applied. After that, 5–10 measurements were made and the mean per each sample batch was calculated. The contact angle was measured immediately after the application of a drop of the measuring liquid (after a few seconds). The measurements were made using a PGX goniometer manufactured by Fibro System (Sweden) and a PG programme for computer image analysis [214].

Surface free energy was determined by the Owens–Wendt method [15, 17, 215–217]. The method assumes that surface free energy (γ_S) of solid is a sum of two components: polar (γ_S^p) and dispersive (γ_S^d), and that they are characterised by the following dependence [17, 217–220]. To determine polar and dispersive components of surface free energy, it is necessary to measure contact angle [17, 214, 215, 218, 220, 221] of surfaces of the materials being tested using two measuring liquids. Measuring liquids used to this aim are liquids whose surface free energy and its polar and dispersive components are known. Distilled water and diiodomethane were used in the test as measuring liquids and their surface free energy γ_L and its components are listed in Table 5.19 (Chapter 5). The components γ_S^d and γ_S^p of the tested materials can be determined using relevant formulas given in the studies [215, 222].

DSC test

NETZSCH Phox PC 200 device was used to DSC test according to PN-EN ISO 1135 Differential scanning calorimetry standard. Test samples were taken from the cured adhesive layer, from samples surfaces submitted two variants of surface treatment (Variant A and Variant B). The samples mass to DCS test was between 8 and 9 mg. The SARTORIUS scales of accuracy ±0.01 mg, with internal calibration and measuring chamber was used. DSC curves were recorded during the heating of the samples at a rate of 10°C/min in the temperature range from room temperature to 400°C.

Microscopic test

Keyence VHX-5000 digital microscope was used to the microscopic tests. The adhesive layer of aluminium alloy adhesive joints (in which adherends were submitted two variants of surface treatment: Variant A and Variant B) after strength test was subjected to microscopic tests.

8.2.2 Results

Surface roughness parameters

The views of surface roughness of tested adherends surfaces after two variants of surface treatment (Variant A and Variant B) were presented in Figs 8.12 and 8.13. The comparison results of surface roughness parameters after two variants of surface treatment are shown in Table 8.6.

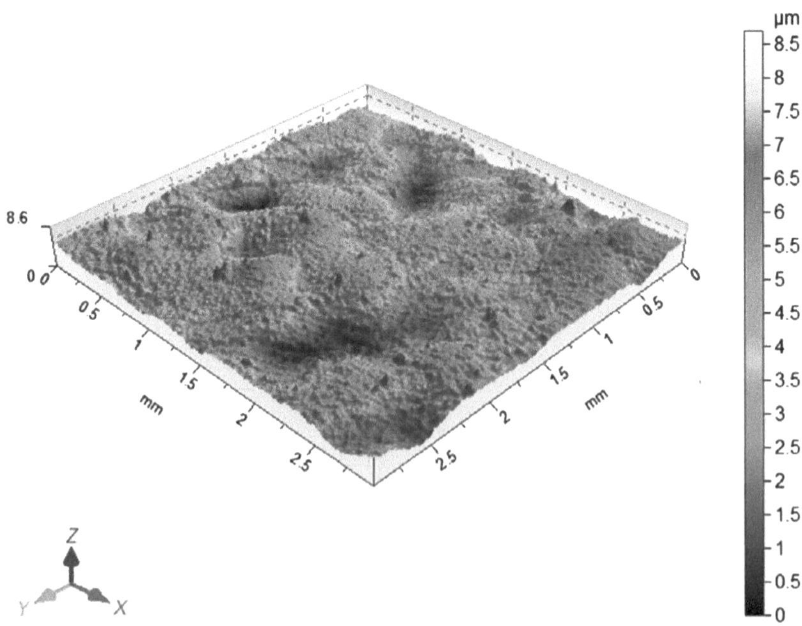

Fig. 8.12 The surface roughness view of samples after degreasing (Variant A).

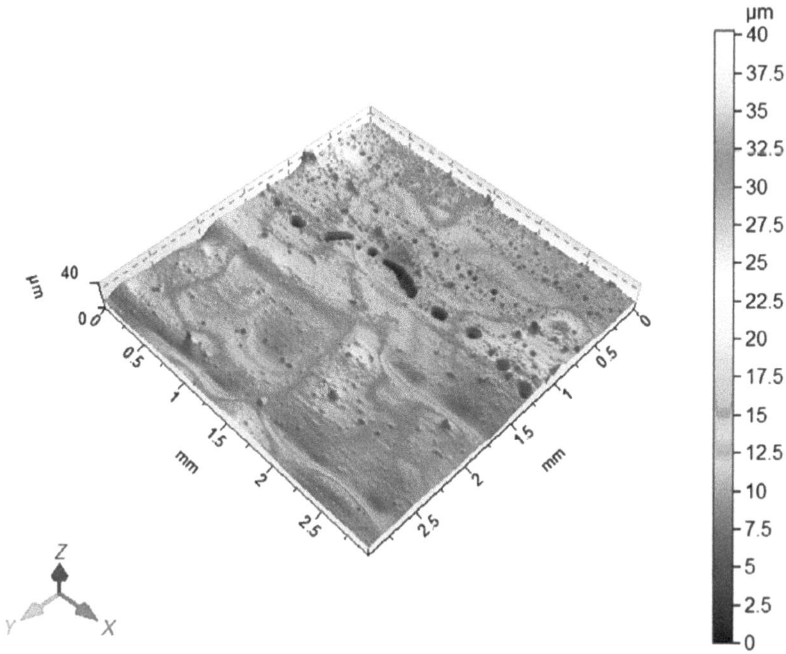

Fig. 8.13 The surface roughness view of samples after degreasing and priming (Variant B) [20].

Table 8.6 Surface roughness parameters of EN AW-5754-H18 aluminum alloy sheet surface after two variants of surface treatment

Variants of surface treatment	Surface roughness parameters (µm)			
	R_a	R_z	R_t	R_{sm}
Variant A	0.34	1.74	2.18	305
Variant B	0.73	4.81	6.53	358

It was observed that the surface roughness of cataphoretic painted EN AW-5754-H18 aluminium alloy sheet, degreased with Teroson FL (Variant A) has lower roughness than the samples after degreasing with Teroson FL and the application of Terostat 8519 P primer (Variant B). On the surface of samples which were treated by Terostat 8519 P primer some residue was observed.

Surface observations were confirmed by results obtained in roughness parameters measurements. All surface roughness parameters for aluminium sheet with surface prepared by degreasing and priming (Variant B) are higher than those obtained in the case when only degreasing was applied.

Strength tests results

The comparison of EN AW-5754-H18 aluminium alloy sheet adhesive joints shear strength with regard to the two surface treatment methods (Variant A and Variant B), after 15 and 64 h of curing was presented in Fig. 8.14.

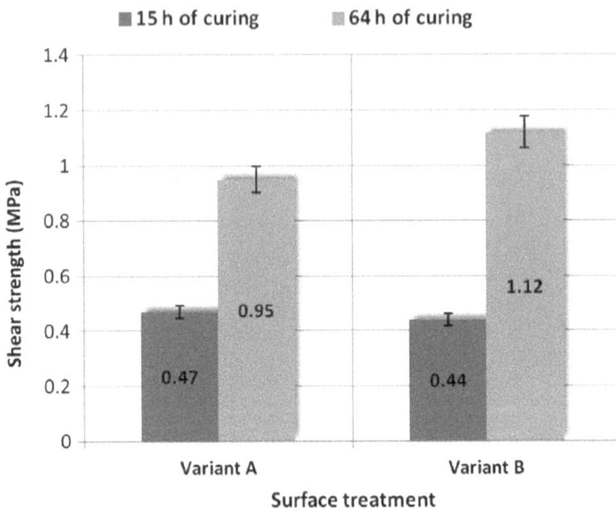

Fig. 8.14 Shear strength of EN AW-5754-H18 aluminium alloy sheet adhesive joints after the application of different surface preparation methods and different curing times.

During the strength tests, it was observed that the adhesive joint is not completely cured inside (the cross-linking process was not completed). Following the strength tests and adhesive bonds failure, the samples of adhesive joints were left to further cure. The temperature of the subsequent curing was ca. 24–26°C, while humidity was ca. 30%.

The results of shear strength of the tested adhesive joints (Fig. 8.14) led to the following observations.

- Curing time significantly affects strength of adhesive joints. A longer curing time leads to a higher strength of adhesive joints compared to those produced at a shorter curing time. For Variant A (degreasing), strength increased by two times, while for Variant B (degreasing and priming), strength increased by over two and a half times.
- A higher strength after 64 h of curing was yielded for Variant B of surface preparation (degreasing followed by priming). The strength obtained for samples after degreasing (Variant A) amounts to ca. 85% of the strength in Variant B (degreasing and priming).

Surface treatment method affects strength depending, however, on curing time. A higher strength was observed for a longer curing time; a shorter curing time does not however affect significantly shear strength of adhesive joints considering the surface preparation method applied. In the work [52] similar results (although there were different combinations of surface treatment) were presented and it was found that the combination of treatments with primers (primer followed by plasma) get the higher failure force that the variant without using primers.

Failure patterns of adhesive joints

An analysis of failure of the tested adhesive joints was performed taking into account the elongation at break of adhesive layer and failure pattern of the joints during

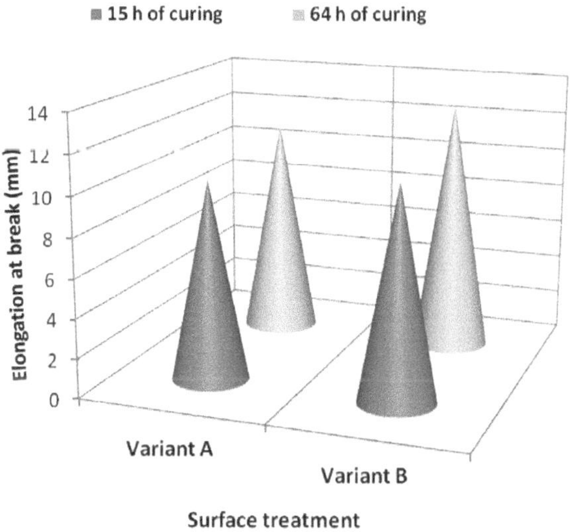

Fig. 8.15 Surface treatment method and time of curing vs adhesive layer elongation in EN AW-5754-H18 aluminium alloy sheet adhesive joints.

strength test. The results of adhesive layer elongation for two surface treatment methods and two time of curing were presented in Fig. 8.15.

Comparing the results (Fig. 8.15) it should be noted that the use of the primer has a positive effect on adhesive layer elongation at break. It was also found that following the 64-h curing, it was observed that the results are more uneven, which might result from the nature of adhesive cross-linking, as it took place over a much greater volume of the adhesive than in the case of 15-h curing.

Visual examination of the failed joints after 15-h curing showed that the adhesive cross-linking process was not completed (it did not take place over the entire volume of the adhesive layer). The adhesive was cured only in the outside layer while in the core it was uncured. It can be claimed that the conditions of adhesive layer curing (humidity and temperature) were inadequate for the adhesive to cure completely over the entire volume of the adhesive layer within 15 h.

Visual valuation of samples of adhesive joints was made on the basis of the standard EN ISO 10365. In accordance with the standard, the failures will be classified to one of the six groups, three of which relate to damage of adherends, while the other to adhesives (adhesives layer). In all tested samples of adhesive joints the cohesive failure (CF) of adhesive layer was noted (Figs 8.16–8.18).

Views in Figs 8.17 and 8.18 presented the sectional view of adhesive layer after failure. The observed variation in adhesive layer thickness confirms the CF mechanism of adhesive. Such failure might indicate a good adhesion between the EN AW-5754-H18 aluminium alloy sheet surface and the polyurethane adhesive, which is the result of proper surface treatment. It can be noticed, that in this case, the more important is the correct selection of the adhesive and curing conditions, including curing time and humidity, than surface treatment.

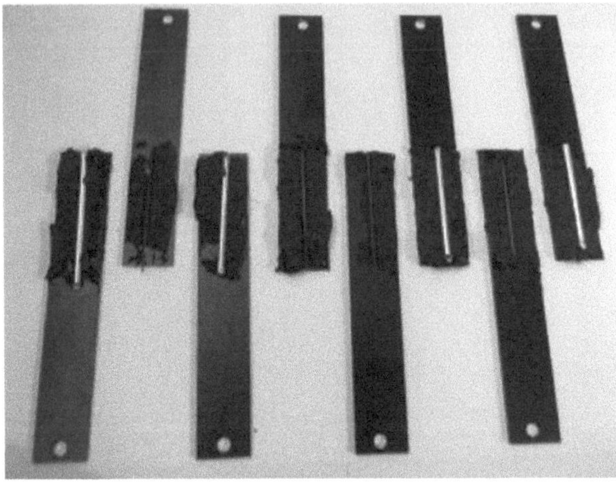

Fig. 8.16 Tested samples of adhesive joints (obtained when applying degreasing and priming) after strength test—cohesive failure (CF)—visual method [20].

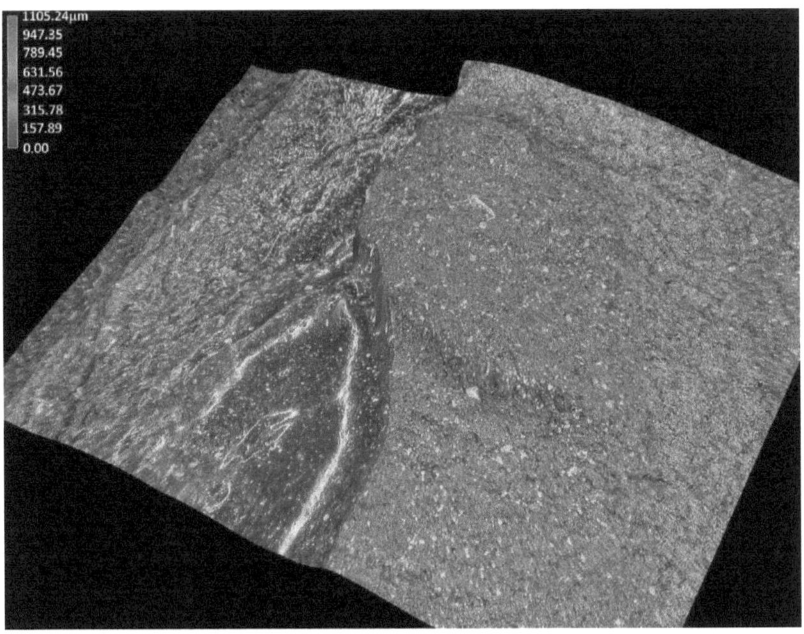

Fig. 8.17 View of failure (CF) of adhesive layer (after Variant A surface treatment) after strength test—microscope method.

The adhesive layer surfaces after failure are given in Figs 8.19 and 8.20.

Based on visual evaluation of the failed joints, it can also be noted that the failed adhesive layers have numerous blowholes (unbonded spots), made by constrained CO_2 generated during the cross-linking of the polyurethane adhesive (Figs 8.19 and 8.20). In this case, the adhesive curing conditions turned out to be the weaker

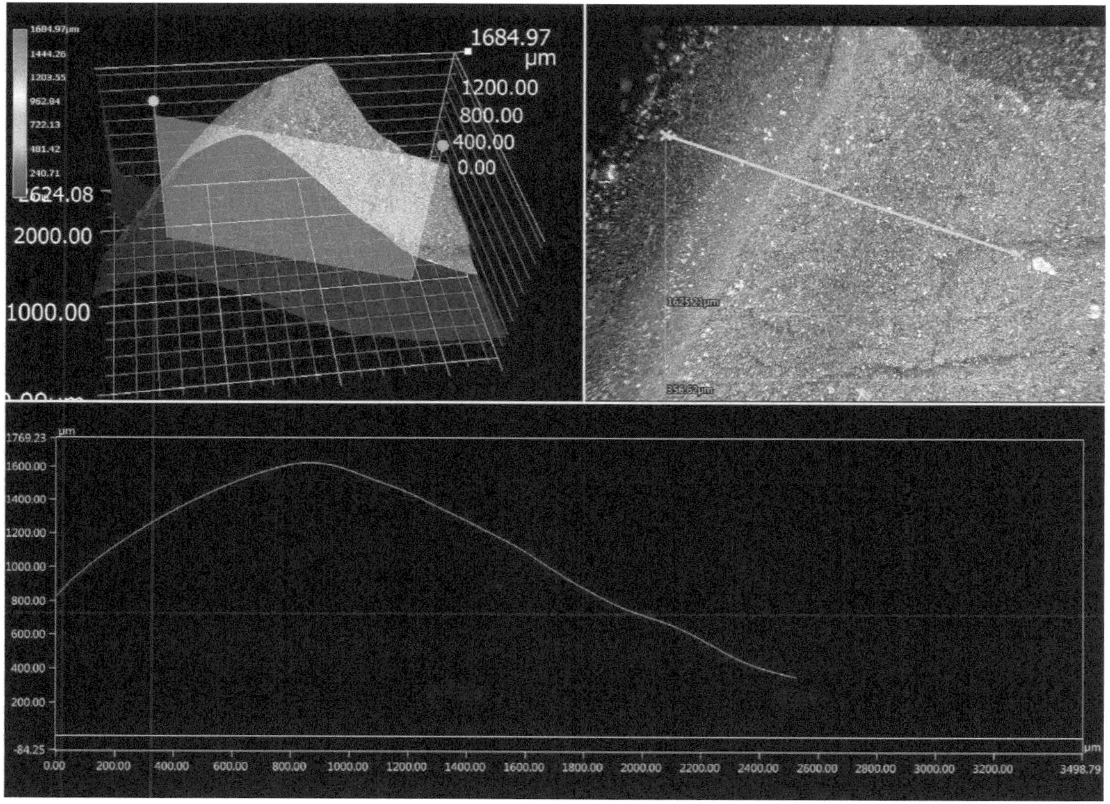

Fig. 8.18 View of failure (CF) of adhesive layer (after variant B surface layer) after strength test—microscope method.

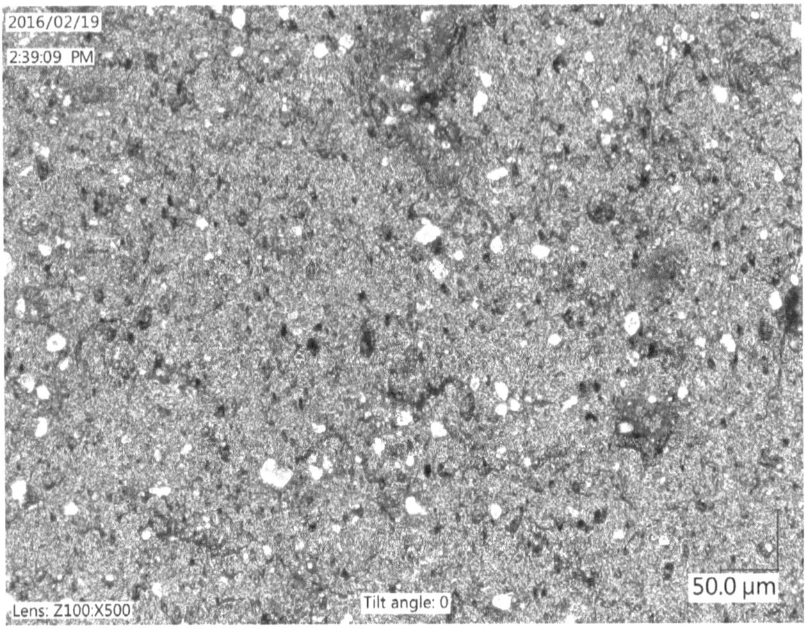

Fig. 8.19 View of adhesive (Variant B surface treatment) after strength test—cohesive failure (CF)—15-h curing time.

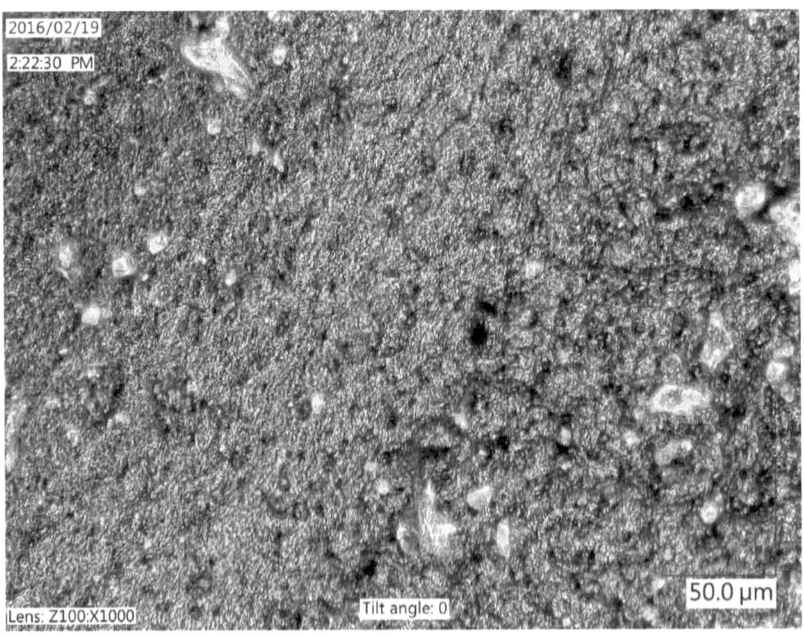

Fig. 8.20 View of adhesive (Variant B surface treatment) after strength test—cohesive failure (CF)—64-h curing time [20].

technological factor, and especially curing time. Furthermore, the results allow to conclude that at a humidity of 33% and curing temperature 24–26°C, the cross-linking of isocyanate groups (NCO) with H_2O groups is still on even after 64 h, because the adhesive layer is bonded again after joint failure and leaving this joint for 48 h to cure.

Surface free energy

Surface free energy and its components results determined based on the contact angle measurement using distilled water and diiodomethane (by Owens–Wendt method) of samples subjected to the tested surface treatment variants were presented in Fig. 8.21.

Based on the results, it can be observed that the surface subjected to cataphoretic painting and priming exhibits better adhesive properties. The values of both surface free energy and its dispersive and polar components are higher than those obtained for the surface subjected only to degreasing. The cataphoretic painting process led to increase in adhesion properties. The surface that was not subjected to painting exhibits the lowest values of γ_S, γ_S^d, and γ_S^p.

Relationships between surface free energy and strength

Considering the results of strength tests (Fig. 8.14) and those of surface free energy and its components (Fig. 8.21) for aluminum alloy sheets, it was observed that a higher value of surface free energy leads to a higher joints strength. It is beneficial to apply priming, as this leads to an increase in surface energy of the adherends, which results in higher strength.

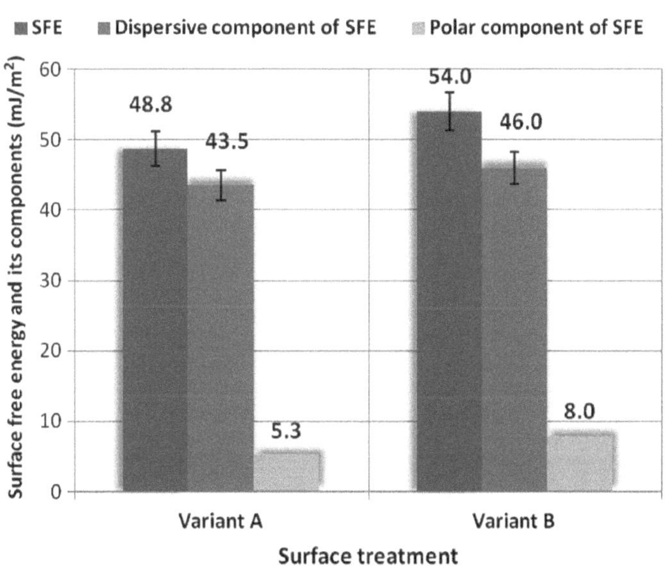

Fig. 8.21 Values of surface free energy γ_S and its components of EN AW-5754-H18 aluminium alloy sheet surface after surface treatment variants.

The priming has contributed to an increase in surface roughness parameters. It can be assumed that, apart from the obvious environmental impacts (chemical bond), perhaps a part surface roughness influences the increase in the strength of the joint (which is also a part of the mechanical adhesion).

In addition, a good adhesion between the surface of aluminium sheet and cataphoretic painting coating was noted and adhesive failure between the surface of the metal and coating painting was not observed. Good adhesion between the joined materials (painted aluminium sheet and painted aluminum sheet with imposed primer) and adhesive was also noted when the nature of the failure was analysed. In all tested samples of adhesive joints the CF of adhesive layer was observed. This can confirm a proper surface treatment.

DSC results

The results of DSC measurements are shown in Fig. 8.22.

DSC curves indicated that at about 250°C the degradation of the adhesive starts. In the range from room temperature to 250°C there is no difference in the DSC curve for both ways of sample surface preparation. On this basis, it appears that the use of the primer during aluminium sheet surface treatment before bonding does not lower the temperature of adhesive decomposition and does not adversely affect the process of adhesive degradation.

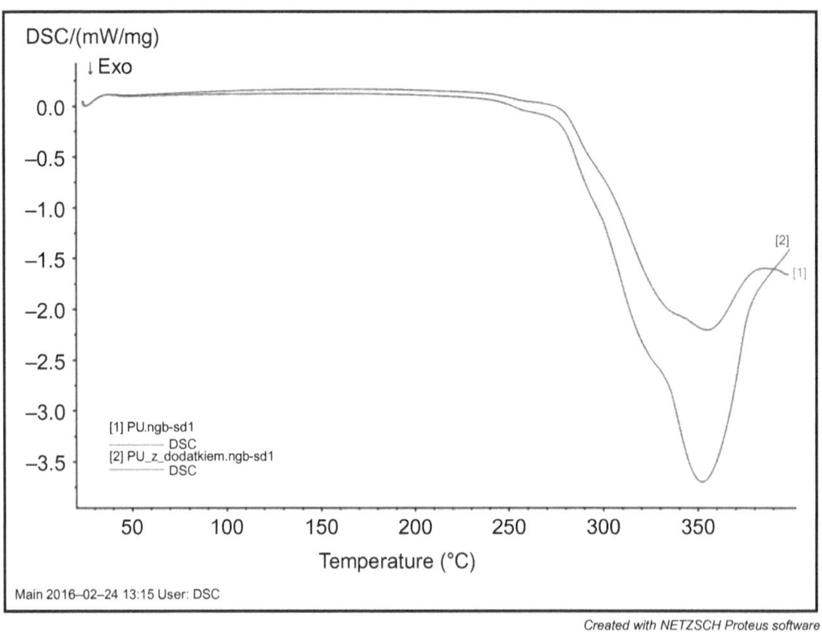

Fig. 8.22 DSC curves of cured adhesives samples, surface preparation method: degreasing (1), degreasing and priming (2) [20].

8.2.3 Conclusions

The application of pro-adhesive agents (primers) leads to obtaining surfaces with better adhesive properties compared to the surfaces subjected only to degreasing. Good adhesive properties are necessary to obtain the required strength of joints, yet other conditions of making adhesive joint, such as the application of a suitable kind of adhesive and curing conditions must be considered too.

Summing up the results of the conducted strength tests, the following observations can be made:

- it is favourable that adhesive layers have a longer curing time at a humidity of 33%, as it leads to a significant increase in strength of the adhesive joints,
- the application of priming prior to application of an adhesive (Variant B) leads to a significant increase in joint strength in compare to the variant in which only degreasing was applied (Variant A),
- a higher elongation was obtained at a longer curing time of the adhesive layers.

Considering the results of adhesive joints strength and values of surface free energy, it was observed that there is the dependence between surface free energy and strength. Namely, the higher the surface free energy, the higher strength of joints is. This confirms the findings given in the literature [31, 77, 223] concerning increased adhesive properties due to the application of priming. Based on the results of experiments, it can be claimed that the curing of polyurethane adhesive layers at humidity 33% has a significant effect on strength of adhesive joints. At such humidity, the curing time of polyurethane adhesive cross-linking must be longer, since adhesives of this type cure under the influence of moisture absorbed from air. Moreover priming during surface treatment for bonding does not adversely affect the process of adhesive degradation. The experiments justify the need to conduct further experimental tests of the presented problem.

References

[1] J. Godzimirski, J. Kozakiewicz, J. Łunarski, W. Zielecki, Konstrukcyjne połączenia klejowe elementów metalowych w budowie maszyn (Structural Adhesive Joint of Metal Elements in Machine Construction), Oficyna Wydawnicza Politechniki Rzeszowskiej, Rzeszów, 1997 (in Polish).

[2] J. Jang, I. Jang, H. Kim, Effect of the mole ratios of silane-modified polyvinylimidazoles on adhesion promotion of the polyimide/copper system, J. Adhes. Sci. Technol. 12 (1998) 323–337.

[3] A. Krysztafkiewicz, Czynniki poprawiające adhezję w układzie napełniacz-polimer, Factors improved adhesion in the filler-polymer system, Chemia Stosowana 4 (1989) 561–584 (in Polish).

[4] K. Uehara, M. Sakurai, Bonding strength of adhesives and surfaces roughness of joined parts, J. Mater. Process. Technol. 127 (2002) 178–181.

[5] L.B. Cohen, Zircoaluminate adhesion promoters, J. Adhes. Sci. Technol. 5 (1991) 439–448.

[6] C. Fotea, C. D'Silva, Adhesion enhancement of chromium tanned heavy-duty leather (Saltz leather) under extreme conditions using photoreagents as surface primers, Int. J. Adhes. Adhes. 25 (2005) 442–449.

[7] T. Vélez-Pagés, J.M. Martin-Martinez, Application of one-components primer to avoid the roughening of leather and increase its adhesion to polyurethane adhesive, Int. J. Adhes. Adhes. 25 (2005) 320–328.
[8] P.H. Winfield, N. Becerra, P. Kaczmar, Investigation of primer/preservative basecoats for timber using vacuum impregnation, Int. J. Adhes. Adhes. 29 (2009) 702–709.
[9] A. Krysztafkiewicz, B. Ragner, M. Maik, Wpływ modyfikacji powierzchni na właściwości fizykochemiczne krzemionki strącanej (The effect of surface modification on the physicochemical properties of precipitated silica), Fizykochemiczne Problemy Mineralurgii 28 (1994) 177–186 (in Polish).
[10] V. Ducháček, A. Kuta, J. Šimek, Roles of alkoxy titanates and zirconates as multifunctional agents for polymers, Polimery 49 (2004) 159–162.
[11] M. Bremont, W. Brockmann, M.F. Guimon, G. Pfister-guillouzo, Improvement of the durability of zinc-cated steel/epoxy bonded joints, J. Adhes. 41 (1993) 147–168.
[12] R.A. Gledhill, S.J. Shaw, D.A. Tod, Durability of adhesive-bonded joints employing organosilane coupling agents, Int. J. Adhes. Adhes. 10 (1990) 192–198.
[13] D.D. Rodrigues, J.G. Broughton, Silane surface modification of boron carbide in epoxy composites, Int. J. Adhes. Adhes. 46 (2013) 62–73.
[14] Y. Xie, C.A.S. Hill, Z. Xiao, H. Militz, C. Mai, Silane coupling agents used for natural fiber/polymer composites: a review, Compos. Part A 41 (2010) 806–819.
[15] M. Żenkiewicz, Adhezja i modyfikowanie warstwy wierzchniej tworzyw wielkocząsteczkowych (Adhesion and Modification of the Surface Layer of Macromolecular Materials), WNT, Warszawa, 2000 (in Polish).
[16] S. Ebnesajjad, Adhesives Technology Handbook, second ed., William Andrew, New York, 2008.
[17] A. Baldan, Adhesion phenomena in bonded joints, Int. J. Adhes. Adhes. 38 (2012) 95–116.
[18] N.A. Darwish, A.B. Shehata, A. Abou-Kandil, A.A.A. El-Megeed, S.N. Lawandy, B.K. Seleh, A novel promoter for enhancing adhesion between natural rubber and brass-plated steel cords, Int. J. Adhes. Adhes. 40 (2013) 135–144.
[19] E. Piva, E.C. Azevedo, A.O. Ogliari, K.J. Pilownic, M.B. Pinto, G.B. Camacho, C.L. Petzhold, F.A. Ogliari, Evaluation of experimental phosphate and sulfur-base primer bonding to metal casting alloys, Int. J. Adhes. Adhes. 58 (2015) 59–62.
[20] A. Rudawska, E. Bociąga, E. Olewnik-Kruszkowska, The effect of primers on adhesive properties and strength of adhesive joints made with polyurethane adhesives, J. Adhes. Sci. Technol. 31 (2017) 327–344.
[21] J.C. Lui, Primer for bonding low surface energy plastics with cyanoacrylate adhesives and bonding method employing same, United States Patent No 5079098, 1992.
[22] G.L. Witucki, A silane primer: chemistry and applications of alkoxy silanes, J. Coatings Technol. 65 (1993) 57–60.
[23] J.A. Bishopp, M.J. Parker, T.A. O'Reilly, The use of statistical experimental design procedures in the development and "robust" manufacture of a water-based, corrosion-inhibiting adhesive primer, Int. J. Adhes. Adhes. 21 (2001) 473–480.
[24] P. He, M. Huang, S. Fisher, C.J. Yue, J. Yang, Effects of primer and annealing treatment on the shear strength between anodized Ti6Al4V and epoxy, Int. J. Adhes. Adhes. 57 (2015) 49–56.
[25] G.M. Pax, J. Bouchet, X. Bulliard, V. Michaud, J.-A.E. Månson, Effect of HBP modified γ-APS on the coating toughness and practical adhesion between galvanized steel and polyethylene, Int. J. Adhes. Adhes. 27 (2007) 539–603.
[26] A.N. Rider, D.R. Arnott, Boiling water and silane pre-treatment of aluminium alloys for durable adhesive bonding, Int. J. Adhes. Adhes. 20 (2000) 209–220.

[27] K. Leena, K.K. Athira, S. Bhuvaneswari, S. Suraj, V. Lakshmana Rao, Effect of surface pre-treatment on surface characteristics and adhesive bond strength of aluminum alloy, Int. J. Adhes. Adhes. 70 (2016) 265–270.
[28] J. Aakkula, O. Saarela, Silane based field level surface treatment methods for aluminium, titanium and steel bonding, Int. J. Adhes. Adhes. 48 (2014) 268–279.
[29] M.-L. Abel, A.N.N. Adams, A.J. Kinloch, S.J. Shaw, J.F. Watts, The effect of surface pretreatment on the cyclic-fatigue characteristics of bonded aluminium-alloy joints, Int. J. Adhes. Adhes. 26 (2006) 50–61.
[30] M. Pantoja, N. Encinas, J. Abenojar, M.A. Martinez, Effect of tetraethoxysilane coating on the improvement of plasma treated polypropylene adhesion, Appl. Surf. Sci. 280 (2013) 850–857.
[31] M. Mohseni, M. Mirabedini, M. Hashemi, G.E. Thompson, Adhesion performance of an epoxy clear coat on aluminum alloy in the presence of vinyl and amino-silane primers, Prog. Org. Coat. 57 (2006) 307–313.
[32] M.-L. Abel, R.D. Allington, R.P. Digby, N. Porritt, S.J. Shaw, J.F. Watts, Understanding the relationship between silane application conditions, bond durability and locus of failure, Int. J. Adhes. Adhes. 26 (2006) 2–15.
[33] F. Awaja, M. Gilbert, G. Kelly, B. Fox, P.J. Pigram, Adhesion of polymers, Prog. Polym. Sci. 34 (2009) 948–968.
[34] P. Walker, Oragnosilanes as adhesion promotores, J. Adhes. Sci. Technol. 5 (1991) 279–305.
[35] N. Brack, A.N. Rider, The influence of mechanical and chemical treatments on the environmental resistance of epoxy adhesive bonds to titanium, Int. J. Adhes. Adhes. 48 (2014) 20–27.
[36] A. Rider, P. Chalkley, Durability of an off-optimum cured aluminium joint, Int. J. Adhes. Adhes. 24 (2004) 95–106.
[37] W. Brockmann, P.L. Geiß, J. Klingen, B. Schröder, Adhesive Bonding. Materials, Applications and Technology, Wiley-VCH Press, Weinheim, 2009.
[38] S.B. Lin, Silicone pressure-sensitive adhesives with selective adhesion characteristics, J. Adhes. Sci. Technol. 10 (1996) 559–571.
[39] W. Polini, L. Sorrentino, Adhesion of a protective coating on a surface of aluminium alloy treated by air cold plasma, Int. J. Adhes. Adhes. 27 (2007) 1–8.
[40] W.M. Rzymski, M. Włóka, T. Grzeżółkowska-Rosiak, Łączenie elastomerów z metalami (Joining elastomers with metals), Polimery 50 (2005) 530–536 (in Polish).
[41] R. Balkova, S. Holcneowa, V. Cech, Testing of adhesives for bonding of polymer composites, Int. J. Adhes. Adhes. 22 (2002) 291–295.
[42] R.M. Podhajny, Comparing surface treatments, Converting 8 (1990) 46–52.
[43] A.N. Rider, Factors influencing the durabilityof epoxyadhesion to silane pretreated aluminium, Int. J. Adhes. Adhes. 26 (2006) 67–78.
[44] A. Stevenson, R. Campion, Durability, in: A.N. Gent (Ed.), Engineering with Rubber. How to Design Rubber Components, second ed., Hanser Publishers Munich, Hanser-Gardner Publications, Inc., Cincinnatti, 2001 (Chapter 7).
[45] V.Y. Lee, Organosilicon Compounds: Theory and Experiment (Synthesis), vol. 1, Academic Press, Elsevier, 2017.
[46] V. Palanivel, D. Zhu, W. van Ooij, Nanoparticle-filled silane films as chromate replacements for aluminium alloys, Prog. Org. Coat. 47 (2003) 384–392.
[47] J. Yang, A. Garton, Primers for adhesive bonding to polyolefins, J. Appl. Polym. Sci. 48 (1993) 359–370.
[48] E.P. Plueddemann, Silane primers for epoxy adhesives, J. Adhes. Sci. Technol. 2 (1988) 179–188.

[49] M.J. Owen, D.E. Wiliams, Surface modification by fluoroalkyl-functional silanes, J. Adhes. Sci. Technol. 5 (1991) 307–320.

[50] V.A. Ogarev, S.L. Selector, Oragnosilicon promoters of adhesion and their influence on the corrosion of metals, Prog. Org. Coat. 21 (1992) 135–187.

[51] I. Gähde, R. Mix, H. Goering, G. Schultz, W. Funke, U. Hermann, Increase in coating wet adhesion stability by highly ordered polymers, J. Adhes. Sci. Technol. 11 (1997) 861–875.

[52] T. Rey, F. Razan, E. Robin, S. Faure, J.-B. Le Cam, G. Chagnon, A. Girard, D. Favier, Mechanical characterization and comparison of different NiTi/silicone rubber interfaces, Int. J. Adhes. Adhes. 48 (2014) 67–74.

[53] C. Xu, K. Ramani, G. Kumar, Thermoplastic adhesive bonding of galvanized steel to polypypropylene composite and its durability, Int. J. Adhes. Adhes. 22 (2002) 187–195.

[54] B. Chwiałkowska, M. Dębski, Badania mas z alkaliczną żywicą rezolową zmierzające do opracowania kompozycji alkaliczna żywica rezolowa i promotor adhezji (Research on masses with alkaline resole resin aimed at developing the composition of alkaline resole resin and adhesion promoter), Solidification of Metals and Alloys 2 (43) (2002) 91–100 (in Polish).

[55] Oranosilicon compounds, https://www.chemicalbook.com/ProductCatalog_EN/1326.htm, 2018 (Accessed October 27, 2018).

[56] J. Chruściel, E. Leśniak, M. Fejdyś, Karbonofunkcyjne silany i polisiloksany. Cz. I. Zastosowania karbofunkcyjnych silanów (Carbofunctional silanes and polysiloxanes. Part I. Applications of carbofunctional silanes), Polimery 53 (2008) 708–716. (in Polish).

[57] J. Chruściel, E. Leśniak, M. Fejdyś, Karbonofunkcyjne silany i polisiloksany. Cz. II. Otrzymywanie i zastosowania karbofunkcyjnych polisiloksanów (Carbofunctional silanes and polysiloxanes. Part II. Preparation and applications of carbofunctional polysiloxanes), Polimery 53 (2008) 817–829 (in Polish).

[58] Y.J. Du, M. Damron, G. Tang, X. Zheng, C.-J. Chu, J.H. Osborne, Inorganic/organic hybrid coatings for aircraft aluminium alloy substrates, Prog. Org. Coat. 41 (2001) 226–232.

[59] F.B. Leal, F.C. Madruga, E.P. Prochnow, G.S. Lima, F.A. Ogliari, E. Piva, R.R. Moraes, Effect of acidic monomer concentration on the dentin bond stability of self-etch adhesives, Int. J. Adhes. Adhes. 31 (2011) 571–574.

[60] P.F. McDonnel, J.K. Kneafsey, Diazabicyclo and triazabicyclo primer compositions an use thereof in bonding non-polar substrates, United States Patent No 4869772, 1989.

[61] J.M. Sykes, Surface treatment for steel, in: D.M. Brewis (Ed.), Surface Analysis and Pretreatment of Plastics and Metals, Applied Science Publishers, London, 1982, pp. 153–174.

[62] A. Rudawska, M.A. Wahab, D. Barta, S. Pukalskas, The effect of technological and structural factors on the strength of polyethylene adhesive joints, in: M.A. Wahab (Ed.), Proceedings of the 7th International Conference on Fracture Fatigue and Wear, FFW 2018, 9-10 July 2018, Ghent University, Belgium, Lecture Notes in Mechanical Engineering, Springer, Singapore, 2019, pp. 241–257.

[63] Plasma, https://www.almukantarat.pl/wiedza/09plazma/, 2018 (Accessed September 10, 2018).

[64] Y. Kusano, H. Mortensen, B. Stenum, S. Goutianos, S. Mitra, A. Ghanbari-Siahkali, P. Kingshott, B.F. Sørensen, H. Bindslev, Atmospheric pressure plasma treatment of glassy carbon for adhesion improvement, Int. J. Adhes. Adhes. 27 (2007) 402–408.

[65] G. Rozovskis, J. Vinkevičius, Plasma surface modification of polyimide for improving adhesion to elekctroless copper coatings, J. Adhes. Sci. Technol. 10 (1996) 399–406.
[66] G.A. Wade, W.J. Cantwell, Adhesive bonding and wettability of plasma treated, glass fiber-reinforced nylon-6,6 composites, J. Mater. Sci. Lett. 19 (2000) 1829–1832.
[67] G.A. Wade, W.J. Cantwell, R.C. Pond, Plasma surface modification of glass fibre-reinforced Nylon-6,6 thermoplastic composites for improved adhesive bonding, Interface Sci. 8 (2000) 363–373.
[68] L.A. Dobrzański, A.D. Dobrzańska-Danikiewicz, Obróbka powierzchni materiałów inżynierskich (Surface Treatment of Engineering Materials), vol. 5, Open Access Library, 2011, pp. 42–367. (Chapter 7) (in Polish). Available from: http://www.openaccesslibrary.com/vol05/7.pdf.
[69] B.R. Striochmeier, Improving the wettability of aluminium foil with oxygen plasma treatments, in: K.L. Mittal (Ed.), Contact Angle, Wettability and Adhesion, 1993, pp. 453–468.
[70] I. De Iorio, C. Leone, L. Nele, V. Tagliaferri, Plasma treatments of polymeric materials and Al alloy for adhesive bonding, J. Mater. Process. Technol. 68 (1997) 179–183.
[71] L. Carrino, G. Moroni, W. Polini, Cold plasma treatment of polypropylene surface: a study on wettability and adhesion, J. Mater. Process. Technol. 121 (2002) 373–382.
[72] R. Stewart, V. Goodship, P. Guild, M. Green, J. Farrow, Investigation and demonstration of the durability of air plasma pre-treatment on polypropylene automotive bumpers, Int. J. Adhes. Adhes. 25 (2005) 93–99.
[73] O. Lunder, B. Olsen, K. Nisancioglu, Pre-treatment of AA6060 aluminium alloy for adhesive bonding, Int. J. Adhes. Adhes. 22 (2002) 143–150.
[74] W. Polini, L. Sorrentino, Improving the wettability of aluminium alloy 2024 by means of cold plasma treatment, Appl. Surf. Sci. 214 (2003) 232–242.
[75] N. Anagreh, L. Dorn, Influence of low-pressure plasma treatment on adhesive bonding between polybutylene terephthalat (PBT) and aluminum, Int. J. Adhes. Adhes. 25 (2005) 165–172.
[76] M. Akram, K.M.B. Jansen, L.J. Ernst, S. Bhowmik, Atmospheric plasma modification of polyimide sheet for joining to titanium with high temperature adhesive, Int. J. Adhes. Adhes. 65 (2016) 63–69.
[77] N. Encinas, J. Abenojar, M.A. Martínez, Development of improved polypropylene adhesive bonding by abrasion and atmospheric plasma surface modifications, Int. J. Adhes. Adhes. 33 (2012) 1–6.
[78] J.H. Ku, I.H. Jung, K.Y. Rhee, S.J. Park, Atmospheric pressure plasma treatment of polypropylene to improve the bonding strength of polypropylene/aluminium composites, Compos. Part B 45 (2013) 1282–1287.
[79] Y. Nakamura, Y. Suzuki, Y. Watanabe, Effect of oxygen plasma etching on adhesion between polyimide films and metal, Thin Solid Films 290 (291) (1996) 357–369.
[80] K. Navaneetha Pandiyaraj, V. Selvarajan, R.R. Deshmukh, C. Gao, Modification of surface properties of polypropylene (PP) film using DC glow discharge air plasma, Appl. Surf. Sci. 255 (2009) 3965–3971.
[81] L. Sorrentino, L. Carrino, Influence of process parameters of oxygen cold plasma treatment on wettability ageing time of 2024 aluminium alloy, Int. J. Adhes. Adhes. 29 (2009) 136–143.
[82] C. Sperandio, J. Bardon, A. Laachachi, H. Aubiert, D. Ruch, Influence of plasma surface treatment on bond strength behaviour of an adhesively bonded aluminium-epoxy system, Int. J. Adhes. Adhes. 30 (2010) 720–728.

[83] Z.Z. You, J.Y. Dong, Effect of oxygen plasma treatment on the surface properties often-doped indium oxide substrates for polymer LEDs, J. Colloid Interface Sci. 300 (2006) 697–703.
[84] T. Kruse, G. Krüger, A. Baalmann, O.-D. Hennemmann, Surface pretreatment of plastic for adhesive bonding, J. Adhes. Sci. Technol. 9 (1995) 1611–1621.
[85] N. De Geyter, R. Morent, C. Leys, L. Gengembre, E. Payen, Treatment of polymer films with a dielectric barrier discharge in air, helium and argon at medium pressure, Surf. Coat. Technol. 201 (2007) 7066–7075.
[86] R.D. Adams, J. Comyn, W.C. Wake, Structural Adhesive Joints in Engineering, second ed., Chapman & Hall, London, 1997.
[87] S.C. Gallo, H. Dong, On the fundamental mechanisms of active screen plasma nitriding, Vacuum 84 (2010) 321–325.
[88] O.J. Kwon, S. Tang, S.W. Myung, N. Lu, H.S. Choi, Surface characteristics of polypropylene firm treated by an atmospheric pressure plasma, Surf. Coat. Technol. 192 (2005) 1–10.
[89] G.S. Sheu, S.S. Shyu, Surface modification of Kevlar 149 fibers by gas plasma treatment. Part II. Improved interfacial adhesion to epoxy resin, J. Adhes. Sci. Technol. 8 (1994) 1027–1042.
[90] R. Morent, N. De Geyter, C. Leys, L. Gengembre, E. Payen, Study of the ageing behaviour of polymer films treated with a dielectric barrier discharge in air, helium and argon at medium pressure, Surf. Coat. Technol. 201 (2007) 7847–7854.
[91] K. Usami, T. Ishijima, H. Toyoda, Rapid plasma treatment of polyimide for improved adhesive and durable copper film deposition, Thin Solid Films 521 (2012) 22–26.
[92] G.Z. Xiao, Effects of solvent on the surface properties of oxygen plasma-treatment polyethylene and polypropylene films, J. Adhes. Sci. Technol. 11 (1997) 655–663.
[93] H. Zhou, F. Li, B. He, J. Wang, B.-d. Sun, Air plasma sprayed thermal barrier coatings on titanium alloy substrates, Surf. Coat. Technol. 201 (2007) 7360–7367.
[94] A. Van Deynse, P. Cools, C. Leys, R. Morent, N. De Geyter, Surface modification of polyethylene in an argon atmospheric pressure plasma jet, Surf. Coat. Technol. 276 (2015) 384–390.
[95] E. Gonzalez II, M.D. Barankin, P.C. Guschl, R.F. Hicks, Surface activation of poly(methyl methacrylate) via remote atmospheric pressure plasma, Plasma Process. Polym. 7 (6) (2010) 482–493.
[96] M.-H. Jung, H.-S. Choi, Surface treatment and characterization of ITO thin film using atmospheric pressure plasma for organic light emitting diodes, J. Colloid Interface Sci. 310 (2007) 550–558.
[97] M. Vishnuvarthanan, N. Rajeswari, Effect of mechanical, barrier and adhesion properties on oxygen plasma surface modified PP, Innov. Food Sci. Emerg. Technol. 30 (2015) 119–126.
[98] D.M. Choi, C.K. Park, K. Cho, C.E. Park, Adhesion improvement of epoxy resin/polyethylene joints by plasma treatment of polyethylene, Polymer 38 (1997) 6243–6249.
[99] A. Van Deynse, P. Cools, C. Leys, N. De Geyter, R. Morent, Surface activation of polyethylene with an argon atmosphericpressure plasma jet: influence of applied power and flow rate, Appl. Surf. Sci. 328 (2015) 269–278.
[100] M. Akram, K.M.B. Jansen, L.J. Ernst, S. Bhowmik, Atmospheric plasma modification of titanium for high temperature adhesive bonding, Int. J. Adhes. Adhes. 31 (2011) 598–604.
[101] N. Encinas, B. Díaz-Benito, J. Abenojar, M.A. Martínez, Extreme durability of wettability changes on polyolefin surfaces by atmospheric pressure plasma torch, Surf. Coat. Technol. 205 (2010) 396–402.

[102] T.S. Wiliams, H. Yu, P.-C. Yeh, J.-M. Yang, R.F. Hicks, Atmospheric pressure plasma effects on the adhesive bonding properties of stainless steel and epoxy composite, J. Compos. Mater. 48 (2) (2014) 219–233.

[103] Y. Li, S. Manolache, Y. Qiu, M. Sarmadi, Effect of atmospheric plasma treatment condition on adhesion of ramie fibres to polypropylene for composite, Appl. Surf. Sci. 364 (2016) 294–301.

[104] U. Lommatzsch, D. Pasedag, A. Baalmann, G. Ellinghorts, H.-E. Wagner, Atmospheric pressure plasma jet treatment of polyethylene surfaces for adhesion improvement, Plasma Process. Polym. 4 (2007) S1041–S1045.

[105] R.F. Hicks, S.E. Babayan, J. Penelon, Q. Truong, D.S.F. Cheng, V.V. Le, J. Ghilarducci, A.G. Hsieh, Atmospheric plasma treatment of polyetheretherketone composites for improved adhesion, in: In SAMPE Fall Technical Conference Proceedings, Global Advances in Materials and Process Engineering, TX, November 6–9, CD-ROM-9 pp., Dallas, 2006.

[106] E. Gonzalez, R.F. Hicks, Surface analysis of polymers treated by remote atmospheric pressure plasma, Langmuir 26 (5) (2010) 3710–3719.

[107] S.J. Park, E.J. Lee, B.J. Kim, A study of atmospheric-pressure CHF3/AR plasma treatment on dielectric characteristics of polyimide films, J. Colloid Interface Sci. 319 (2008) 365–369.

[108] E. Panousis, C. Franck, J.-F. Loiseau, C. Guimon, Titanium alloy surface treatment using an atmospheric plasma jet in nitrogen pulsed discharge conditions, Surf. Coat. Technol. 201 (2007) 7292–7302.

[109] S. Bhowmik, P.K. Ghosh, S. Ray, K. Barthwal, Surface modification of high density polyethylene and polypropylene by DC glow discharge and adhesive bonding to steel, J. Adhes. Sci. Technol. 12 (1998) 1181–1204.

[110] N. Shirai, H. Shito, S. Ibuka, S. Ishii, DC corona and glow discharges generated along the intersecting axial miniature gas flows under atmospheric pressure air, Appl. Phys. Express 2 (2009) 076001.

[111] M. Pascual, R. Sanchis, L. Sánchez, D. García, R. Balart, Surface modification of low density polyethylene (LDPE) film using corona discharge plasma for technological applications, J. Adhes. Sci. Technol. 22 (2008) 1425–1442.

[112] K. Navaneetha Pandiyaraj, V. Selvarajan, R.R. Deshmukh, C. Gao, Adhesive properties of polypropylene (PP) and polyethylene terephthlate (PET) film surfaces treated by DC glow discharge plasma, Vacuum 83 (2009) 332–339.

[113] P. Zheng, K. Liu, J. wang, Y. Dai, B. Yu, X. Zhou, H. Hao, Y. Luo, Surface modification of polyimide (PI) film using water cathode atmospheric pressure glow discharge plasma, Appl. Surf. Sci. 259 (2012) 494–500.

[114] N.Y. Cui, N.M.D. Brown, Modification of the surface properties of polypropylene (PP) film using an air dielectric barrier discharge plasma, Appl. Surf. Sci. 253 (2007) 3690–3695.

[115] J. Niu, D. Liu, Y. Wu, Large-area and uniform surface modification of polymers by barrier discharge plasmas, Surf. Coat. Technol. 205 (2011) 3434–3437.

[116] F. Caiazzo, P. Canonico, R. Nigro, V. Tagliaferri, Electrode discharge for plasma surface treatment of polymeric materials, J. Mater. Process. Technol. 58 (1996) 96–99.

[117] F.M. El-Hossary, N.Z. Negm, S. Khalil, M. Raaif, Surface modification of titanium by radio frequency plasma nitriding, Thin Solid Films 497 (2006) 196–202.

[118] A. Nihlstrand, T. Hjertberg, H.P. Schreiber, J.E. Klemberg-Sapieha, Plasma treatment and adhesion properties of a rubber-modified polypropylene, J. Adhes. Sci. Technol. 10 (1996) 651–675.

[119] Y. Takagi, Y. Gunjo, H. Toyoda, H. Sugai, Rapid treatment of polyimide film surface using high-density microwave plasma for enhancement of Cu layer adhesion, Vacuum 83 (2009) 501–505.
[120] M. Strobel, C.S. Lyons, K.L. MIttal (Eds.), Plasma Surface Modification Of Polymers: Relevance to Adhesion, VSP BV, Ultrecht, 1994.
[121] C. Mandolfino, E. Lertora, S. Genna, C. Leone, C. Gambaro, Effect of laser and plasma surface cleaning on mechanical properties of adhesive bonded joints, Procedia CIRP 33 (2015) 458–463.
[122] W. Petasch, B. Kegel, H. Schmid, K. Lendenmann, H.U. Keller, Low pressure plasma cleaning: a process for precision cleaning applications, Surf. Coat. Technol. 97 (1997) 176–181.
[123] L. Sorrentino, L. Carrino, 2024 Aluminium alloy wettability and superficial cleaning improvement by air cold plasma treatment, J. Mater. Process. Technol. 209 (2009) 1400–1409.
[124] E.M. Liston, L. Martinu, M.R. Wertheimer, Plasma surface modification of polymers for improved adhesion: a critical review, J. Adhes. Sci. Technol. 7 (1993) 1091–1127.
[125] D.M. Brewis, I. Mathieson, M. Wolfensberger, Treatment of low energy surfaces for adhesive bonding, Int. J. Adhes. Adhes. 15 (1995) 87–90.
[126] G.S. Oehrlein, Surface processes in low pressure plasmas, Surf. Sci. 386 (1997) 222–230.
[127] M. Noeske, J. Degenhardt, S. Strudthoff, U. Lommatzsch, Plasma jet treatment of five polymers at atmospheric pressure: surface modification and relevance for adhesion, Int. J. Adhes. Adhes. 24 (2004) 171–177.
[128] P. Molitor, V. Barron, T. Young, Surface treatment of titanium for adhesive bonding to polymer composites: a review, Int. J. Adhes. Adhes. 21 (2001) 129–136.
[129] R. Dorai, M.J. Kushner, A model for plasma modification of polypropylene using atmospheric pressure discharges, J. Phys. D. Appl. Phys. 36 (2003) 666–685.
[130] W. Leahy, V. Barron, M. Buggy, T. Young, A. Mas, F. Schue, T. McCabe, M. Bridge, Plasma surface treatment of aerospace materials for enhanced adhesive bonding, J. Adhes. 77 (2001) 215–249.
[131] J.W. Chin, J.P. Wightman, Adhesion to plasma-modified LaRC-TPI II. Effect of plasma treatment on peel strength, J. Adhes. 41 (1993) 23–34.
[132] C. Dartevelle, E. McAlpine, G.E. Thompson, M.R. Alexanderc, Low pressure plasma treatment for improving the strength and durability of adhesively bonded aluminium joints, Surf. Coat. Technol. 173 (2003) 249–258.
[133] C. Mandolfino, E. Lertora, C. Gambaro, Effect of cold plasma treatment on surface roughness and bonding strength of polymeric substrates, Key Eng. Mater. 611–612 (2014) 1484–1493.
[134] M.-H. Jung, H.-S. Choi, Surface treatment and characterization of ITO thin films using atmospheric pressure plasma for organic light emitting diodes, J. Colloid Interface Sci. 310 (2007) 550–558.
[135] E. Tamio, R. Lakshmi, N. Hiroaki, K. Satoru, N. Yoshinobu, K. Kazuhiro, Composite engineering—direct bonding of plastic PET films by plasma irradiation, Procedia Eng. 171 (2017) 88–103.
[136] A. Baalmann, K.D. Vissing, E. Born, A. Gross, Surface treatment of polyetheretherketone (PEEK) composites by plasma activation, J. Adhes. 46 (1994) 57–66.
[137] N. Anagreh, L.D.C. Bilke-Krause, Low-pressure plasma pretreatment of polyphenylene sulfide (PPS) surfaces for adhesive bonding, Int. J. Adhes. Adhes. 28 (2008) 16–22.
[138] H. Lim, Y. Lee, S. Han, J. Cho, K.J. Kim, Surface treatment and characterization of PMMA, PHEMA, and PHPMA, J. Vac. Sci. Technol. A 19 (4) (2001) 1490–1496.

[139] K.G. Kostov, T.M.C. Nishime, A.H.R. Castro, A. Toth, L.R.O. Hein, Surface modification of polymeric materials by cold atmospheric plasma jet, Appl. Surf. Sci. 314 (2014) 367–375.
[140] F. Rezaei, M.D. Dickey, M. Bourham, P.J. Hauser, Surface modification of PET film via a large area atmospheric pressure plasma: an optical analysis of the plasma and surface characterization of the polymer film, Surf. Coat. Technol. 309 (2017) 371–381.
[141] S. Zanini, P. Massimi, M. Mietta, E. Grimoldi, C. Riccardi, Plasma treatments of PET meshes for fuel-water separation applications, J. Colloid Interface Sci. 322 (2008) 566–571.
[142] A.B. Meddeb, Z. Ounaies, M. Lanagan, Enhancement of electrical properties of polyimide films by plasma treatment, Chem. Phys. Lett. 649 (2016) 111–114.
[143] K.N. Katnam, A.J. Comer, W.F. Stanley, M. Buggy, A.R. Ellingboe, T.M. Young, Characterising pre-preg and non-crimp-fabric composite single-lap bonded joints, Int. J. Adhes. Adhes. 31 (2011) 679–686.
[144] J.K. Kim, D.G. Lee, Characteristics of plasma surface treated composite adhesive joints at high environmental temperature, Compos. Struct. 57 (2002) 37–46.
[145] M.R. Gude, S.G. Prolongo, A. Ureña, Hydrothermal ageing of adhesive joints with nanoreinforced adhesives and different surface treatments of carbon fibre/epoxy substrates, Int. J. Adhes. Adhes. 40 (2013) 179–187.
[146] D. Gallant, V. Savard, New surface treatment technologies for adhesive bonding of lightweight aluminum-polypropylene hybrid joints in semi-structural applications, SAE Int. J. Mater. Manuf. 4 (1) (2011) 314–327. Available from: https://doi.org/10.4271/2011-01-0217.
[147] E. Panousis, F. Clément, J.-F. Loiseau, N. Spyrou, B. Held, J. Larrieu, E. Lecoq, C. Guimon, Titanium alloy surface treatment using an atmospheric plasma jet in nitrogen pulsed discharge conditions, Surf. Coat. Technol. 201 (2007) 7292–7302.
[148] B. Kegel, H. Schmid, Low-pressure plasma cleaning of metallic surfaces on industrial scale, Surf. Coat. Technol. 112 (1999) 63–66.
[149] T.J. Lin, J.A. Antonelli, D.J. Yang, H.K. Yasuda, F.T. Wang, Plasma treatment of automotive steel for corrosion protection-a dry energetic process for coatings, Prog. Org. Coat. 31 (1997) 351–361.
[150] C. Alves Jr., J.A. Lima, V. Hajek, J.B. Marimon da Cunha, C.A. dos Santos, Effect of cooling rate on properties of plasma nitrided AISI 1010 steel, Surf. Coat. Technol. 201 (2007) 7566–7573.
[151] E. Portolan, I.J.R. Baumvol, C.A. Figueroa, Electronic structure and mechanical properties of plasma nitrided ferrous alloys, Appl. Surf. Sci. 255 (2009) 6661–6665.
[152] K. Lin, X. Li, H. Dong, P. Guo, D. Gu, Nitrogen mass transfer and surface layer formation during the active screen plasma nitriding of austenitic stainless steels, Vacuum 148 (2018) 224–229.
[153] L.A. Espitia, H. Dong, X.-Y. Li, C.E. Pinedo, A.P. Tschiptschin, Scratch test of active screen low temperature plasma nitride AISI410 martensitic stainless steel, Wear 376-377 (2017) 30–36.
[154] M.C.S. Duarte, C. Godoy, J.C. Avelar-Batista Wilson, Analysis of sliding wear tests of plasma processed AISI 316L steel, Surf. Coat. Technol. 260 (2014) 316–325.
[155] Y. Li, Y. He, J.J. Xiu, W. Wang, Y.J. Zhu, B. Hu, Wear and corrosion properties of AISI 420 martensitic stainless steel treated by active screen plasma nitriding, Surf. Coat. Technol. 329 (2017) 184–192.
[156] Y. Hoshiyama, R. Mizobata, H. Miyake, Mechanical properties of austenitic stainless steel treated by active screen plasma nitriding, Surf. Coat. Technol. 307 (2016) 1041–1044.

[157] J. Custódio, J. Broughton, H. Cruz, P. Winfield, Activation of timber surface by flame and corona treatments to improve adhesion, Int. J. Adhes. Adhes. 29 (2009) 167–172.
[158] M. Żenkiewicz, Wyładowania koronowe w powietrzu jako metoda modyfikowania warstwy wierzchniej materiałów polimerowych (Corona discharge in the air as a method of modifying the surface layer of polymeric materials), Polimery 53 (2008) 3–13 (in Polish).
[159] A.V. Pocius, Adhesion and Adhesive Technology: An Introduction, second ed., Carl Hanser Gardner Verlag, Munchen, 2002.
[160] K.L. Mittal, A. Pizzi (Eds.), Adhesion Promotion Techniques, Technological Applications, first ed., CRC Press, 1999.
[161] A. Pizzi, K.L. Mittal (Eds.), Handbook of Adhesives Technology, third ed., CRC Press, 2017.
[162] J. Gołębiewski, M. Żenkiewicz, Influence of corona treatment energy and storage time on the changes of the contact angle of activated polysteryrene film, Polimery 48 (2003) 134–137.
[163] H. Moghadamzadeh, H. Rahimi, M. Asadollahzadeh, A.R. Hemmati, Surface treatment of wood polymer composites for adhesive bonding, Int. J. Adhes. Adhes. 31 (2011) 816–821.
[164] M.D. Green, F.J. Guild, R.D. Adams, Characterization and comparison of industrially pre-treated homopolymer polypropylene, HF 135M, Int. J. Adhes. Adhes. 22 (2002) 81–90.
[165] J.N. Strobel, M. Strobel, C.S. Lyons, C. Dunatov, S.J. Perron, Aging of air-corona-treated polypropylene film, J. Adhes. Sci. Technol. 5 (1991) 119–130.
[166] L.J. Gerenser, J.F. Elman, M.G. Mason, J.M. Pochan, ESCA studies of corona-discharge-treated polyethylene surfaces by use of gas-phase derivatization, Polymer 26 (1985) 1162–1166.
[167] A.J. Kinloch, Adhesion and Adhesives, Chapman and Hall, New York, 1987.
[168] Aktywowanie powierzchni (Surface activation) (in Polish). https://www.aktywacjapowierzchni.pl, 2018 (Accessed August 17, 2018).
[169] C.D. Volpe, A. Deimichei, T. Riccò, A multiliquid approach to the surface free energy determination of flame-treated surfaces of rubber-toughened polypropylene, J. Adhes. Sci. Technol. 12 (1998) 1141–1180.
[170] A.J. Kinloch, G.K.A. Kdokian, On the calculation of dispersion and polar force components of the surface free energy, J. Adhes. 34 (1991) 41–44.
[171] M. Żenkiewicz, Płomieniowe modyfikowanie warstwy wierzchniej wytworów tworzywowych (Flame modification of the surface of plastics articles), Polimery 45 (2000) 81–88 (in Polish).
[172] I. Sutherland, D.M. Brewis, R. Health, E. Sheng, Modification of polypropylene surface by flame treatment, Surf. Interface Anal. 17 (1991) 507–510.
[173] E.G. Baburaj, D. Starikov, J. Evans, G.A. Shafeev, A. Bensaoula, Enhancement of adhesive joint strength by laser surface modification, Int. J. Adhes. Adhes. 27 (2007) 268–276.
[174] R. Rechner, I. Jansen, E. Beyer, Influence on the strength and aging resistance of aluminium joints by laser pre-treatment and surface modification, Int. J. Adhes. Adhes. 30 (2010) 595–601.
[175] M. Żenkiewicz, P. Rytlewski, J. Czupryńska, J. Polański, T. Karasiewicz, W. Engelgard, Contact angle and surface free energy of electron-beam irradiated polymer composites, Polimery 53 (2008) 446–451.
[176] M. Żenkiewicz, Effect of electron radiation dose on oxidation of surface layers of polyethylene, polypropylene and poly(ethylene terephthalate) films, J. Adhes. Sci. Technol. 18 (2004) 607–616.

[177] H.C. Man, M. Li, T.M. Yue, Surface treatment of thermoplastic composite with an excimer laser, Int. J. Adhes. Adhes. 18 (1998) 151–157.
[178] Q. Bénard, M. Fois, M. Grisel, P. Laurens, Surface treatment of carbon/epoxy composites with an excimer laser beam, Int. J. Adhes. Adhes. 26 (2006) 534–549.
[179] C.D. Wagner, W.M. Riggs, L.E. Davis, J.F. Moulder, G.E. Muilenberg (Eds.), Handbook of X-Ray Photoelectron Spectroscopy, Perkin-Elmer Corporation, Eden Prairie, MN, 1979, pp. 38–76.
[180] L. Lavisse, J.M. Jouvard, L. Imhoff, O. Heintz, J. Korntheuer, C. Langlade, S. Bourgeois, M.C.M. de Lucas, Pulsed laser growth and characterization of thin films on titanium substrates, Appl. Surf. Sci. 253 (2007) 8226–8230.
[181] C. Spadaro, C. Sunseri, C. Dispenza, Laser surface treatments for adhesion improvement of aluminium alloys structural joints, Radiat. Phys. Chem. 76 (2007) 1441–1446.
[182] M. Rotel, J. Zahavi, S. Tamir, A. Buchman, H. Dodiuk, Pre-bonding technology based on excimer laser surface treatment, Appl. Surf. Sci. 154–155 (2000) 610–616.
[183] K. Arakawa, T. Seguchi, Y. Watanabe, N. Hayakawa, Radiation-induced oxidation of polyethylene, ethylene–butene copolymer, and ethylene–propylene copolymer, J. Polym. Sci. Polym. Chem. 20 (1982) 2681–2692.
[184] F. Shen, H.A. McKellop, R. Salovey, Irradiation of chemically crosslinked ultrahigh molecular weight polyethylene, J. Polym. Sci. B Polym. Phys. 34 (1996) 1063–1077.
[185] W. Urbaniak-Domagala, Pretreatment of polypropylene films for the creation of thin polymer layers, part 1: the use of chemical, electrochemical, and UV methods, J. Appl. Polym. Sci. 122 (2011) 2071–2080.
[186] J. Haubrich, K. Schulze, J. Hausmann, Laser treatment of titanium surfaces in different atmospheres for improved adhesion properties, in: International Conference Euro Hybrid Materials and Structures April 10–11, Stade, Germany, pp. 23–41, 2014. Available from: https://core.ac.uk/download/pdf/31013268.pdf.
[187] A.P. del Pino, P. Serra, J.L. Morenza, Oxidation of titanium through Nd:YAG laser irradiation, Appl. Surf. Sci. 197 (2002) 887–890.
[188] G.W. Critchlow, D.M. Brewis, D.C. Emmony, C.A. Cottam, Initial investigation into the effectiveness of CO2-laser treatment of aluminium for adhesive bonding, Int. J. Adhes. Adhes. 15 (1995) 233–236.
[189] G.W. Critchlow, C.A. Cottam, D.M. Brewis, D.C. Emmony, Further studies into the effectiveness of CO2-laser treatment of metals for adhesive bonding, Int. J. Adhes. Adhes. 17 (1997) 143–150.
[190] H. Dodiuk, A. Buchman, S. Kenig, M. Rotel, J. Zahavi, T.J. Reinhart, Preadhesion laser treatment of aluminum surfaces, J. Adhes. 41 (1993) 93–112.
[191] P. Pou, A. Riveiro, J. del Val, R. Comesaña, J. Penide, F. Arias-González, R. Soto, F. Lusquiños, J. Pou, Laser surface texturing of titanium for bioengineering applications, Procedia Manuf. 13 (2017) 694–701.
[192] M. Momcilovic, D. Milovanovic, M. Bokorov, M. Trtica, Surface modification of the titanium implant using TEA CO2 laser pulses in controllable gas atmospheres—comparative study, Appl. Surf. Sci. 258 (2012) 2741–2748.
[193] F. Torrent, L. Lavisse, P. Berger, J.-M. Jouvard, H. Andrzejewski, G. Pillon, S. Bourgeois, M.C. Marco de Lucas, Wavelength influence on nitrogen insertion into titanium by nanosecond pulsed laser irradiation in air, Appl. Surf. Sci. 278 (2013) 245–249.
[194] W.D. Müller, K. Seliger, J. Meyer, J.L. Gilbert, The improvement of adhesion of polymer on titanium surface after treatment with tea-CO2 laser irradiation, J. Mater. Sci. Mater. Med. 5 (1994) 692–694.

[195] C.-C. Yang, Y.-C. Lin, C.-C. Yang, Y.-H. Lin, K.-C. Huang, K.-m. Lin, W.-T. Hsiao, Laser-induced coloring of titanium alloy using ultraviolet nanosecond pulses scanning technology, J. Alloys Compd. 715 (2017) 349–361.

[196] D.P. Adams, R.D. Murphy, D.J. Saiz, D.A. Hirschfeld, M.A. Rodriguez, P.G. Kotula, B.H. Jared, Nanosecond pulsed laser irradiation of titanium: oxide growth and effects on underlying metal, Surf. Coat. Technol. 248 (2014) 38–45.

[197] F.L. Palmieri, K.A. Watson, G. Morales, T. Williams, R. Hicks, C.J. Wohl, J.W. Hopkins, J.W. Connell, Laser ablative surface treatment for enhanced bonding of Ti-6Al-4V alloy, ACS Appl. Mater. Interfaces 5 (2013) 1254–1261.

[198] M.A. Belcher, M.S. List, C.J. Wohl, S. Ghose, K.A. Watson, J.W. Hopkins, J.W. Connell, Laser surface preparation for adhesive bonding of Ti-6Al-4V, 2018. Available from: https://ntrs.nasa.gov/. (Accessed September 10, 2018).

[199] R.C.P. Wong, A.P. Hoult, J.K. Kim, T.X. Yu, Improvement of adhesive bonding in aluminium alloys using a laser surface texturing process, J. Mater. Process. Technol. 63 (1997) 579–584.

[200] M. Langer, R. Rechner, M. Thieme, I. Jansen, E. Beyer, Surface analytical characterisation of Nd:YAG-laser pre-treated $AlMg_3$ as a preparation for bonding, Solid State Sci. 14 (2012) 926–935.

[201] G. Rotella, M. Alfano, T. Schiefer, I. Jansen, Evaluation of mechanical and laser surface pre-treatments on the strength of adhesive bonded steel joints for automotive industry, J. Adhes. Sci. Technol. 30 (2016) 747–758.

[202] T. Köckritz, T. Schiefer, I. Jansen, E. Beyer, Improving the bond strength at hybrid-yarn textile thermoplastic composites for high-technology applications by laser radiation, Int. J. Adhes. Adhes. 46 (2013) 85–94.

[203] X.M. Zhang, T.M. Yue, H.C. Man, Enhancement of ceramic-to-metal adhesive bonding by excimer laser surface treatment, Mater. Lett. 30 (1997) 327–332.

[204] A.J. Antończak, Ł. Skowroński, M. Trzciński, V.V. Kinzhybalo, Ł.K. Łazarek, K.M. Abramski, Laser-induced oxidation of titanium substrate: Analysis of thephysicochemical structure of the surface and sub-surface layers, Appl. Surf. Sci. 325 (2015) 217–226.

[205] M. Imai, L.G. Xu, K. Ametani, M. Tutiya, γ-Irradiation of polypropylene in vacuum and in air, J. Polym. Sci. A Polym. Chem. 27 (1989) 1763–1773.

[206] A. Shojaei, R. Fathi, N. Sheikh, Adhesion modification of polyethylenes for metallization using radiation-induced grafting of vinyl monomers, Surf. Coat. Technol. 201 (2007) 7519–7529.

[207] V. Haddadi-Asl, R.P. Burford, Radiation graft modification of ethylene-propylene rubber—II. Effect of additives, Radiat. Phys. Chem. 45 (1995) 191–198.

[208] V. Haddadi-Asl, R.P. Burford, Radiation graft modification of ethylene-propylene rubber—III. Effect on water uptake, wettability and biocompatibility, Radiat. Phys. Chem. 47 (1996) 907–912.

[209] J. Zhao, G. Geuskens, Surface modification of polymers VI. Thermal and radiochemical grafting of acrylamide on polyethylene and polystyrene, Eur. Polym. J. 35 (1999) 2115–2123.

[210] E.M. El-Nesr, Gamma radiation induced graft copolymerization of acrylamide onto EPDM blended with PE, Polym. Adv. Technol. 13 (2002) 626–635.

[211] Mechanical properties of EN AW-5754 aluminum alloy. https://www.makeitfrom.com/material-properties/5754-H18-Aluminum, 2018 (Accessed September 12, 2018).

[212] A. Rudawska, M.A. Wahab, The effect of cataphoretic and powder coatings on the strength and failure modes of EN AW-5754 aluminium alloy adhesive joints, Int. J. Adhes. Adhes. 89 (2019) 40–50.

[213] Technical data sheet, http://www.chryssafidis.com/files/el/data_sheets/965032A.pdf, 2018 (Accessed September 16, 2018).
[214] A. Rudawska, I. Danczak, M. Müller, P. Valasek, The effect of sandblasting on surface properties for adhesion, Int. J. Adhes. Adhes. 70 (2016) 176–190.
[215] A. Rudawska, E. Jacniacka, Analysis of determining surface free energy uncertainty with the Owens-Wendt method, Int. J. Adhes. Adhes. 29 (2009) 451–457.
[216] P. Dalet, E. Papon, J.-J. Villenave, Surface free energy of polymeric materials: relevancy of conventional contact angle data analyses, J. Adhes. Sci. Technol. 13 (1999) 857–870.
[217] J. Comyn, Diffusion and adhesion, in: J. Comyn (Ed.), Polymer Permeability, Chapman and Hall, London, 1985, pp. 177–215 (Chapter 5).
[218] S. Ahadian, M. Mohseni, S. Morawian, Ranking proposed models for attaining surface free energy of powders using contact angle measurements, Int. J. Adhes. Adhes. 29 (2009) 458–469.
[219] V.S. Ajaev, T. Gambaryan-Roisman, P. Stephan, Static and dynamic contact angles of evaporating liquids on heated surfaces, J. Colloid Interface Sci. 342 (2010) 550–558.
[220] C.J. van Oss, Interfacial forces in aqueous media, Dekker, New York, 1994.
[221] A. Rudawska, E. Jacniacka, Evaluating uncertainty of surface free energy measurement by the van Oss-Chaudhury-Good method, Int. J. Adhes. Adhes. 82 (2018) 139–145.
[222] A. Rudawska, Adhesive properties, in: V. Kazmiruk (Ed.), Scanning Electron Microscope, InTech Publisher, Rijeka, 2012, pp. 101–126 (Chapter 7).
[223] I.A. Ashcroft, J. Comyn, S. Tellwright, Adhesives for automotive windscreen replacement: kinetics of cure and bond strength, Int. J. Adhes. Adhes. 29 (2009) 155–166.

Further reading

[224] Primers, www.loctite-kleje.pl, 2018 (Accessed September 10, 2018).
[225] B.N. Zand, M. Mahdavian, Evaluation of the effect of vinyltrimethoxysilane on corrosion resistance and adhesion strength of epoxy coated AA1050, Electrochim. Acta 52 (2007) 6438–6442.
[226] Koronowanie (Corona treatment) (in Polish). https://www.technograph.pl, 2018 (Accessed August 20, 2018).
[227] Koronowanie (Corona treatment) (in Polish). https://www.transwest.pl, 2018 (Accessed August 20, 2018).

Assessment of surface preparation for the bonding/adhesive technology

9

9.1 Effect of surface roughness on wettability, contact angle, surface free energy, and strength

The roughening surfaces prior to bonding enhance the adhesive joints strength [1–7] An increase of adherend roughness should lead to an increase of effective area for the bond [8–12]. However, it is reported that excess roughness decreases the ability for adhesive penetration, increasing void formation and therefore introducing localized stress concentration [11, 13–15]. Indeed, a decrease of strength is often found when the adherend surface is too rough. The value of critical roughness depends on many parameters, such as, e.g. the roughening pretreatment applied, the type of adherend, the type of adhesive, the geometry of joint, and the stresses applied. It was also found that roughening process may introduce physicochemical changes, which affect wettability and surface free energy (SFE) [16–25]. The surface roughness can affect the spreading of the adhesive, either because the adhesive cannot penetrate the adherend or because it gels before it completes the penetration. The relationship between roughness and adhesion is not very simple. Optimum surface profile varies from one adhesive to another, and depends on the type of stress applied [26]. The various surface roughness parameters were used to characterise the surface topography.

Several researches have compared the various surface roughness parameters with SFE or wetting or contact angle [9, 10, 13, 21–24, 27–29]. Although some research [29] suggested that surface roughness was relatively unimportant during contact angle measurement.

The effect of surface roughness on wettability have been studied by many researchers using surface roughness factors such as the Wenzel roughness factor [13, 30]. The Wenzel equation [30, 31] describes the effect of surface roughness on the contact angle. It is generally accepted that if $R_a < 0.5\,\mu m$, then the impact of roughness on the contact angle is negligible. According to the assumption of the Wenzel equation, increasing the R_a value can improve (when $\theta < 90$ degrees) or worsen (when $\theta > 90$ degrees) the surface wetting conditions.

Wenzel [32] proposed a parameter 'r' to characterise a surface as follows [30]:

$$r = \frac{\text{total surface area}}{\text{apparent geometric area}}$$

and assumed that the value of r increased as the roughening increased. It is generally believed that the apparent contact angle decreases with increasing roughness values [33]. Surface roughness is one of the major causes of the thermodynamic hysteresis of the contact angle [30, 34–42].

Surface Treatment in Bonding Technology. https://doi.org/10.1016/B978-0-12-817010-6.00009-6

Sommers and Jacobi [20] investigated the behaviour of water droplets on aluminium surface with parallel grooves tens of microns in width and depth. They noticed that water droplets placed on a micro-grooved aluminium surface using a micro-syringe exhibited significantly increased apparent contact angles, and the droplet volume at incipient sliding was reduced by more than 50% compared to droplets on a surface without micro-grooves.

The effect of density and height of peaks on adhesion fracture strength is discussed by Uehara and Sakurai [43].

Wu [44] and Brewis and Briggs [45] suggested that the effect of roughness at angles less than 90 degrees is to produce lower contact angles, better wetting, and thus improved printability. In effect, a rough surface behaves as if it consists of a larger number of tiny flat segments oriented at all possible angles to the surface [46].

Xu et al. [47] estimated the effect of the substrate roughness on the liquid droplet spreading. They demonstrated the experimental process of determining the mobility of the contact line via a droplet spreading on a steel substrate and the results exhibited that the mobility of value is lower for a rougher surface.

The effect of grit blasting on surface properties for adhesion is presented by Harris and Beevers [48]. They presented the effect of surface roughness parameter R_q on SFE. They proved that, in general, smoother grit-blasted surfaces display higher surface energy values.

Some issue of surface pretreatment on aluminium surface is investigated by Prolongo and Urena [14]. They emphasised that chemical and chemical treatment generated different structure of aluminium surface. All aluminium surfaces etched by chemical treatments presented the characteristic pits generated by etching treatment; in contrast, the surface of those subjected to mechanical abrasion is more compact. The oxide layer formed through chemical etching is more porous than the natural alumina film. In principle, the generation of high porous oxide layer could imply a high adhesive strength, increasing the contact area and providing some degree of mechanical interlocking and 'keying' with the adhesive [48].

Shahid and Hashim [49] examined the influence of surface roughness of a steel adherend on cleavage strength. They studied the surface roughness parameters R_a and R_{lo} and also root mean square slope, R_{dq}, after polishing and grit blasting. They defined the relationship between average cleavage strength and R_a value of the steel adhered surface. The cleavage strength appears to increases linearly with the R_a value.

Hay et al. [8] presented some information of the influence of surface roughness on the wettability. They depicted a theoretical investigation of the effect of surface roughness on fluid spreading. The analytical solution implies the existence of a critical contact angle that is a function of roughness geometry, below which fluid will spread and above which fluid will resist spreading.

Serro et al. [21] proved that adhesion forces measured in partially wetting liquids depend strongly on the surface topography, as a consequence of the existence of nanobubbles close to topographic accidents.

Hitchcock et al. [24] also observed that, within certain limits, roughening a substrate usually causes its wettability to decrease. Some researcher suggested that the peaks, ridges, and asperities from barriers which restrict the spreading of the droplet.

Shanahan [39] reported that heterogeneities such as surface roughness could produce a positive influence on wetting characteristics of a drop by 'pulling' the liquid mass towards the abnormality but the negative effect of peaks and asperities on wetting characteristics could was observed.

Birkeman [23] suggested that wetting was independent of roughness on grooved and ridged surfaces and ridged surfaces and implied that as a drop front moves a ridge the true contact remains constant.

Packham [13] reported that the extent of contact angle between a liquid and a rough surface depends on the details of the topography. For example penetration into an idealised pore, occurs until the back pressure of trapped air equals the capillary driving pressure.

da Silva et al. [16] investigated the influence of macroscopic state of the aluminium surface (prepared as several patterns) on the strength of adhesive joints. The patterns applied to the specimens consisted of a series of grooves, which were applied with 0, 45, or 90 degrees orientations. The main conclusion was that the patterns can increase the joint strength of nontreated aluminium surfaces in the case of the brittle adhesive.

De Bruyne [50] studied the shapes of pits into which the adhesive would not penetrate the contact angle with the adherends walls was greater than a certain value.

Khrulev [19] presented the alternative analysis for describing the influence of surface irregularities on the interaction between the adhesive and the adherend and assumed that the continuous layer of adhesive that withstood shear would be less due to the penetration of the adhesive into the surface irregularities.

Islam et al. [28] studied the several techniques for surface preparation of mild steel (especially mechanical treatment methods). The surface profile apparent surface energy parameters such as surface free, work of adhesion energy, Lifshitz-van der Waals acid-base components were measured from these trials connection made with these properties to the metal to polymer coating bond strength and bond failure mode.

Iqbal et al. [51] investigated the effects of surface morphology on adhesion properties of two different based fibre composite materials before and after atmospheric pressure plasma treatment and hand sanding.

The influence of surface roughness on the interaction between the adhesive and adherend have been also investigated by numerous researchers [11, 17, 27, 48–54].

Arrowsmith [52] investigated the effect of surface topography on peel adhesion and claimed that not only the average surface roughness but also the particular geometry of the surface topography had influence on the peel strength.

Keisler and Lataillade [27] studied the effect of average roughness, the width of valleys, and the dominance of valleys and peaks on the strength of shear lap joints.

The influence of various surface roughness parameters (obtained after various surface treatment methods) on the wettability and adhesive joints strength were also investigated by numerous researchers [43, 55–57].

The Bresson et al. [55] used the arithmetic mean roughness R_a to assessment the surface roughness of aluminium substrates after various surface pretreatment (sandblasting and silane application). Surface roughness was found to be crucial and increased roughness (due to control of grit size, nozzle/substrate distance, and time of sandblasting) was found to be beneficial. Moreover they notice that the use of silane improved surface wettability and overall strength.

Jiang et al. [56] measured the average surface roughness R_a after sandblasting of pure titanium. They underlined that the average roughness R_a of the sandblasted surface was about 1.1 μm, which was larger than that of the untreated samples (0.3 μm).

Moghadamzadeh et al. [57] used the R_a and R_z (also S_a and S_z) surface roughness parameters to characterise the surface topography after surface treatment. They emphasised that surface roughness could be a key factor that contributes to mechanical interlocking phenomenon.

9.2 Assessment and degrees of surface treatment

The surface preparation assessment (Fig. 9.1) may include [31, 43, 58]:

- surface preparation grades,
- assessment of the degree of removing impurities from the surface, and
- determination of surface roughness.

The aforementioned types of assessment concern the preparation of surfaces prior to application of paint coats; however, they are similarly applicable in the case of surface preparation for adhesive joining. International standards containing guidelines are detailed in the following groups of standards (Fig. 9.2):

- ISO 8501—visual assessment of surface cleanliness,
- ISO 8502—tests for the assessment of surface cleanliness, and
- ISO 8503—Surface roughness characteristics of steel substrates.

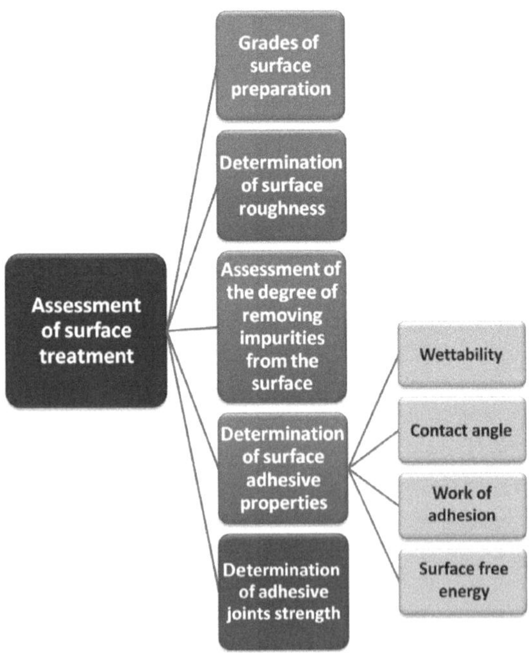

Fig. 9.1 Type of surface treatment assessment.

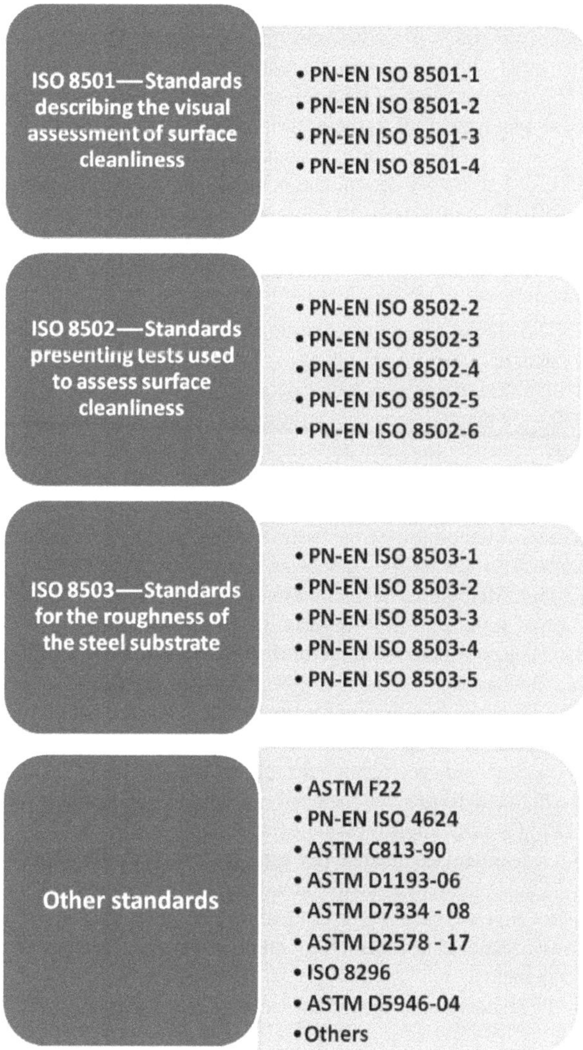

Fig. 9.2 Type of standards groups used to assessment of surface treatment.

Among these groups of standards, the following can be distinguished:

- PN-EN ISO 8501-1. Preparation of steel substrates before application of paints and related products—Visual assessment of surface cleanliness—Part 1: Rust grades and preparation grades of uncoated steel substrates and of steel substrates after overall removal of previous coatings,
- PN-EN ISO 8501-2. Preparation of steel substrates before application of paints and related products—Visual assessment of surface cleanliness—Part 2: Preparation grades of previously coated steel substrates after localised removal of previous coatings,

- PN-EN ISO 8501-3. Preparation of steel substrates before application of paints and related products—Visual assessment of surface cleanliness—Part 3: Preparation grades of welds, edges, and other areas with surface imperfections,
- PN-EN ISO 8501-4. Preparation of steel substrates before application of paints and related products—Visual assessment of surface cleanliness—Part 4: Initial surface conditions, preparation grades and flash rust grades in connection with high-pressure water jetting,
- PN-EN ISO 8502-2. Laboratory determination of chloride on cleaned surfaces,
- PN-EN ISO 8502-3. Assessment of dust on steel surfaces prepared for painting (pressure-sensitive tape method),
- PN-EN ISO 8502-4. Guidance on the estimation of the probability of condensation prior to paint application,
- PN-EN ISO 8502-5. Measurement of chloride on steel surfaces prepared for painting (ion detection tube method,
- PN-EN ISO 8502-6. Extraction of soluble contaminants for analysis—The Bresle method,
- PN-EN ISO 8503-1. Specifications and definitions for ISO surface profile comparators for the assessment of abrasive blast-cleaned surfaces,
- PN-EN ISO 8503-2. Method for the grading of surface profile of abrasive blast-cleaned steel—Comparator procedure,
- PN-EN ISO 8503-3. Method for the calibration of ISO surface profile comparators and for the determination of surface profile—Focusing microscope procedure,
- PN-EN ISO 8503-4. Method for the calibration of ISO surface profile comparators and for the determination of surface profile—Stylus instrument procedure,
- PN-EN ISO 8503-5 Replica tape method for the determination of the surface profile,
- ASTM F22-02. Standard test method for hydrophobic surface films by the water-break test,
- ASTM C813-90. Standard test method for hydrophobic contamination on glass by contact angle measurement,
- ASTM D1193-06. Standard specification for reagent water,
- ASTM D7334-08. Standard practice for surface wettability of coatings, substrates, and pigments by advancing contact angle measurement,
- ASTM D2578-17. Standard test method for wetting tension of polyethylene and polypropylene films,
- ISO 8296. Plastics film and sheeting determination of wetting tension,
- ASTM D 5946-04. Standard test method for corona-treated polymer films using water contact angle measurements,
- ASTM D 724-99. Standard test method for surface wettability of paper (Angle-of-contact method), and
- Other standards.

9.2.1 Surface preparation grades according to PN-EN ISO 8501-1

Standards PN-EN ISO 8501-1 and PN-EN ISO 8501-2 describe the surface preparation grades in the visual assessment of rust grades and grades of steel substrate preparation for painting. The International Standard PN-EN ISO 8501-1 concerns uncoated surfaces, whereas PN-EN ISO 8501-2 previously painted steel substrates whose previous coatings have been locally removed. The methods and grades of surface preparation may well apply to bonding of substrates which are assembled for the first time, or rejoined (and therefore obtained after the disassembly), in case of which residual adhesives are found on the surface. The information on PN-EN ISO 8501-1 is presented below.

The said standard determines four rust grades of the surface:

- A—steel surface covered with strongly adhering scale on almost 100% of surface, and little if any rust,
- B—steel surface, which has begun to rust and from which the mill scale has begun to scale,
- C—steel surface on which the mill scale has rusted away or from which it can be scraped, but with slight pitting visible under normal vision, and
- D—steel surface on which the mill scale has rusted away and on which general pitting is visible under normal vision.

The surface is assessed visually in scattered light, eyesight corrected to normal vision, by comparing it to a specific photographic reference given in the standards.

The PN-EN ISO 8501-1 standard gives the degrees of surface preparation using the following methods (described in Chapters 4, 5, and 8):

- abrasive blasting (the standard does not distinguish between dry and wet abrasive treatment methods),
- cleaning methods with hand and power tools, and
- flame treatment method.

Tables 9.1–9.3 show the surface preparation grades along with their description.

Table 9.1 Degrees of surface preparation according to PN-EN ISO 8501-1: abrasive blast cleaning

Grade	Description
Sa 1—Light blast-cleaning	When viewed without magnification, the surface is free from visible oil, grease, and dirt, and from poorly adhering mill scale, rust, paint coatings, and foreign matter Grades: B Sa 1, C Sa 1, D Sa 1
Sa 2—Thorough blast-cleaning	When viewed without magnification, the surface is free from visible oil, grease, and dirt, and from most of the mill scale, rust, paint coatings, and foreign matter. Any residual contamination should be firmly adhering Grades: B Sa 2, C Sa 2, D Sa 2
Sa 2½—Very thorough blast-cleaning	When viewed without magnification, the surface is free from visible oil, grease, and dirt, and from mill scale, rust, paint coatings, and foreign matter. Any remaining traces of contamination show only slight stains in the form of spots or stripes Grades: A Sa 2½, B Sa 2½, C Sa 2½, D Sa 2½
Sa 3—Blast cleaning to visually clean steel	When viewed without magnification, the surface is free from visible oil, grease, and dirt, and from mill scale, rust, paint coatings, and foreign matter. The surface should have a uniform metallic colour Grades: A Sa 3, B Sa 3, C Sa 3, D Sa 3

Source: Own work based on PN-EN ISO 8501-1.

Table 9.2 Surface preparation grades according to PN-EN ISO 8501-1: thorough hand and power tool cleaning

Grade	Description
St 1	When viewed without magnification, the surface is free from visible oil, grease, and dirt, and from poorly adhering mill scale, rust, paint coatings, and foreign matter Grades: B St 2, C St 2, D St 2
St 2	When viewed without magnification, the surface is free from visible oil, grease, and dirt, and from poorly adhering mill scale, rust, paint coatings, and foreign matter. The surface is cleaned better than for St 1 Grades: B St 2, C St 2, D St 2
St 3	When viewed without magnification, the surface is free from visible oil, grease, and dirt, and from poorly adhering mill scale, rust, paint coatings, and foreign matter. The surface is cleaner than in the case of St 2, metallic surface is locally visible Grades: B St 3, C St 3, D St 3

Source: Own work based on PN-EN ISO 8501-1.

Table 9.3 Surface preparation grades according to PN-EN ISO 8501-1: flame cleaning

Grade	Description
FI	When viewed without magnification, the surface is free from visible oil, grease, and dirt, and from poorly adhering mill scale, rust, paint coatings, and foreign matter. Any residual rust should be visible in the form of surface discolouring (shades) Grades: A FI, B FI, C FI, D FI

Source: Own work based on PN-EN ISO 8501-1.

The International Standard PN-EN ISO 8501-1 is supplemented with photographs of the initial surface condition and its condition following particular stages of surface preparation.

9.2.2 Surface preparation grades according to PN-EN ISO 8501-4

The International Standard PN-EN ISO 8501-4 details surface preparation grades in the case of high-pressure water jetting treatment. Currently, there exists no European Standard for surfaces preparation with wet abrasive methods. Should the need arise, the standards prepared by International 'Slurry blasting Standards' can be used instead [59]. High-pressure water jetting applied to steel substrates typically leads to the emergence of flash rust on the surface of cleaned elements, in which case the following are assessed:

- initial surface condition,
- surface appearance after cleaning, and
- appearance of the surface with flash rust.

Surface preparation grades given by PN-EN ISO 8501-4 are presented in Table 9.4.

Table 9.5 shows rust grades after surface preparation by means of high-pressure water jetting.

The following initial surface conditions are distinguished:

- DC A—a surface where the paint coating system has degraded to an extent similar to that illustrated by ISO 4628-3, grade Ri3,

Table 9.4 Surface appearance after cleaning according to PN-EN ISO 8501-4

Grade	Description
Wa 1—Light high-pressure water jetting	When viewed without magnification, the surface is free from visible oil and grease, loose paint or rust, and other foreign matter
Wa 2—Thorough high-pressure water jetting	When viewed without magnification, the surface is from visible oil, grease and dirt and most of the rust, previous paint coatings and other foreign matter. Any residual contamination is randomly dispersed and can consist of firmly adherent coatings, firmly adherent foreign matter and stains of previously existent rust
Wa 2½—Very thorough high-pressure water jetting	When viewed without magnification, the surface is free from all visible rust, oil, grease, dirt, previous paint coatings and, except for slight traces, all other foreign matter. Discolouration of the surface can be present. The gray or brown/black discolouration observed on pitted and corroded steel cannot be removed by further water jetting

Source: Own work based on PN-EN ISO 8501-4.

Table 9.5 Flash rust grades according to PN-EN ISO 8501-4

Grade	Description
L—Light flash rust	A surface which, when viewed without magnification, exhibits small quantities of a yellow/brown rust layer through which the steel substrate can be seen. The rust can be evenly distributed or present in patches, but it will be tightly adherent. It is not easily removed by gentle wiping with a cloth
M—Medium flash rust	A surface which, when viewed without magnification, exhibits a layer of yellow/brown rust that obscures the original steel surface. The rust layer can be evenly distributed or present in patches, but it will be reasonably well adherent. It will lightly mark a cloth that is gently wiped over the surface
H—Heavy flash rust	A surface which, when viewed without magnification, exhibits a layer of red-yellow/brown rust that obscures the original steel surface and is loosely adherent. The rust layer can be evenly distributed or present in patches and it will readily mark a cloth that is gently wiped over the surface

Source: Own work based on PN-EN ISO 8501-4.

- DC B—a surface where the paint coating system has degraded to an extent similar to that illustrated by ISO 4628-3, grade Ri4,
- DC C—a surface which has degraded to a major extent, as illustrated by ISO 4628-3, grade Ri5, or when completely degraded as illustrated by ISO 8501-1, rust grade C,
- DP I—an iron oxide epoxy prefabrication (shop) primer surface that has degraded, and
- DP Z—a zinc silicate prefabrication (shop) primer surface that has degraded.

9.2.3 Determination of water-soluble salts on cleaned surfaces, according to PN-EN ISO 8502-2

Ionic contaminants on the surface of substrates prepared for adhesive joining and bonding processes (paint or adhesive coating) may cause subcoating corrosion, blistering and joint delamination. When ionic residues are found on the surface following proper treatment methods, they need to be subjected to assessment. PN-EN ISO 8502-2 standard provides a description of several methods for removing ionic contaminants from the surface for quantitative analysis. Methods that can be used in laboratory and field conditions are:

- the swabbing method,
- the Bresle method, and
- the conductometric method.

The most frequently used method for the determination of the total amount of all water-soluble ionic contaminants on the surface is the conductometric method. It consists of measuring the conductivity of the solution obtained from the surface by means of a conductometer.

9.2.4 Assessment of dust on steel surfaces according to PN-EN ISO 8502-3

The assessment of the level of dust contaminants on the surface is carried out in accordance with the PN-EN ISO 8502-3 standard, which consists of:

- applying the 150 mm-long adhesive tape firmly onto the surface of a substrate by repeatedly pressing it with one's finger or a roller,
- assessing the quantity of dust on the tape by referencing them with the pictures given in the standards. The assessment should be made against a contrasting background for better visibility of removed contaminants.

Concentration density standards (included in the PN-EN ISO 8502-3 standard) are shown in Fig. 9.3, and the classification of dust particle size is given in Table 9.6.

9.2.5 Assessment of oil contamination on surfaces (according to ASTM F22 standard)

The term *oil contamination* embraces a number of compounds of various chemical structure, which is why it is impossible to simply label them based on their chemical structure. The previously referenced group of standards concerning surface

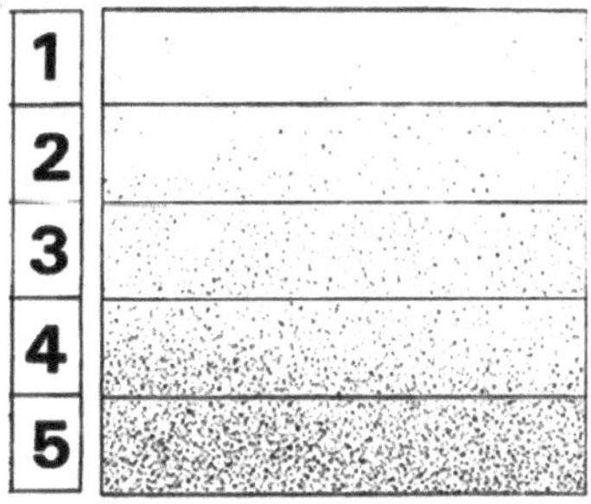

Fig. 9.3 Dust quantity ratings according to PN-EN ISO 8502-3.

Table 9.6 Classification of dust size classes according to PN-EN ISO 8502-3

Class	Dust size
0	Particles not visible under ×10 magnification
1	Particles hardy visible under ×10 magnification (usually less than 50 μm in diameter)
2	Particles visible with normal vision (particle diameter between 50 and 100 μm)
3	Particles clearly visible with normal vision (up to 0.5 mm in diameter)
4	Particles between 0.5 and 2.5 mm in diameter

Source: Own work based on PN-EN ISO 8502-3.

contamination, PN-EN ISO 8502, fails to address the assessment of oil contamination. The only common feature of oil contaminants is their hydrophobic character, which is manifested by the fact that water on greasy surfaces occurs in drops. This property is the basis of the ASTM F22 standard, according to which oil contaminants are determined on vertical surfaces by spraying the surface with water and assessing after 10 s the behaviour of water: whether it forms drops or a continuous water film, as well as whether:

- water covers 50% of the surface,
- water covers 20%–45% of the surface, and
- water covers up to 15% of the surface.

The visual test procedure constitutes either a preliminary assessment of the cleanliness of the treated surface or is implemented prior to the selection of a particular surface treatment method. With less accuracy, this method can also be used on horizontal surfaces.

In addition, the subsequent sections of this work shall present several methods, which may also be applied in workshop conditions that are used when assessing the surface preparation for adhesive joining and other adhesive processes (including the aforementioned paint coatings and working with steel substrates). The selection of particular methods depends on various factors, including interalia size and shape of the surface, or workshop conditions.

9.3 Determination of surface roughness

The surface of objects is the boundary separating them from the environment (from another object, substance, or space). The total of all surface irregularities is what is understood as the surface texture [58, 60–66]. Surface roughness is a measure of the texture of a surface. It is quantified by the vertical deviations of a real surface from its ideal form. If these deviations are great, the surface is rough, if they are small, the surface is smooth [67]. The surface texture is examined in cross sections of the surface on a perpendicular plane (two-dimensional analysis) or by evaluating the selected surface area (three-dimensional analysis of the surface topography) [68, 69].

Different surface preparation methods (particularly the mechanical methods) modify the surface and thus allow obtaining different surface roughness as well as stereometric structure [68, 70–73]. The surface roughness is also directly related to the wettability of the surfaces prepared for adhesive joining (as described in Sections 9.1 and 9.4).

Surface roughness is determined by trails left by cutting edge of tool. Method of surface measurement and evaluation is normalised by international ISO standards, based on profile method, evaluates the size of the surface profile—line originated from cutting the actual surface by defined area [69].

Terms, definitions and parameters of surface character are described in PN-EN ISO 4287 'Geometrical Product Specifications (GPS)—Surface texture: Profile method—Terms, definitions and surface texture parameters' [65].

The standard in question lists methods that enable expressing the surface roughness of profiles in numerical terms, by means of basic parameters, such as the following [60, 64, 68, 74, 75]:

- arithmetic mean height of the roughness profile R_a,
- maximum height of the profile R_z (or ten-point mean roughness R_z, according to the obsolete PN-87/M-04256 standard),
- maximum profile valley depth R_m,
- mean spacing of profile irregularities S_m,
- mean spacing of local peaks of the profiles,

and others.

The R_a parameter is the most commonly used and widely considered as the basic roughness parameter; according to the PN-EN ISO 4287 standard it is defined as: arithmetical mean deviation of the assessed profile (Fig. 9.4) within a sampling length l.

The R_a parameter describes the distance between two parallel lines limiting the surface area of the profile area that is filled with material and lies above the mean line m and profile areas without material lying below the mean line (Fig. 9.4). This is the height of the space around the middle line m created after the 'levelling' of hills and valleys. The R_a parameter, determined in micrometers (μm), provides a better reflection of roughness on larger surfaces, as it eliminates the influence of individual, irregular peaks or valleys.

Although R_a is a useful average, it does not differentiate between peaks and valleys. Very different profiles can have the same R_a value. Although Ra values are commonly

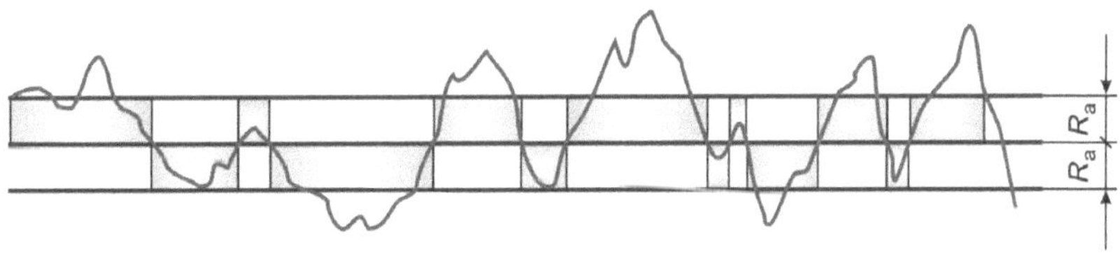

Fig. 9.4 Graphic interpretation of the R_a parameter.
Source: Own work based on B. Nowicki, Struktura geometryczna powierzchni. Chropowatość i falistość powierzchni (Geometric structure of the surface. Surface roughness and waviness), WNT, Warszawa, 1991 (in Polish).

used to describe surfaces, the limits of this indicator should not be forgotten [67].

Arithmetical mean height indicates the average of the absolute value along the sampling length. When dealing with the roughness profile, R_a is referred to as Arithmetic mean roughness while W_a is referred to as Arithmetic mean waviness for the waviness profile.

The R_z parameter (*maximum height of the profile*) is the sum of the height of the largest profile peak R_p and the largest profile valley depth R_v within the sampling length l_n (Fig. 9.5).

The surface roughness assessment employs both vertical (amplitude) and horizontal (spacing) parameters. In addition, the assessment may include the parameters of the roughness profile (2D) or be based on the stereometric characteristics of the area roughness (3D), which are included in ISO 25178 and ISO 16610.

Among the amplitude parameters, [67, 68, 74] there are:

- maximum profile peak height of the profile R_p,
- maximum profile valley depth of the profile R_v,
- arithmetic mean height of the roughness profile R_a (Fig. 9.4),
- root mean square (RMS) roughness R_q,
- maximum height of the profile R_z (Fig. 9.5), and
- total height of the profile (between the largest peak height and the largest valley depth) R_t.

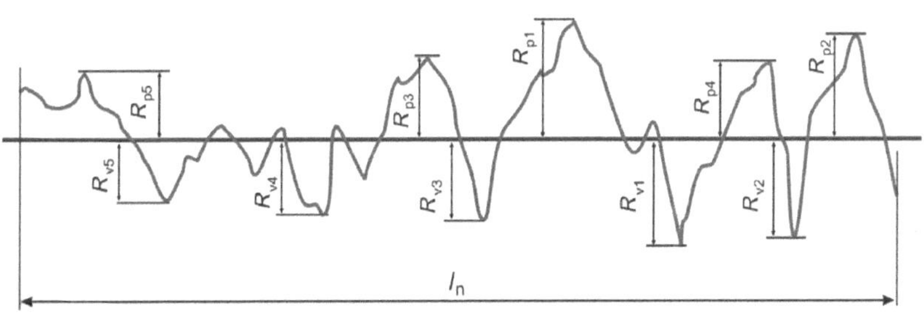

Fig. 9.5 Sketch of the surface profile for determination of R_z.
Source: Own work based on E.S. Gadelmawla, M.M. Koura, T.M.A. Maksoud, I.M. Elewa, H.H. Soliman, Roughness parameters, J. Mater. Process. Technol. 123 (2002) 133–145; B. Nowicki, Struktura geometryczna powierzchni. Chropowatość i falistość powierzchni (Geometric structure of the surface. Surface roughness and waviness), WNT, Warszawa, 1991 (in Polish).

Spacing parameters include:

- mean peak width of the profile RS_m and
- material component of the profile $R_{mr}(c)$.

The total height of the roughness profile R_t is the sum (vertical distance) of the height of the highest peak of the profile R_p and the largest depth of the profile valley R_v within the sampling length ($R_t = R_p + R_v$)—Fig. 9.6A. The R_t parameter is defined on the sampling length larger than the elementary segment; therefore, all profiles will always exhibit the dependence: $R_t \geq R_z$ [60] (Fig. 9.6B) [69].

The material component curve of the profile contains information regarding the shape of the profile, even of extreme valley depths. This curve allows describing the variations in profile properties that change with its depth. The parameters that have

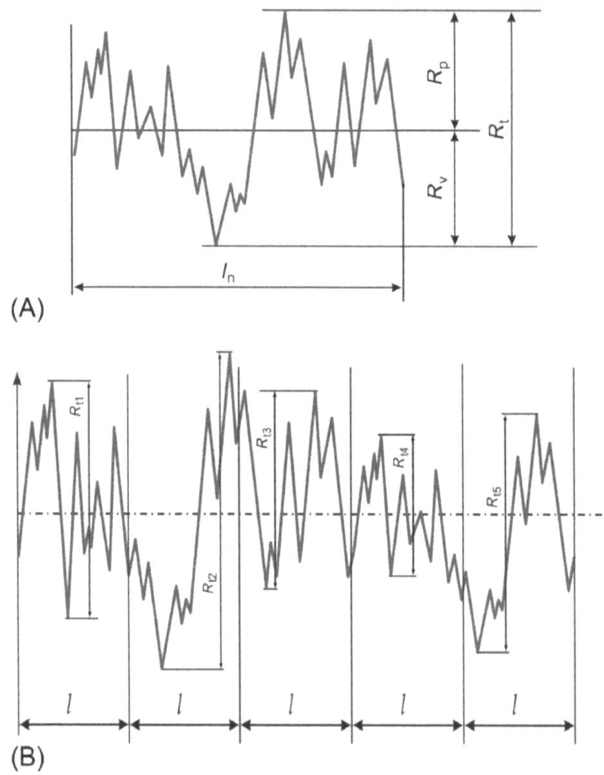

Fig. 9.6 Graphical representation of the R_t: (A) l_n—sampling length, R_p—maximum profile peak height of the profile, R_v—maximum profile valley depth of the profile, (B) example of assessment length divided into five samples.
Source: Own work based on E.S. Gadelmawla, M.M. Koura, T.M.A. Maksoud, I.M. Elewa, H.H. Soliman, Roughness parameters, J. Mater. Process. Technol. 123 (2002) 133–145; B. Nowicki, Struktura geometryczna powierzchni. Chropowatość i falistość powierzchni (Geometric structure of the surface. Surface roughness and waviness), WNT, Warszawa, 1991 (in Polish).

been determined based on the data provided by the bearing ratio curve include the following [60, 67–69]:

- R_k—core roughness depth,
- R_{pk}—reduced peak height,
- R_{vk}—reduced depth of valleys,
- M_{r1}—material ratio of peaks, and
- M_{r2}—material ratio of valleys.

These parameters are predominantly employed to provide a clear distinction between surfaces with exhibiting an identical value of R_a, which is possible due to the fact that they do not only describe the height of roughness, but also its shape. Therefore, it seems that the analysis of the suitability of a surface for bonding could include the material component curve of the profile, as well as the parameters characterising it. A graphic interpretation of the parameters listed above is presented in Fig. 9.7.

Reduced peak height R_{pk} is determined from a measure of the peak height above the core roughness. This parameter characterises the behaviour during lapping of sliding and pitch surfaces. A small R_{pk} value reflects high abrasion resistance [60, 74].

Reduced depth of valleys R_{vk} is the average valley depth below the core roughness of the profile. It is a measure of the ability of the surface to retain oil [60, 71], by means of exemplification—high R_{vk} values are required by lubricating surfaces. This parameter could presumably apply to filling the depths of valleys by the adhesive when producing adhesive joints. However, due to differences in the viscosity of adhesives, the width of the valley is equally relevant.

Surface roughness parameters are determined with surface profilers of different designs and sizes. The measuring element is a sharp sliding gauge head, which moves at a constant speed on the analysed surface and collects data and sends it for analysis in the electronic system.

In addition to the determination of surface roughness parameters through contact measurement, the surface roughness assessment may also be based on the series of standards PN-EN ISO 8503, and in particular, the surface profile comparator method

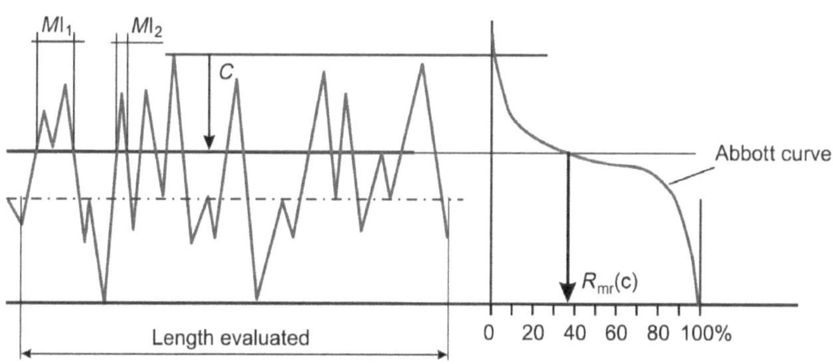

Fig. 9.7 Graphical representation of the material ratio curve (Abbott-Firestone curve). Source: Own work based on Z. Humienny, P.H. Osanna, M. Tamre, A. Weckenmann, L. Blunt, W. Jakubiec, Specyfikacja geometrii wyrobów (GPS) (Product Geometry Specification (GPS)), WNT, Warszawa, 2004 (in Polish).

Fig. 9.8 Surface profile comparator [76].

described by PN-EN ISO 8503-2. This assessment follows one of the two standards: for surfaces obtained by abrasive blasting treatment with metallic or mineral grit (G) or abrasive blasting with metallic shot (S). Fig. 9.8 presents the reference profile comparator in the form of a flat plate with four segments (i.e. a coupon) on which the reference surface profiles are prepared. The roughness profile is subsequently defined as: fine, medium, or coarse [60, 71, 74].

Profile assessment based on the reference profile is as follows:

- fine profile—corresponds to segment 1 or between 1 and 2, but excluding 2;
- intermediate profile (medium)—corresponds to segment 2 or between 2 and 3, but excluding 3;
- coarse profile—corresponds to segment 3 or between 3 and 4, but excluding 4.

Numerical values of roughness parameters corresponding to reference profiles are shown in Table 9.7. The value in μm presented in Table 9.7 is the value of the parameter R_t (R_t—the largest spacing between peak and valley within one elementary section, determined using a roughness contact device).

Table 9.7 Nominal values for surface roughness parameters [76]

	Comparator G		**Comparator S**	
Segment	**Nominal reading (μm)**	**Tolerance (μm)**	**Nominal reading (μm)**	**Tolerance (μm)**
1	25	3	25	3
2	60	10	40	5
3	100	15	70	10
4	150	20	100	15

9.4 Determination of adhesive properties of adherends surface

9.4.1 Wettability

One of the ways to determine the suitability of the selected method for the preparation of a given substrate surface for bonding is to assess the surface wettability. Wettability is an extremely important factor of high importance to adhesive joining and adhesive bonding and in various technological processes because it significantly affects the adhesion phenomenon, increasing adhesion or contributing to its reduction [5, 23, 24, 77–81]. According to Iqbal et al. [51], Eckert [82], and Duncan et al. [83] materials with better surface wetting properties ensure better adhesive bonding.

The wettability is a very important engineering property [84–86]. Wettability notably affects performance properties like icing/de-icing, soiling and cleanability, printing, and adhesion [46, 84]. One parameter for the characterisation of wettability is the static contact angle, whereby small contact angles refer to good wettability and vice versa [87].

Wetting is a surface phenomenon consisting in substitution of the surface of the solid and the liquid with a boundary surface, characterised by certain tension (σ), which results from the difference in the surface tension between the solid and liquid in the gaseous medium [5, 87–90]. The environmental atmosphere is assumed to be filled with saturated vapour of liquid [81]. The measure of surface wettability is the contact angle θ (also referred to as the contact angle or the critical angle) formed between the wetted surface of the solid and the tangent to the wetting liquid surface (to the meniscus of the wetting liquid), at the contact point of the liquid and the solid surface (Fig. 9.9) [35, 40, 41, 90–94]. The amount of wetting depends on the energies of the interfaces involved [91].

Contact angle represent different degrees of wetting such that [4, 46, 78, 91, 93, 95–98]:

- $\theta = 0$ degrees (spreading)—the case when the contact angle $\theta = 0$ degrees, indicates that the liquid spreads over the surface evenly and, furthermore, represents complete wetting of a solid surface by a liquid.
- $\theta = 0–90$ degrees (nonspreading)—when wetting the surface of a solid, the contact angle value will be lower than 90 degrees,
- $\theta = 90–180$ degrees (nonwetting),
- $\theta = 180$ degrees (repellency)—if the contact angle $\theta = 180$ degrees, then the result is absolute nonwetting.

The scientific literature analyses various aspects regarding the surface wetting, which account for the differences in size and relationships (concerned with the contact angle) [13, 20, 23, 41, 94, 99–110]. In order for the liquid to wet the surface of the solid favourably, its surface tension should be lower than the surface tension of the liquid [31].

The method for wettability determination consists in depositing a thin layer of the measuring liquid onto the test surface, which is followed by an immediate measurement of the time after which the layer of applied liquid will separate into single drops.

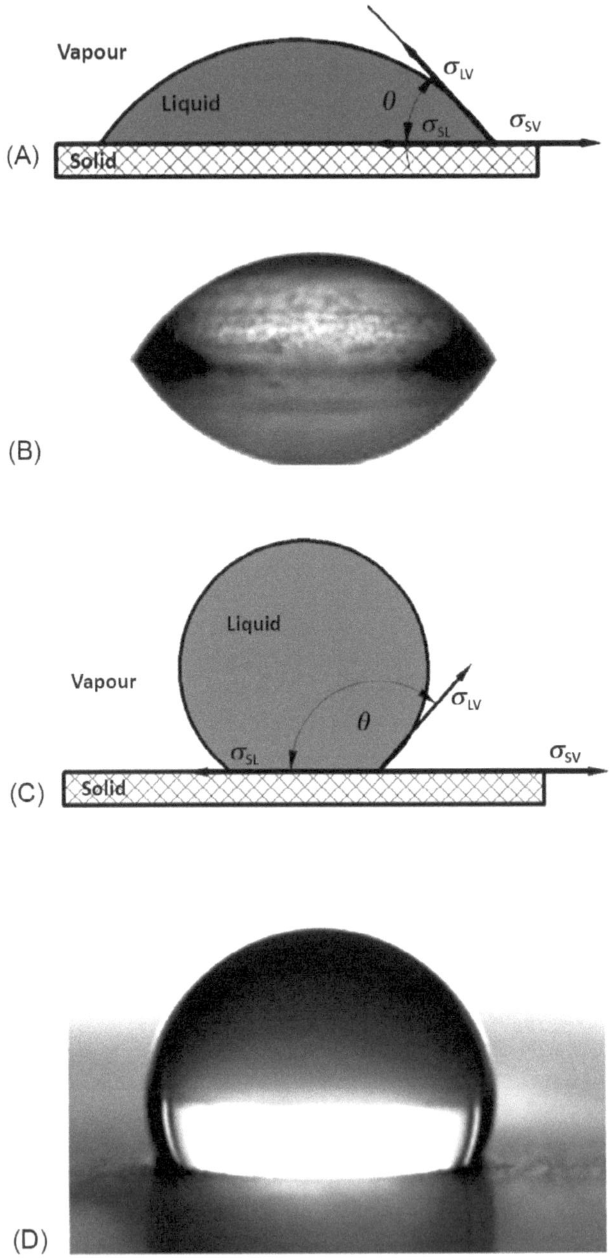

Fig. 9.9 Wetting of a solid surface by the liquid in the case of: (A) and (B) good wettability $\theta < 90$ degrees, (C) and (D) poor wettability $\theta > 90$ degrees.

The separation of the liquid layer depends both on the value of the SFE of the measuring liquid and on the value of the SFE of the tested surface.

The following guidelines should be taken into account when applying the measuring liquid [29, 31, 80, 110]:

- the thickness of the applied measuring liquid should approx. amount to 12 μm,
- the liquid can be applied in the form of a band with a width of approximately 10 mm, and
- if the applied liquid layer maintains its shape without clustering to single drops for a period longer than 2 s, the operation should be repeated on another surface using a measuring liquid with a higher SFE value.

The executed activities should be carried out until the layer of the measuring liquid is separated into single drops in the time closest to 2 s. In order to obtain the correct result, these measurements should be repeated at least three times using the same measuring liquid. It is then assumed that the wettability of the substrate surface is approximately equal to the SFE of the measuring liquid used [31]. However, the method is connected with a certain difficulty when determining wettability—namely the phenomenon of the narrowing of the measuring liquid band without the separation of this band into single drops (Fig. 9.10). Initially, a constant width of the stenosis persists, and after exceeding certain value γ_L takes on an irregular shape or narrows [21].

One of the workshop methods for controlling the surface wettability after the surface treatment for adhesive joining is the sessile drop method [31, 39, 47, 111–116]. It consists in depositing a drop of water (or a few drops of water) on prepared surfaces and observing the behaviour of the drop(s) on the surface [33, 117]. If the drop assumes a spherical shape, it means that the surface is not sufficiently prepared and requires further processing. However, if a drop of water dissolves, it may indicate that

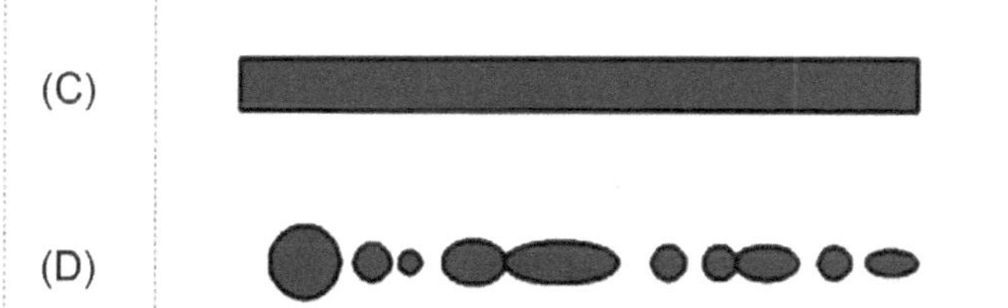

Fig. 9.10 Images of the measuring liquid on the surface of the material during the wettability test: (A) uniform distribution of liquid ($\gamma_S > \gamma_L$), (B) unequal distribution of liquid, (C) narrowed liquid band, and (D) liquid in the form of individual drops ($\gamma_L \geq \gamma_S$).
Source: Own work based on M. Żenkiewicz, Adhezja i modyfikowanie warstwy wierzchniej tworzyw wielkocząsteczkowych (Adhesion and Modification of the Surface Layer of Macromolecular Materials), WNT, Warszawa, 2000 (in Polish).

the surface prepared for adhesive processes is sufficiently prepared. The advantages of this method are ease and speed of execution and the fact that it does not require the use of special equipment. However, this method has some limitations, among which the following are typically listed [31]:

- it is not applicable in surface preparation quality control of certain structural materials, e.g. aluminium alloys after anodising, or magnesium and its alloys,
- different degrees of water hardness affect the surface tension and may cause incorrect results,
- in certain cases, distilled water does not produce reliable results, due to, e.g. the influence of material factors, as well as other external factors (temperature and humidity),
- it may be necessary to use different sets of measuring liquids, which significantly extends the time of such control and may contribute to difficulties in interpreting the obtained results.

In addition, the contact angle measurements are used to assess wettability. The measuring liquids used to determine the surface wettability should be of high chemical purity and should be free from any contaminants. The measuring liquids most often have values of the SFE γ_L from 22.6 to 73.0 mJ/m^2. In the case of industrial research, liquids with 30–56 mJ/m^2 are most commonly used [31].

9.4.2 Contact angle

From fundamental point of view, the contact angle is dependent on surface energy and it characterises the surface wettability [91]. Contact angle measurement is easily performed by establishing the tangent (angle) of liquid drop with a slid surface at the base. The attractiveness of using contact angles θ to estimate the solid–vapour and solid–liquid interfacial tensions is due to the relative ease with which contact angle can be measured on suitable prepared solid surfaces [28, 33, 34, 83, 90, 114, 118, 119].

The contact angle depends on a lot of surface parameters namely, on topology, chemical composition, and temperature. Contact angle, measured on sample surfaces, is very often given as an integral, or statistically average, number. This procedure may not cover the real wettability characteristics of surfaces [84].

Formation of the angle liquid drop to a solid surface is determined through the energetic situation. Drop formation is connected with the formation of additional surfaces, either at the solid or a liquid [84, 120].

The review of contact angle measurement and contact angle interpretation was presented by Kwok and Neumann [90].

There is a wide range of methods for the determination of the contact angle [20, 93, 114, 121–126], and the most commonly used methods are [12, 23, 29, 37, 80, 85, 86, 110, 113, 122, 127–129]:

- captive bubble method,
- geometric methods (e.g. sessile drop method),
- method of capillary liquid growth on a sample of the tested material (including the Wilhelmy method and Washburn method), and
- direct measurement method.

Alghunaim et al. [86] presented the summary of some common methods for determination the wettability of powders and presented a detailed review of the different techniques that measure, directly or indirectly, the contact angle that a liquid would form on powders.

The characteristics of the technique for evaluating the wettability of powdered foods which is the capillary rise method in which the contact angle is usually estimated using the classical Washburn model and the new mathematical model for the analysis of capillary rise date were showed by Hammes et al. [129].

Galet et al. [128] presented the study of the Washburn method for determining the contact angle of liquids with powders. The study indicated that the capillary rise of liquids in a packed bed of particles was more complex than would indicate the simple theory based on capillary rise of liquids in a tube system.

Shang et al. [112] compared the different methods of determination contact angle of colloids typical for soils and sediments, like: static sessile drop, dynamic sessile drop, Wilhelmy plate, thin layer wicking, and column wicking.

Schuster et al. [124] used the sessile drop method to measure the contact angle between solids (Ti6Al4V and AISI 316 stainless steel) and liquids.

Schmitt and Heib [114] underlined that drop shape analysis is one of the most important and frequently used methods to characterize surfaces in the scientific and industrial communities.

Nowadays, the method of direct contact measurement is a method of direct measurement using special devices called goniometers or contact angle analysers [31, 40, 112, 113, 129].

Captive bubble method

In the captive bubble method, the examined surface of the plate-shaped substrate is immersed in the measuring liquid in such a way that its lower plane is perpendicular to the direction of the buoyancy force (Fig. 9.11).

A bubble of air is introduced with a syringe directly under the plate and settles on its lower surface. The contact angle is determined from the shape of the bubble [31].

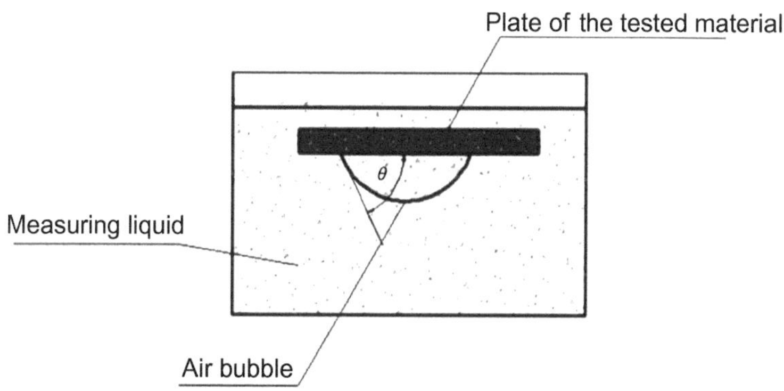

Fig. 9.11 Contact angle measurement by means of the captive bubble method.
Source: Own work based on M. Żenkiewicz, Adhezja i modyfikowanie warstwy wierzchniej tworzyw wielkocząsteczkowych (Adhesion and Modification of the Surface Layer of Macromolecular Materials), WNT, Warszawa, 2000 (in Polish).

The advantage of this method the value of the contact angle reading is unaffected by gravity.

Geometric method

Determination of the contact angle with the geometric method (sessile drop method) accounts for different assumptions and models of the measuring liquid drops. The models of liquid drops are classified into [29, 31, 90, 112, 113, 130]:

- the spherical droplet model (Fig. 9.12),
- the drop curve model (the Bashforth method), and
- the ellipsoidal droplet model.

In the case of a spherical model of a measuring liquid droplet, gravitational effects that contribute to its distortion are not taken into account. The diameter of the droplet is given by the following formulas (9.1), (9.2) [31]:

$$D_0 \le \left(\frac{2\gamma_L}{\rho g}\right)^{0.5} \tag{9.1}$$

$$D_0 < 0.45\left(\frac{\gamma_L}{\rho}\right)^{0.5} \tag{9.2}$$

where:

D_0—diameter of the measuring droplet,
γ_L—the SFE of the measuring liquid,
ρ—density of the measuring liquid, and
g—force of gravity.

It should be noted that when water that is applied as a measuring liquid in the procedure of the geometric method with a spherical droplet model, the droplet diameter should be less than 3.86 mm. Given the spherical shape of the measuring liquid (Fig. 9.13), the following dependencies are shown: for $\theta < 90$ degrees (Eq. 9.3) and for $\theta > 90$ degrees (Eq. 9.4), which are used to determine the value of the contact angle θ [31].

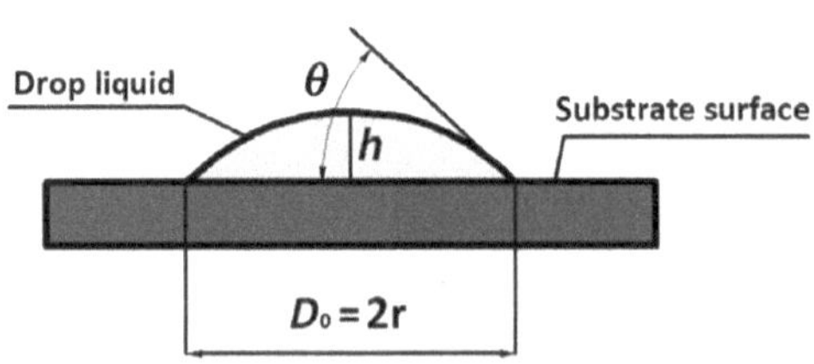

Fig. 9.12 Spherical model of the measuring droplet: D_0—diameter of the measuring droplet on the surface of solid, h—drop height.
Source: Own work based on M. Żenkiewicz, Adhezja i modyfikowanie warstwy wierzchniej tworzyw wielkocząsteczkowych (Adhesion and Modification of the Surface Layer of Macromolecular Materials), WNT, Warszawa, 2000 (in Polish).

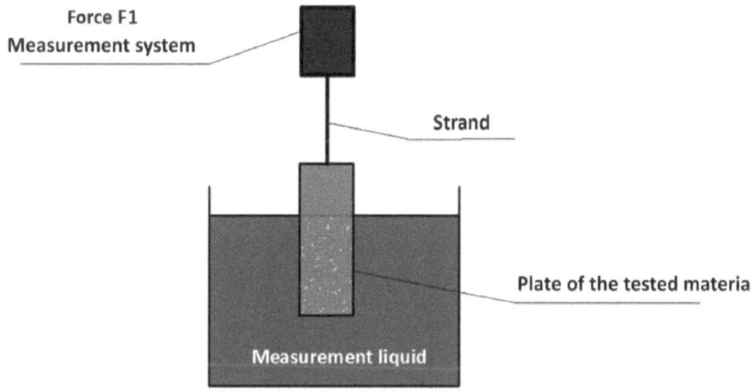

Fig. 9.13 The contact angle measurement system used in the Wilhelmy plate method, *h*—immersion depth of the measuring plates.
Source: Own work based on M. Żenkiewicz, Adhezja i modyfikowanie warstwy wierzchniej tworzyw wielkocząsteczkowych (Adhesion and Modification of the Surface Layer of Macromolecular Materials), WNT, Warszawa, 2000 (in Polish); M. Zielecka, Methods of contact angle measurement as a tool for characterization of wettability of polymers, Polimery 49 (2004) 327–332; A. Depalo, A.C. Santomaso, Wetting dynamics and contact angle of powders studied through capillary rise experiments, Colloids Surf. A Physicochem. Eng. Asp. 436 (2013) 371–379; A. Alghunaim, S. Kirdponpattara, B.-M. Z. Newby, Techniques for determining contact angle and wettability of powders, Powder Technol. 287 (2016) 201–215.

$$\theta = \arccos \frac{1-\left(\frac{h}{r}\right)^2}{1+\left(\frac{h}{r}\right)^2} \tag{9.3}$$

$$\theta = 90\text{degrees} + \arccos \frac{2hr}{h^2+r^2} \tag{9.4}$$

The use of direct contact angle measurement in the case of small contact angles is burdened with error, which however may be reduced by using considerable image enlargements.

Żenkiewicz [31] presents a table with values of the *h*/*r* quotient, which enables the determination of the contact angle value, applying the assumptions of the spherical model together with relevant relationships. In this work, other models of the measuring liquid droplet used in the geometrical method are characterised.

Wilhelmy plate method

The Wilhemy method is based on the measurements of the force which is needed to overcome the resistance of the liquid as the solid plate with known wetted length is immersed in or withdrawn from the liquid of known surface tension. The precision of the contact angle measurements by Wilhemy method is higher as compared the sessile drop method [80]. Packham [13], Zielecka [80], Alghunaim et al. [86], Shang et al. [112], and Della Volpe and Siboni [122] described and use the Wilhemy method to contact angle measurements.

In the first step the measuring plate is immersing and removing from the measuring liquid. Next, the sample, which is typically in the shape of plate or fibre, is suspended by a wire on the arm of the force measurement system (Fig. 9.13).

The proper instrument registers the force at the point of withdrawing and immersing the plate (F_1). The buoyant force F_2 (9.5) acts on the plate

$$F_2 = (\rho_1 - \rho_2) gSh \quad (9.5)$$

where:

ρ_1—plate density,
ρ_2—measuring liquid density,
g—gravitational force,
S—cross-sectional area of the plate, and
h—immersion depth of the measuring plates.

The force of gravity F_3 acts on the plate in the opposite direction (9.6):

$$F_3 = mg \quad (9.6)$$

where: m—mass of plates and force F_4 of interactions at the boundary of three phases, connected with the surface energy of the measuring liquid γ_L and perimeter of the plate (9.7):

$$F_4 = \gamma_L l\cos\theta \quad (9.7)$$

By measuring F_l, and given the dimensions, density, immersion depth, and surface tension of the liquid, $\cos\theta$ can be calculated from the relationship (9.8)

$$F_1 + F_2 = F_3 + F_4 \quad (9.8)$$

The Wilhelmy plate method is used to determine the advancing contact angle θ_A when immersing the plate and the receding contact angle θ_R while removing the plate from the liquid. The difference between θ_A and θ_R is the hysteresis of the contact angle (Fig. 9.14) [34, 41, 42, 78, 96, 120, 131–134]. In real conditions, the receding contact angle is always smaller than the advancing contact angle, and the value of the equilibrium angle θ (Eq. 9.9) [34] takes intermediate values

$$\theta_R < \theta < \theta_A \quad (9.9)$$

Extreme care is advised when performing the contact angle measurements—they ought to be taken in accordance with the established guidelines given below—so that the Young's equation is fulfilled [8, 24, 25, 28, 29, 34, 35, 38, 40, 46, 78, 113, 122, 125]:

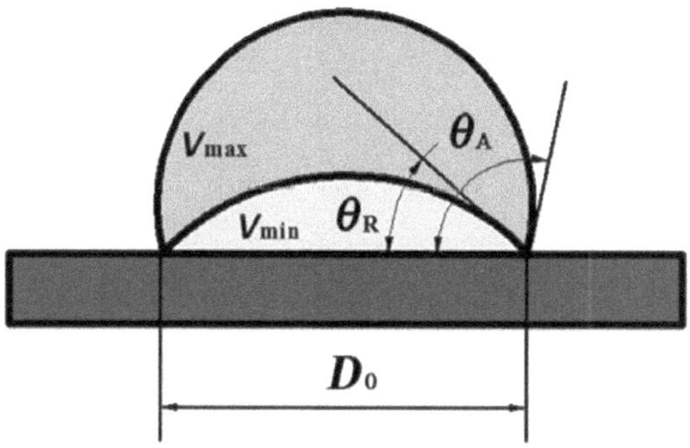

Fig. 9.14 Contact angles at constant diameter D (D=const): θ_A—advancing contact angle, θ_R—receding contact angle, V_{max}—liquid drop of maximum volume, V_{min}—liquid drop of minimum volume.
Source: Own work based on M. Żenkiewicz, Adhezja i modyfikowanie warstwy wierzchniej tworzyw wielkocząsteczkowych (Adhesion and Modification of the Surface Layer of Macromolecular Materials), WNT, Warszawa, 2000 (in Polish).

- the tested surface should be flat and smooth,
- the smallest possible volume of the measuring liquid droplet should be applied, i.e. below <2μL (or even <0.2 μL), in order to minimise the spread of drops as the effect of gravitational forces, in which case the contact angle measurement is lower than in reality,
- once the drop is set, the measurement time should be as short as possible to minimise the effects of gravitational forces and evaporation,
- the analysed surfaces must be clean and free from contamination,
- tests should be carried out at a constant temperature with low, strictly-defined humidity (the use of air-conditioning chambers is widely recommended),
- standard measuring liquids of identified surface tension are highly recommended, and
- during the measurement, the sample should remain stationary so that no distortion of the measuring droplet could result from vibration.

Factors that significantly influence the correctness of measurements of the contact angle [20, 34, 41, 46, 96, 102, 103, 113, 116, 135, 136] are the following:

Drop size and symmetry

The literature in the field [20, 31, 80, 113] indicates that measurements of the contact angle should be performed on two sides of a drop: since the surface of the sample is heterogeneous, the arithmetic mean of the measurements of at least nine different drops should be taken as the result of the test. For a given series of measurements, it is necessary to observe the principle that the time which elapses from the deposition of the drop to the contact angle reading potentially the shortest possible and identical for all drops in a given measurement series—this reduces the risk of misreading as the effect of the drop's reaction with the substrate or drop size change resulting from evaporation. The data collected in the literature [31], indicate that the droplet volume should be in the range from 28 to 0.5 mm^3. Schuster et al. [113] found that drop asymmetry is a very important parameter that must be considered in CA measurements.

Vafaei and Podowski [106] investigated the relationship between liquid droplet size and contact angle.

Temperature and humidity

Up to approximately 80°C, however, temperature changes within the range cause little deviation in the SFE readings as natural temperature oscillations that may occur during laboratory tests have no significant effect on the measurement of the contact angle of plastics [31, 113].

The Young's modulus (surface stiffness)

The surface of the material on which drops of the measuring liquid are deposited must be sufficiently stiff, and therefore its longitudinal elasticity coefficient must be greater than 10 kPa—in order to prevent the drop-shape distortion due to its weight [137].

Surface roughness (affecting thermodynamic hysteresis)

It is generally accepted that if $R_a < 0.5\,\mu m$, then the impact of roughness on the contact angle is negligible [40]. The surface roughness causes the drop to exhibit different metastable states. Scientists have established experimentally a significant effect of the shape and the density of surface irregularities [8–10, 31, 35, 41, 47, 93, 132]. Sometimes a droplet on a rough surface traps air between the liquid and solid and this issue was presented by Yamamoto and Ogata [134].

Surface homogeneity (and surface layer homogeneity) both physical and chemical and heterogeneity (the second source of thermodynamic hysteresis)

The contact angle measurement result is strongly affected by migration of auxiliary components and diverse supramolecular structure and also heterogeneity of the surface as a result of the formation of functional groups of various sizes and different character [33, 34, 38–40, 88, 91, 95, 104, 138].

Surface contamination

Surface contamination contributes to significant differences in the obtained contact angle values [31].

The type of measuring liquid

Penetration of the measuring liquid into the gaps in the surface layer, as well as penetration into the intermolecular areas of the material (swelling phenomenon)—are the main causes of dynamic hysteresis [31, 47]. Furthermore, the molar volume of the measuring liquid is important: the increase in the molar volume retards and limits the processes of liquid penetration into the structure. Water, due to its small molar volume, easily penetrates the structure of certain materials; therefore, it is important to choose the right measuring liquid so that it does not react with the components of the material under analysis [40, 120, 122].

These factors hinder measurements of the contact angle, contribute to obscuring the interpretation of obtained results and are the cause of various metastable states of the drop itself [20]. The effect of these phenomena is the hysteresis of the contact angle [29, 36, 40, 120, 122, 131, 133]. Learning the hysteresis phenomenon is of great practical importance, as it, for instance, is the key component in the calculation of the SFE [100, 110, 124, 135, 139].

The contact angle hysteresis has two components: thermodynamic and dynamic [31, 34–36, 41, 89, 131]. The first one is affected by the roughness and heterogeneity of the surface, which leads to the drop occurring in different metastable states, which in turn translate to discrepancies in the contact angle measurement results. The second component depends on time and its source is found in such phenomena as:

- chemical interaction of liquid-material,
- penetration of the measuring liquid into the pores, and
- particle reorganisation on the surface.

9.4.3 Work of adhesion

The adhesion work W_a describes the work required to create a surface unit due to separating the measuring liquid and tested material. W_a can be defined by the following Eq. (9.10) [22, 30, 89, 100, 118, 140–142]:

$$W_a = \gamma_L (1 + \cos\theta) \tag{9.10}$$

The work of adhesion between the solid and the liquid can be described by means of the Dupré equation (9.11) [13, 30, 31]:

$$W_a = \gamma_{SV} + \gamma_{LV} - \gamma_{SL} \tag{9.11}$$

Owens and Wendt proposed the following form of the work of adhesion between interacting solid and liquid (Eq. 9.12) [30, 31]:

$$W_a = 2\left(\gamma_S^d \gamma_{LV}^d\right)^{0.5} + 2\left(\gamma_S^p \gamma_{LV}^p\right)^{0.5} \tag{9.12}$$

According to the thermodynamic theory of adhesion, it is a fundamental condition of adhesion that a solid be thoroughly wet by a liquid. The criterion for estimating the wetting process is defined as the magnitude of the coefficient S [13, 30, 78] described by the following Eq. (9.13):

$$S = W_a - 2\gamma_L \tag{9.13}$$

where W_a is the work of adhesion and γ_L is the SFE of a wetting liquid.

According to this theory, wetting can be considered thorough when coefficient $S > 0$.

Bearing in mind thermodynamic principles, it can be assumed that wetting is possible when the SFE of a wetting liquid is higher than the SFE of the material being wetted. Moreover, the more thorough the wetting, the higher the difference between the SFE and the material subjected to wetting.

9.4.4 Surface free energy

It is a widely shared conviction among numerous researchers [12, 30, 32, 103, 108, 109, 138, 143–152] that adhesive properties can be determined by means of the SFE. The SFE is one of the thermodynamic quantities describing the state of atomic equilibrium in the surface layer of materials [30, 31]. It reflects a specific state of imbalance in intermolecular interactions that occurs at the interface between two different media (e.g. solid–liquid) [12].

The SFE of materials depends on many factors [13, 31, 130, 142], including the surface geometry [9, 23, 48], certain physical properties of materials, as well as the properties of the physical adsorption phase on the solid surface, produced as a result of various types of production processes [108, 153, 154] or surface treatments [121, 146, 155–157].

The Young's equation (9.14), which combines a measurable geometric parameter, e.g. the contact angle, with three thermodynamic indicators, and explains the interaction properties in the contact at the interface [5, 34, 41, 78, 109, 158] lies at the heart of both the wetting theory and of the methods for the calculation of the SFE with the contact angle measurements. The equation in question was derived from the equilibrium of forces, which represent surface tensions at the interface between three phases—solid, liquid, and gas [130]:

$$\sigma_{SV} = \sigma_{SL} + \sigma_{LV} \cdot \cos\theta_Y \tag{9.14}$$

where

σ_{SV}—surface tension at the solid–gas interface,
σ_{SL}—surface tension at the solid–liquid interface,
σ_{LV}—surface tension at the liquid–gas interface,
θ_Y—equilibrium contact angle.

The Young's equation (9.14) can also be derived from the energy balance of the three-phase balance, which then is given as Eq. (9.15) [31, 40, 78, 89–91, 140–142, 148, 159, 160]

$$\gamma_{LV} \cdot \cos\theta_Y = \gamma_{SV} - \gamma_{SL} \tag{9.15}$$

where

γ_{LV}—SFE at the liquid–gas interface,
γ_{SV}—SFE at the solid–gas interface,
γ_{SL}—SFE at the solid–liquid interface,
θ_Y—equilibrium contact angle.

Table 9.8 The surface free energy and its components of selected measuring liquids—the Owens–Wendt method [31, 102, 143]

Measuring liquid	Surface free energy and its components (mJ/m^2)		
	γ_L	γ_L^d	γ_L^p
Distilled water	72.8	21.8	51.0
Diiodomethane	50.8	48.5	2.3

Table 9.9 The surface free energy and its components of selected measuring liquids—the van Oss–Good method [31, 78, 163]

Measuring liquid	Surface free energy and its components (mJ/m^2)				
	γ_L	γ_L^{LW}	γ_L^{AB}	γ_L^+	γ_L^-
Distilled water	72.8	21.8	51.0	25.0	25.0
Glycerol	64.0	34.0	30.0	3.9	57.4
Formamide	58.0	39.0	19.0	28.9	39.6
Diiodomethane	50.8	50.0	0	0	0
α-Bromonaphthalene	44.4	43.5	0	0	0

Determination of the SFE may be performed by means of a number of methods, e.g. the approaches of Fowkes, Owens–Wendt, van Oss–Good (also known as van Oss–Chaudhury–Good), Zisman, Neumann, or Wu [41, 110, 113, 122, 139, 144, 146, 161, 162]. The basis for many methods of determining the SFE from appropriate relationships are the measurements of the contact angle with specific measurement liquids of identified SFE and its components according to the method in use (Tables 9.8 and 9.9) [30, 39, 108, 142].

The SFE, obtained from contact angle measurements, depended greatly on the calculation method [142]. The historical background of solid SFE calculation has been reported in detail by Correia et al. [164] and the important features were summarised by Michalski et al. [142].

Below are presented two selected methods, which are most widely applied in the determination of the SFE of various structural materials. The theoretical foundations of these methods are most profoundly described and explained in the scientific literature [4, 30, 31, 130, 142, 150, 152, 165–168].

The Owens–Wendt method (O–W)

The Owens–Wendt method (sometimes referred to as Kaelble–Owens–Wendt Method) is a frequently applied method for determining the SFE of, e.g. polymers and metals [108, 150, 169]. This method consists in determining the dispersive and polar components of the SFE based on Berthelot principle [31], which assumes that interaction between molecules of two bodies in their surface layers equals the geometric mean of the cohesion work between the molecules of each body.

This method assumes that the SFE (γ_S) is a sum of two components: polar (γ_S^p) and dispersive (γ_S^d), and that there exists a relationship (9.16) between the three quantities [46, 78, 110, 146]:

$$\gamma_S = \gamma_S^d + \gamma_S^p \tag{9.16}$$

The polar component is regarded to be the sum of components derived from such intermolecular interactions as: polar, hydrogen, induction, and acid-base, with the exception of dispersive interactions. Dispersive interactions constitute the dispersive component of the SFE.

Comparing Eqs (9.10), (9.12), the following Eq. (9.17) [13, 30, 48, 110, 140] is obtained:

$$\gamma_{LV}\left(1+\cos\theta\right) = 2\left(\gamma_S^d\gamma_{LV}^d\right)^{0.5} + 2\left(\gamma_S^p\gamma_{LV}^p\right)^{0.5} \tag{9.17}$$

This equation allows determining the SFE of a solid and its SFE components.

In order to determine the polar and the dispersive components of the SFE, the measurements of the contact angle of the analysed samples need to be conducted with two measuring liquids. The SFE of the measuring liquids used in the test is known, including its polar and dispersive components [29, 102, 123, 143, 170]—typically, one of the liquids is nonpolar and the other is bipolar. Most frequently, the tests are carried out with distilled water as the polar liquid and diiodomethane as the nonpolar one.

The SFE γ_S is obtained from the adjusted dependence describing the dispersive component of the SFE (Eq. 9.18) and the polar component of the SFE (Eq. 9.19) [123, 171]

$$\left(\gamma_S^d\right)^{0.5} = \frac{\gamma_d\left(\cos\theta_d+1\right) - \sqrt{\frac{\gamma_d^p}{\gamma_w^p}}\gamma_w\left(\cos\theta_w+1\right)}{2\left(\sqrt{\gamma_d^d} - \sqrt{\gamma_d^p\frac{\gamma_w^d}{\gamma_w^p}}\right)} \tag{9.18}$$

$$\left(\gamma_S^p\right)^{0.5} = \frac{\gamma_w\left(\cos\theta_w+1\right) - 2\sqrt{\gamma_S^d\gamma_w^d}}{2\sqrt{\gamma_w^p}} \tag{9.19}$$

where

γ_S^d—the dispersive component of the test material SFE,
γ_S^p—the polar component of the test material SFE,
γ_d—the SFE of diiodomethane,
γ_d^d—the dispersive component of the SFE of diiodomethane,
γ_d^p—the polar component of the SFE of diiodomethane,
γ_w—the SFE of water,
γ_w^d—the dispersive component of the SFE of water,
γ_w^p—the polar component of the SFE of diiodomethane,
θ_d—the contact angle of diiodomethane,
θ_w—the contact angle of water.

Certain issues, both theoretical and related to the practical use of the Owens-Wendt method to determine SFE, have been included in many works [31, 102, 108, 112, 161, 172–174].

The van Oss–Chaudhury–Good method (OCG)

The van Oss–Chaudhury–Good method is one of the indirect methods for the determination of the SFE of solids. In this method, the SFE is the sum of two components (9.20), one of which γ_i^{LW} is connected with long-range interactions (dispersion, polar, and inductive interactions), whereas the other component γ_i^{AB} is associated with the acid-base interactions (interactions related to the formation of Lewis acid–base bonds) [40, 143, 148, 152, 175–177].

$$\gamma_I = \gamma_i^{LW} + \gamma_i^{AB} \tag{9.20}$$

In addition, the γ_i^{AB} component described by means of the equation for bipolar compounds (showing properties of both Lewis acids and bases, e.g. water) (Eq. 9.21) [13, 31, 40, 78]:

$$\gamma_i^{AB} = 2\left(\gamma_i^{+}\gamma_i^{-}\right)^{0.5} \tag{9.21}$$

where

γ_i^{+}—Lewis acid SFE component,
γ_i^{-}—Lewis base SFE component,
index *i*—subsequent measuring solids or liquids.

Determining the SFE of test materials will consist in measuring their surfaces contact angle with three different measuring liquids and calculating the γ_S of the system of three equations: (9.22) [40, 140, 152]:

$$\left(\gamma_S^{LW}\gamma_{Li}^{LW}\right)^{0.5} + \left(\gamma_{S_s}^{+}\gamma_{Li}^{-}\right)^{0.5} + \left(\gamma_S^{-}\gamma_S^{+}\right)^{0.5} = \gamma_{Li}\left(1+\cos\theta_i\right)/2 \tag{9.22}$$

where $i = 1,2,3$.

Measuring the contact angle requires the application of two polar and one nonpolar liquid; nevertheless, solving the Eq. (9.22) requires additional information regarding particular values for the applied measuring liquids. Polar liquids applied in tests are water, glycerol, formamide or ethylene glycol, and nonpolar liquids (not showing properties of either Lewis acids or bases) diiodomethane or α-bromonaphthalene. A detailed description of this method is provided in the publications [31, 99, 148, 152, 174, 176, 178, 179].

9.5 Determination of strength of adhesive joints

Strength of adhesive bonds (including adhesive joints) is one of the factors enabling the assessment of surface preparation [16, 57, 149, 180–183]. Nevertheless, this is an assessment of not only the surface preparation but also of the strength of the adhesive bond itself as well as of the entire bonding technology [12, 14, 17, 28, 33, 54, 56, 57, 184].

The strength of adhesive joints depends on many technological, constructional, material, and operational factors, among which the most prominent are [1, 2, 4, 43, 53, 106, 130, 184–207]:

- stress distribution in the joint,
- mechanical (such as the Young's modulus or the Poisson's ratio) and physical properties of substrates,
- energy properties of the substrate surface,
- adhesive properties and mechanical properties of the bond,
- surface treatment,
- cure conditions,
- type of loads acting on the joint,
- the extent and nature of the deformation of adherends,
- the joint failure mode,
- time, temperature, and humidity conditions of joint operation, and
- other.

Variety of test methods for evaluation of adhesively bonded joints have been developed and established, including ISO (International Organization of Standardization—http://www.iso.org), EN (European Committee for Standardization—http://www.cen.eu) and/or ASTM (American Society for testing Materials—http://astm.org) standards. The spectra of mechanical test for adhesive joint can be subdivided by the type of mechanical stress to be created in the adhesive joints [4]. Four types of static tests of adhesive joint strength are typically recommended by relevant ISO, EN, and ASTM standards:

- shear tests,
- tensile tests,
- peel tests, and
- cleavage and wedge tests.

However, other norms specify different methods for the determination of shear, cleavage, and peel strength of adhesive joints. With regard to surface preparation wedge and cleavage tests [208] are common test methods applied to assess the quality of surface treatments and provide an estimate of long-term durability [4]. During the tests, the breaking force is measured at the specified rate of load build-up. Shear, peel, and bending tests determine the break strength of adhesive joints [2–7, 12, 25, 55, 77, 191, 197, 209]. A commonly used method for testing the strength of adhesive joints is the determination of shear strength, as these joint types offer the highest load-carrying capacity [2, 31].

The subsequent paragraphs present an overview of selected methods for the determination of the strength of adhesive joints, which have been the basis for tests found in numerous publications [1, 2, 4, 40, 67, 197, 207]. The test categories by the type of mechanical stress were described by Adams [1, 2], Cagle [3], Brockmann et al. [4], Crocombe and Adams [210], and others. The characteristics of the adhesion strength (including adhesive joint strength) were presented by Awaja et al. [130].

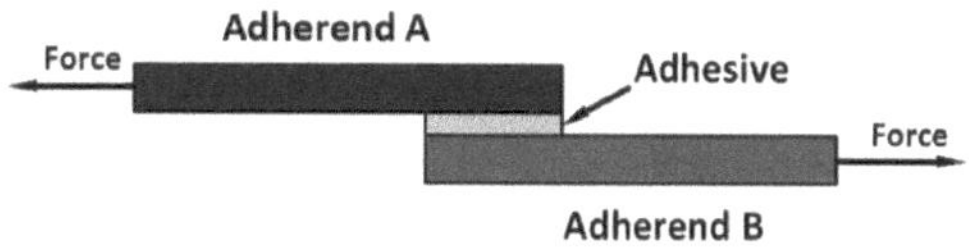

Fig. 9.15 Schematic representation of the shear test.

The terminology covers the principal terms relating to methods of mechanical testing of solids are presented in ASTM E6-15e1 standard [211] to encourage uniformity of terminology in product specifications. In this standard the general definitions are restricted and interpreted, when necessary, to make them particularly applicable and practicable for use in standards requiring or relating to mechanical tests.

9.5.1 Shear test

The shear test (Fig. 9.15) is one of the most commonly used strength tests for the determination of the shear strength of adhesive joints [17, 130, 149]. The true shear strength of an adhesive can be determined only if normal stresses are entirely absent. These conditions can be approached under special conditions, but not in single-lap specimens made with thin adherends, which are normally used in manufacturing and the most standard specimens [4]. The basic geometry of a thick adherends shear specimen and the description of the test of this type of adhesive joints are presented by Brockmann et al. [4].

With certain approximation, it represents the service life of a properly designed adhesive joint; however, it should be noted that it does not enable the quantitative comparison of the strength properties of adhesives of various modulus of rigidity. In addition, during the shear test, the bending moment, which causes the peel and deformation of the adherends and affects the final strength of adhesive joints, occurs as a result of the loss of the co-axiality of the joint (the adherends are not aligned in one plane) [2, 4].

The commonly used ASTM standard shear tests on adhesive include [4]:

- ASTM E229: Standard method of test for shear strength and shear modulus of structural adhesive,
- ASTM D1002: Standard method of test for strength properties of adhesives in shear by tension loading (metal-to-metal),
- ASTM D3165: Strength properties of adhesives in shear by tension loading of single-lap joint laminated assemblies,
- ASTM D5656: Thick-adherend metal lap shear joints for determination of stress-strain behaviour of adhesives in shear by tension loading,
- ASTM D3983: Standard test method for measuring strength and shear modulus of nonrigid adhesives by thick-adherend tensile-lap specimen.
- ASTM D905-08: Standard test method for strength properties of adhesive bonds in shear by compression loading,
- ASTM E229-97. Standard test method for shear strength and shear modulus of structural adhesives,
- DIN EN 1465:2009. Adhesives. Determination of tensile lap-shear strength of bonded assemblies,
- Others.

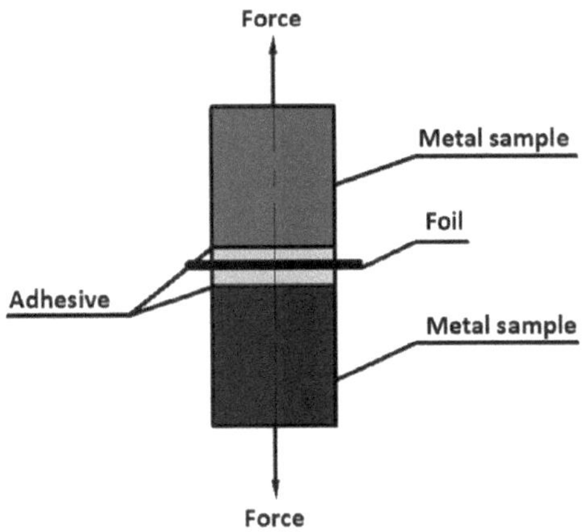

Fig. 9.16 Schematic representation of the pull-off test.

9.5.2 Pull-off test

The pull-off test for adhesion assessment in bonded and other adhesive joints (as per the PN-EN ISO 4624 standard [212] and ASTM 4541-02 standard [213]) consists in subjecting the bonded dolly assemblies to a destructive force (Fig. 9.16).

A characteristic feature of this method is the uneven stress distribution along the specimen's diameter: the highest stress occurs on the edges of the specimen, and the smallest in its central zone. During the test, the assembly is subjected to loads that lead to its destruction, during which the measurement of the force required to break it is taken and the maximum stress value is obtained.

The following factors influence the accuracy of the pull-off test [1, 2, 4, 31]:

- proper surface treatment of the flat face of the adherends—the surface of specimens should be prepared with utmost care to rid any contamination as well as to maintain the desired sample geometry (coaxial alignment of the dollies),
- the bending moment and the torque moment of the test sample should be eliminated,
- maintain uniform bond thickness over the entire surface,
- test speed (travel of the testing machine clamps) at a uniform rate throughout the entire test, and
- shape and dimensions of the cylindrical samples, especially the diameter to height ratio.

9.5.3 Peel tests

Peel tests are quite similar to cleavage tests, except that at last one of adherends is prepared from a flexible material which could be plastically deformed during the measurement [4].

First, there are the roller peel testing methods for bonded assemblies (and other adhesive joints) using centrifugal force, the peel-off is performed at an angle of 90 degrees, 2 × 90 degrees (T-peel test), or 180 degrees, and the joint is destroyed by the use of the floating roller or the climbing drum peel method. In the case with the methods mentioned above, their results depend mainly on the angle and the speed of peel: obviously, the increase in the angle and the speed of peel translates to increased peel action performance. The floating roller peel and the climbing drum peel methods provide a constant angle of peel throughout the test. Peel tests enable acquiring data on the resistance of welds to crack propagation.

90 degrees peel test

This 90 degrees peel strength determination test (in accordance with PN EN 28510-1 [214], EN 1939 [215], ASTM D3330/D3330M-4 [216], and others [217]) applies to the adhesive joint assembly of two adherends, of which at least one is flexible. The drawback of the method is that the typical tensile testing apparatus does not guarantee maintaining the constant 90 degrees peel angle. Therefore, if a constant angle of precisely 90 degrees is among the requirements of a given solution, a roller peel device [214] is employed instead. The determination of the peel strength at an angle of 90 degrees is particularly useful in the case of substrates exhibiting lower flexibility, whose peel strength cannot be determined at an angle of 180 degrees, for the risk of cracking, breaking or delamination [1, 4, 31, 130].

Awaja et al. [130] described the peel test and emphasised that the peel test is an excellent of an adhesion test whose values are only useful in a relative sense.

Crocombe and Adams [210] investigated the adhesive peel test and only elastic behaviour was considered. In this work both noncracked and cracked configurations were analysed, representing initial and continuous failure of the peel test. They emphasised that, inter alia, failure (propagation) showed to occur at a critical applied bending moment for a particular adherend and adhesive, independent of peel angle. Moreover, the strength measured by the peel test is not proportional to the actual strength of the adhesive, a small increase in the adhesive strength causing a much larger increase in the applied peel load.

2 × 90 degrees peel test (T-peel test)

The peel method for adhesion strength testing of joints (in accordance with PN EN ISO 11339 [218] ASTM D1876 [219]) consists in measuring the breaking force of two flexible specimens connected with a T-shaped adhesive joint (Fig. 9.17); the angle between the joint axis and the applied force axis is not determined and the elements are separated at an approximately constant speed, which is applied at the uncoated end of the joint so that the delamination would occur gradually along the joint axis [2, 4, 31, 163]. This method is applicable not only to metal substrates and may well be used for flexible adherends, whose bending at an angle of <90 degrees does not cause noticeable cracks or bends.

The assembled joints are fixed in the testing machine clamps and subjected to a tensile force, applied at a constant test speed, typically 100 mm/min. During the test, the pull-off force is measured, and the result of the measurement is the average pull-off force over the length of 150 mm of the test sample [31].

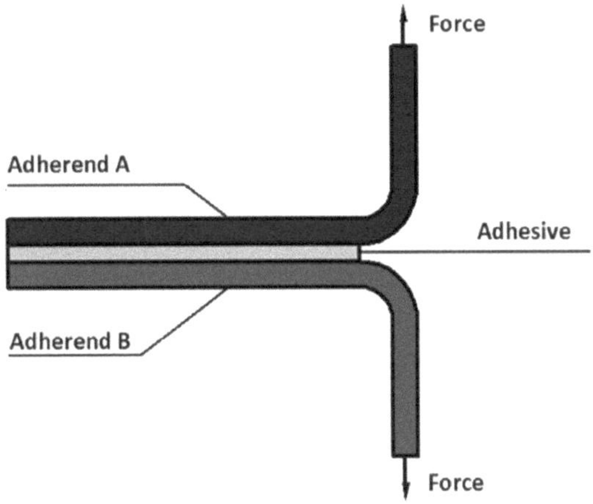

Fig. 9.17 Schematic representation of the 2×90 degrees peel test.

180 degrees peel test

The 180 degrees peel test (in accordance with PN-EN ISO 8510 [220] and ASTM D903 [221]) is applicable to testing bonded joints (and other adhesive joints) of rigid and flexible adherends, which can be bent to the angle of 180 degrees without causing cracks or fractures (Fig. 9.18) [4, 31].

The joint is assembled of the metal or polymeric element—the rigid component, and a thin polymeric material or a polymeric film—the flexible component. This method assesses the adhesive strength of both the adhesive and of paint coating to the polymer film.

Floating roller peel method

The floating roller peel method (PN-EN 1464 [222] and ASTM D3167 [223]) is designed to determine the peel strength of adhesive joints between a rigid and a flexible adherend, under specific conditions of specimen preparation and strict measurement procedure (Fig. 9.19) [2, 4]. The method, in contrast to other peeling tests mentioned above, enables carrying out the test at a bending angle of the flexible element by selecting the appropriate roller diameter, which is the specimen bending point. It is also possible to select the diameter of the roll to adjust the location of the cohesive failure in the joint. In the case a small-diameter roll, the cohesive failure occurs at the flexible element. However, if a larger-diameter roll with employed for testing, the failure will be located towards the centre of the bond. The test is performed at a constant speed of 152 mm/min [31].

It ought to be highlighted that the floating roller peel method ensures more repeatable measurement results than other peel tests [31].

Climbing drum peel method

The climbing drum peel method is applicable to test adhesive joint assemblies of a rigid and a flexible element (Fig. 9.20).

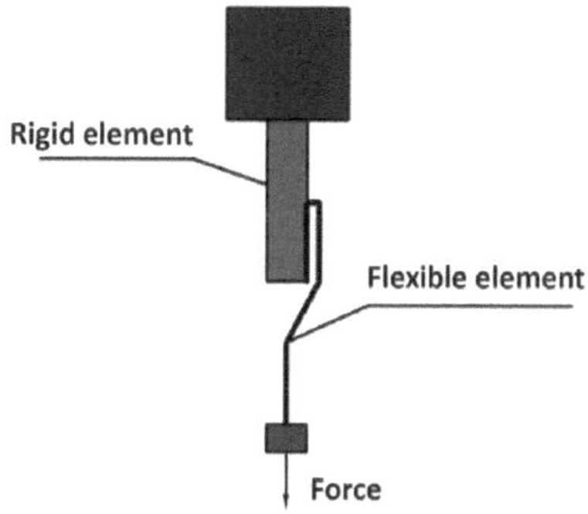

Fig. 9.18 Schematic representation of the 180 degrees peel test.
Source: Own work based on M. Żenkiewicz, Adhezja i modyfikowanie warstwy wierzchniej tworzyw wielkocząsteczkowych (Adhesion and Modification of the Surface Layer of Macromolecular Materials), WNT, Warszawa, 2000 (in Polish).

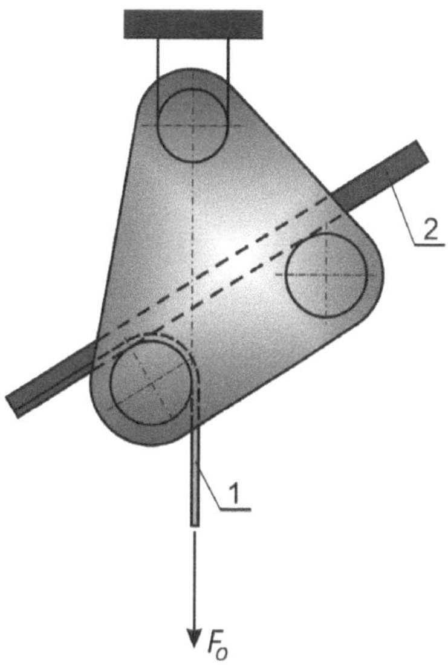

Fig. 9.19 Schematic representation of the floating roller peel method.
Source: Own work based on W. Brockmann, P.L. Geiß, J. Klingen, B. Schröder, Adhesive Bonding. Materials, Applications and Technology, Wiley-VCH Press, Weinheim, 2009; K.L. DeVries, D.O. Adams, Mechanical testing of adhesive joints, in: D.A. Dillard, A.V. Pocius (Eds.), The Mechanics of Adhesion, Adhesion Science and Engineering-1, Elsevier Science B.V., Amsterdam, 2002.

During the tests, the bending angle of the flexible element is constant. The measurements are carried out using a device equipped with a drum weighing 3.6 kg and an external radius of 51 ± 0.13 mm. The drum is equipped with two flanges, to which the tapes transmitting the force are attached. The radius of these flanges together with half the thickness of the strip is greater by 12.7 ± 0.13 mm than the radius of the drum. In addition, the drum is provided with a fixture mechanism that secures the end of the

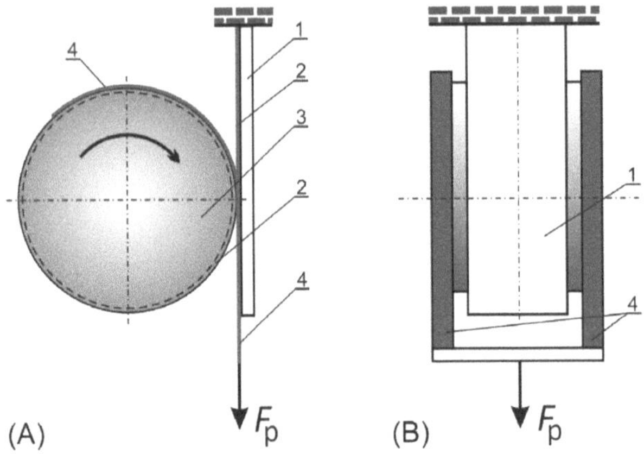

Fig. 9.20 Schematic representation of the climbing drum peel testing fixture: (A) front view, (B) side view, (1) rigid layer, (2) flexible layer, (3) drum, (4) external force conveyor tapes, F_p—external force.
Source: Own work based on M. Żenkiewicz, Adhezja i modyfikowanie warstwy wierzchniej tworzyw wielkocząsteczkowych (Adhesion and Modification of the Surface Layer of Macromolecular Materials), WNT, Warszawa, 2000 (in Polish).

flexible joint component. The external force applies the torque which in turn causes the drum to roll on the specimen in a specific direction, simultaneously detaching the flexible adherend from the rigid body [31].

The breadth of the information provided in this Chapter exemplifies the wide variety of methods for the determination of the strength of bonds and bonded joints, including adhesive joints. The selection of the presented tests for the assessment of adhesive strength depends on many factors; however, the design of the joint and the type of substrates building the assembly are important beyond comparison. It should also be noted that the adhesive strength determined using different methods may show certain discrepancies and therefore in many cases cannot be compared. It is for this very reason that prior to testing, a thorough analysis of the properties of the analysed joint is due.

References

[1] R.D. Adams (Ed.), Adhesive Bonding. Science, Technology and Applications, Woodhead Publishing Limited, Cambridge, 2010.
[2] R.D. Adams, J. Comyn, W.C. Wake, Structural adhesive joints in engineering, second ed., Chapman & Hall, London, 1997.
[3] C.V. Cagle, Handbook of Adhesive Bonding, McGraw-Hill, New York, 1973.
[4] W. Brockmann, P.L. Geiß, J. Klingen, B. Schröder, Adhesive Bonding. Materials, Applications and Technology, Wiley-VCH Press, Weinheim, 2009.
[5] L.-H. Lee (Ed.), Adhesive Bonding, first ed., Springer Science + Business Media, New York, 1991.
[6] A. Pizzi, K.L. Mittal (Eds.), Handbook of Adhesive Technology, third ed., CRC Press, Boca Raton, FL, 2008.

[7] A.V. Pocius, Adhesion and Adhesives Technology: An Introduction, Carl Hanser, Munich, 2002.
[8] K.M. Hay, M.I. Dragila, J. Liburdy, Theoretical model for the wetting of a rough surface, J. Colloid Interface Sci. 325 (2008) 472–477.
[9] R.D. Hazlett, On surface roughness effects in wetting phenomena, J. Adhes. Sci. Technol. 6 (1992) 625–633.
[10] R.D. Hazlett, On surface roughness effects in wetting phenomena, in: K.L. Mittal (Ed.), Contact Angle, Wettability and Adhesion, VSP, Zeist, 1993, pp. 173–181.
[11] S.G. Prolongo, G. del Rosario, A. Ureña, Study of the effect of substrate roughness on adhesive joints by SEM image analysis, J. Adhes. Sci. Technol. 20 (2006) 457–470.
[12] S. Ebnesajjad, Adhesives Technology Handbook, second ed., William Andrew, New York, 2008.
[13] D.E. Packham, Surface energy, surface topography and adhesion, Int. J. Adhes. Adhes. 23 (2003) 437–448.
[14] S.G. Prolongo, A. Ureña, Effect of surface pre-treatment on the adhesive strength of epoxy-aluminium joints, Int. J. Adhes. Adhes. 29 (2009) 23–31.
[15] D.E. Packham (Ed.), Handbook of Adhesion, Longman, New York, 1992.
[16] L.F.M. da Silva, N.M.A.J. Ferreira, V. Richter-Trummer, E.A.S. Marques, Effect of grooves on the strength of adhesively bonded joints, Int. J. Adhes. Adhes. 30 (2010) 735–743.
[17] L.F.M. da Silva, R.J.C. Carbas, G.W. Critchlow, M.A.V. Figueiredo, K. Brown, Effect of material, geometry, surface treatment and environment on the shear strength of single lap joints, Int. J. Adhes. Adhes. 29 (2009) 621–632.
[18] G.W. Critchlow, D.M. Brewis, Review of surface pretreatments for alluminium alloys, Int. J. Adhes. Adhes. 16 (1996) 255–275.
[19] W.M. Khrulev, Surface roughness and rheological properties of adhesives as factors determining optimal thickness of glue line, Polym. Mech. (Mekhanika Polimerov) 6 (1965) 103–107.
[20] A.D. Sommers, A.M. Jacobi, Wetting phenomena on micro-grooved aluminium surfaces and modelling of the critical droplet size, J. Colloid Interface Sci. 328 (2008) 402–411.
[21] A.P. Serro, R. Colaço, B. Saramago, Adhesion forces in liquid media: effect of surface topography and wettability, J. Colloid Interface Sci. 325 (2008) 573–579.
[22] A. Rudawska, I. Danczak, M. Müller, P. Valasek, The effect of sandblasting on surface properties for adhesion, Int. J. Adhes. Adhes. 70 (2016) 176–190.
[23] J.J. Bikerman, Surface roughness and contact angles, J. Phys. Colloid Chem. 54 (1950) 653–658.
[24] S.J. Hitchcock, N.T. Caroll, M.G. Nicholas, Some effects of substrate roughness on wettability, J. Mater. Sci. 16 (1981) 714–732.
[25] A.J. Kinloch, Adhesion and Adhesives, Chapman & Hall, New York, 1987.
[26] J.M. Sykes, Surface treatment for steel, in: D.M. Brewis (Ed.), Surface Analysis and Pretreatment of Plastics and Metals, Applied Science Publishers, London, 1982, pp. 153–174.
[27] C. Keisler, J.L. Lataillade, The effect of substrate roughness characteristics on wettability and on the mechanical properties of adhesive joints loaded at high strain rates, J. Adhes. Sci. Technol. 9 (1995) 395–411.
[28] M.S. Islam, L. Tong, P.J. Falzon, Influence of metal surface preparation on its surface profile, contact angle, surface energy and adhesion with glass fibre prepreg, Int. J. Adhes. Adhes. 51 (2014) 32–41.
[29] C.W. Extrand, Y. Kumagai, An experimental study of contact angle hysteresis, J. Colloid Interface Sci. 191 (1997) 378–383.

[30] A. Baldan, Adhesion phenomena in bonded joints, Int. J. Adhes. Adhes. 38 (2012) 95–116.
[31] M. Żenkiewicz, Adhezja i modyfikowanie warstwy wierzchniej tworzyw wielkocząsteczkowych (Adhesion and Modification of the Surface Layer of Macromolecular Materials), WNT, Warszawa, 2000 (in Polish).
[32] R.N. Wenzel, Resistance of solid surfaces to wetting by water, Ind. Eng. Chem. 28 (1936) 988–994.
[33] W.C. Wake, Adhesion and the Formulation of Adhesives, second ed., Applied Science Publishers Ltd, Essex/New York, 1982.
[34] S. Vedantam, M.V. Panchagnula, Constitutive modeling of contact angle hysteresis, J. Colloid Interface Sci. 321 (2008) 393–400.
[35] I.S. Bayer, C.M. Megaridis, J. Zhang, D. Gamota, A. Biswas, Analysis and surface energy estimation of various polymeric surfaces using contact angle hysteresis, J. Adhes. Sci. Technol. 21 (2007) 1439–1467.
[36] V.S. Ajaev, T. Gambaryan-Roisman, P. Stephan, Static and dynamic contact angles of evaporating liquids on heated surfaces, J. Colloid Interface Sci. 342 (2010) 550–558.
[37] Y.L. Chen, C.A. Helm, J.N. Israelachvili, Molecular mechanisms associated with adhesion and contact angle hysteresis of monolayer surfaces, J. Phys. Chem. 95 (1991) 10736–10747.
[38] A.W. Neumann, R.J. Good, Thermodynamics of contact angles. I. Heterogeneous solid surfaces, J. Colloid Interface Sci. 38 (1972) 341–358.
[39] M.E.R. Shanahan, Sessile drops on heterogeneous surfaces: static and dynamic behavior, in: K.W. Allen (Ed.), Adhesion 15, 28th Annual Conference on Adhesion and Adhesives, City University, Elsevier, London, Amsterdam, pp. 116–130, 1990.
[40] R.S. Faibish, W. Yoshida, Y. Cohen, Contact angle study on polymer-grafted silicon wafers, J. Colloid Interface Sci. 256 (2002) 341–350.
[41] E. Chibowski, F. González-Caballero, Interpretation of contact angle hysteresis, J. Adhes. Sci. Technol. 7 (1993) 1195–1209.
[42] R.J. Good, A Thermodynamic derivation of Wenzel's modification of Young's equation for contact angles: together with a theory of hysteresis, J. Am. Chem. Soc. 74 (1952) 5041–5042.
[43] K. Uehara, M. Sakurai, Bonding strength of adhesives and surfaces roughness of joined parts, J. Mater. Process. Technol. 127 (2002) 178–181.
[44] S. Wu, Polymer Interface and Adhesion, CRS Press, 1982, pp. 1–26. (Chapter 1).
[45] D.M. Brewis, D. Briggs (Eds.), Industrial Adhesion Problems, Orbital Press, 1985, pp. 57–59 (Chapter 3.3).
[46] S.A. McCarthy, Dynamic contact angle analysis and its application to paste PVC products, Polimery 43 (1998) 314–319.
[47] L. Xu, H. Fan, C. Yang, W.M. Huang, Contact line mobility in liquid droplet spreading on rough surface, J. Colloid Interface Sci. 323 (2008) 126–132.
[48] A.F. Harris, A. Beevers, The effect of grit-blasting on the surface properties for adhesion, Int. J. Adhes. Adhes. 19 (1999) 445–452.
[49] M. Shahid, S.A. Hashim, Effect of surface roughness on the strength of cleavage joint, Int. J. Adhes. Adhes. 22 (2002) 235–244.
[50] N.A. De Bruyne, The adhesive properties of epoxy resins, J. Appl. Chem. 6 (1956) 303–310.
[51] H.M.S. Iqbal, S. Bhowmik, R. Benedictus, Study on the effect of surface morphology on adhesion properties of polybenzimidazole adhesive bonded composite joints, Int. J. Adhes. Adhes. 72 (2017) 43–50.

[52] D.J. Arrowsmith, Adhesion of electroformed copper and nickel to plastic laminates, Trans. Inst. Met. Finish. 48 (1970) 88–92.
[53] A. Rudawska, Selected aspects of the effect of mechanical treatment on surface roughness and adhesive joint strength of steel sheets, Int. J. Adhes. Adhes. 50 (2014) 235–243.
[54] D.N. States, K.L. deVries, Geometric factors impacting adhesive lap joint strength and design, Int. J. Adhes. Adhes. 26 (2012) 89–107.
[55] G. Bresson, J. Jumel, M.E.R. Shanahan, P. Serin, Strength of adhesively bonded joints under mixed axial and shear loading, Int. J. Adhes. Adhes. 35 (2012) 27–35.
[56] X. Jiang, X. Qiang, M.H. Kolstein, F.S.K. Bijlaard, Experimental investigation on mechanical behavior of FRP-to-steel adhesively-bonded joint under combined loading—Part 2: After hygrothermal ageing, Compos. Struct. 125 (2015) 687–697.
[57] H. Moghadamzadeh, H. Rahimi, M. Asadollahzadeh, A.R. Hemmati, Surface treatment of wood polymer composites for adhesive bonding, Int. J. Adhes. Adhes. 31 (2011) 816–821.
[58] PN-EN ISO 3274. Specyfikacja geometrii wyrobów (GPS). Struktura geometryczna powierzchni: Metoda profilowa. Charakterystyki nominalne przyrządów stykowych (z ostrzem odwzorowującym) Product geometry specification (GPS). Geometric structure of the surface: Profile method. Nominal characteristics of contact devices (with a mating blade) (in Polish).
[59] International Slurry blasting Standards, http://www.materialsperformance.com/articles/coating-linings/2016/03/nace-sspc-publish-joint-wet-abrasive-blast-standards/, 2018. (Accessed November 15, 2018).
[60] Z. Humienny, P.H. Osanna, M. Tamre, A. Weckenmann, L. Blunt, W. Jakubiec, Specyfikacja geometrii wyrobów (GPS) (Product Geometry Specification (GPS)), WNT, Warszawa, 2004 (in Polish).
[61] PN-EN ISO 16610-1:2015-08. Specyfikacje geometrii wyrobów (GPS). Filtrowanie. Część 1: Przegląd i podstawowe pojęcia.
[62] PN-EN ISO 25178-1. Specyfikacje geometrii wyrobów (GPS). Struktura geometryczna powierzchni: Przestrzenna. Część 1: Oznaczanie struktury geometrycznej powierzchni (Specifications of product geometry (GPS). Geometric structure of the surface: Spatial. Part 1: Determination of the geometric structure of the surface) (in Polish).
[63] PN-EN ISO 25178. Specyfikacje geometrii wyrobów (GPS). Struktura geometryczna powierzchni: Przestrzenna. Część 2: Terminy, definicje i parametry struktury geometrycznej powierzchni Specifications of product geometry (GPS). Geometric structure of the surface: Spatial. Part 2: Terms, definitions and parameters of the geometric structure of the surface (in Polish).
[64] PN-EN ISO 25178-601:2010. Specyfikacje geometrii wyrobów (GPS). Struktura geometryczna powierzchni: Przestrzenna. Część 601: Charakterystyki nominalne przyrządów stykowych (z ostrzem odwzorowującym) (Specifications of product geometry (GPS). Geometric structure of the surface: Spatial. Part 601: Nominal characteristics of contact devices (with a mating blade)) (in Polish).
[65] PN-EN ISO 4287. Specyfikacje geometrii wyrobów. Struktura geometryczna powierzchni: metoda profilowa. Terminy, definicje i parametry struktury geometrycznej powierzchni (Specifications of product geometry. Geometric structure of the surface: profile method. Terms, definitions and parameters of the geometric structure of the surface) (in Polish).
[66] PN-EN ISO 4288. Specyfikacje geometrii wyrobów (GPS). Struktura geometryczna powierzchni: Metoda profilowa. Zasady i procedury oceny struktury geometrycznej powierzchni (Specifications of product geometry (GPS). Geometric structure of the surface: Profile method. Principles and procedures for the assessment of the geometric structure of the surface) (in Polish).

[67] Roughness measurements of stainless steel surfaces, http://www.worldstainless.org/Files/issf/non-image-files/PDF/Euro_Inox/RoughnessMeasurement_EN.pdf, 2018. (Accessed November 10, 2018).
[68] W. Grzesik, Wpływ topografii powierzchni na właściwości eksploatacyjne części maszyn (The influence of surface topography on the operational properties of machine parts), Mechanik 8–9 (2015) 587–593 (in Polish).
[69] E.S. Gadelmawla, M.M. Koura, T.M.A. Maksoud, I.M. Elewa, H.H. Soliman, Roughness parameters, J. Mater. Process. Technol. 123 (2002) 133–145.
[70] Z. Lipa, D. Tomaníčková, Choosing the most appropriate mathematical model to approximate the Abbott curve, https://pdfs.semanticscholar.org/226d/88946eaf7de79dd4a9df871b0f78780460d9.pdf/, 2018. (Accessed November 10, 2018).
[71] A.A. Adamczak, Pomiary geometryczne powierzchni (Geometric Measurements of the Surface), WNT, Warszawa, 2008 (in Polish).
[72] A. Rudawska, E. Jacniacka, Ocena przydatności parametrów chropowatości powierzchni do określenia wytrzymałości połączeń klejowych (Evaluation of the usefulness of surface roughness parameters to determine the strength of adhesive joint), Przegląd Mechaniczny (2005) 119–122. 9S. (in Polish).
[73] M. Sedlacek, B. Podgornik, J. Vizitin, Influence of surface preparation on roughness parameters, friction and wear, Wear 266 (2009) 482–487.
[74] B. Nowicki, Struktura geometryczna powierzchni. Chropowatość i falistość powierzchni (Geometric structure of the surface. Surface roughness and waviness), WNT, Warszawa, 1991 (in Polish).
[75] R. Bartos, A. Gorog, Determination of parameter Ra for the methods of finishing machining, http://www.mmscience.eu/archives/MM_Science_201307.pdf/, 2018. (Accessed November 10, 2018).
[76] A. Rudawska, Przygotowanie powierzchni do klejenia wybranych materiałów konstrukcyjnych (Surface treatment of selected contructional materials for bonding), Wydawnictwa Uczelniane, Lublin, 2017 (in Polish).
[77] K.L. DeVries, D.O. Adams, Mechanical testing of adhesive joints, in: D.A. Dillard, A.V. Pocius (Eds.), The Mechanics of Adhesion, Adhesion Science and Engineering-1, Elsevier Science B.V, Amsterdam, 2002.
[78] L.H. Lee, Roles of molecular interactions in adhesion, adsorption, contact angle and wettability, J. Adhes. Sci. Technol. 7 (1993) 538–634.
[79] K.L. Mittal, Interfacial chemistry and adhesion: recent developments and prospects, Pure Appl. Chem. 52 (1980) 1295–1305.
[80] M. Zielecka, Methods of contact angle measurement as a tool for characterization of wettability of polymers, Polimery 49 (2004) 327–332.
[81] K. Katoh, Contact angle and surface tension measurement, in: S. Hartland (Ed.), Surface and Interfacial Tension, Measurement, Theory and Applications, CRC Press, Boca Raton, FL, 2004.
[82] W. Eckert, Improvement of adhesion on polymer film, foil and paperboard by flame treatment, in: Proceedings of TAPPI European PLACE Conference, 2003.
[83] B. Duncan, R. Mera, D. Leatherdale, M. Taylor, R. Musgrove, Techniques for characterising the wetting, coating and spreading of adhesives on surfaces, NPL Report, DEPC MRP 020, 2005.
[84] A.W. Momber, A.G. Muehlhan, M. Irmer, N. Glück, A note on the distribution of stationary contact angles on organic coatings, Int. J. Adhes. Adhes. 72 (2017) 1–5.

[85] A. Depalo, A.C. Santomaso, Wetting dynamics and contact angle of powders studied through capillary rise experiments, Colloids Surf. A Physicochem. Eng. Asp. 436 (2013) 371–379.
[86] A. Alghunaim, S. Kirdponpattara, B.-M.Z. Newby, Techniques for determining contact angle and wettability of powders, Powder Technol. 287 (2016) 201–215.
[87] Y. Yuan, T.R. Lee, Contact angle and wetting properties, Springer Series, Surface Science, vol. 51, 2013, pp. 3–34.
[88] J. Comyn, Diffusion and adhesion, in: J. Comyn (Ed.), Polymer Permeability, Chapman and Hall, London, 1985, pp. 177–215 (Chapter 5).
[89] D. Kwok, A. Neumann, Contact angle interpretation in term of solid surface tension, Colloids Surf. A Physicochem. Eng. Asp. 161 (2000) 31–48.
[90] D. Kwok, A.W. Neumann, Contact angle measurement and contact angle interpretation, Adv. Colloid Interf. Sci. 81 (1999) 167–249.
[91] J. Chinnam, D. Das, R. Vajjha, J. Satti, Measurements of the contact angle of nanofluids and development of a new correlation, Int. Commun. Heat Mass Transf. 62 (2015) 1–12.
[92] J. Comyn, Contact angels and adhesive bonding, Int. J. Adhes. Adhes. 12 (1992) 145–149.
[93] P.G. De Gennes, Wetting: statics and dynamics, Rev. Mod. Phys. 57 (1984) 827–863.
[94] M.G. Norton, The influence of contact angle, wettability, and reactivity on the development of indirect-bonded metallizations for aluminum nitride, J. Adhes. Sci. Technol. 6 (1992) 635–651.
[95] F.M. Fowkes, D.C. Mc Carthy, M.A. Mostafa, Contac angles and the equilibrium spreading pressures of liquids on hydrophobic solids, J. Colloid Interface Sci. 78 (1980) 200–206.
[96] C.W. Extrand, Uncertainty in contact angle measurements from the tangent method, J. Adhes. Sci. Technol. 30 (2016) 1597–1601.
[97] S. Ebnesajjad, A.H. Landrock, Adhesives Technology Handbook, William Andrew, London, 1985.
[98] E. Lugscheider, K. Bobzin, The influence on surface free energy of PVD-coatings, Surf. Coat. Technol. 142–144 (2001) 755–760.
[99] F. Barberis, M. Capurro, Wetting in the nanoscale: a continuum mechanics approach, J. Colloid Interface Sci. 326 (2008) 201–210.
[100] P. Dalet, E. Papon, J.-J. Villenave, Surface free energy of polymeric materials: relevancy of conventional contact angle data analyses, J. Adhes. Sci. Technol. 13 (1999) 857–870.
[101] X. Qin, W.V. Chang, The role of interfacial free energy in wettability, solubility, and solvent crazing of some polymeric solids, J. Adhes. Sci. Technol. 10 (1996) 963–987.
[102] A. Rudawska, E. Jacniacka, Analysis of determining surface free energy uncertainty with the Owens-Wendt method, Int. J. Adhes. Adhes. 29 (2009) 451–457.
[103] A. Rudawska, E. Jacniacka, Evaluating uncertainty of surface free energy measurement by the van Oss-Chaudhury-Good method, Int. J. Adhes. Adhes. 82 (2018) 139–145.
[104] D. Shaw, Colloid and Surface Chemistry, fourth ed., Butterworth-Heinemann, 1992, pp. 151–159. (Chapter 6).
[105] C.J. Van Oss, L. Ju, M.K. Chaudhury, R.J. Good, Estimation of polar parameters of the surface tension of liquids by contact angle measurements on gels, J. Colloid Interface Sci. 128 (1989) 313–319.
[106] S. Vafaei, M.Z. Podowski, Analysis of the relationship between liquid droplet size and contact angle, Adv. Colloid Interf. Sci. 113 (2005) 133–146.
[107] M. Morra, E. Occhiello, F. Garbassi, Knowledge about polymer surfaces from contact angle measurements, Adv. Colloid Interf. Sci. 32 (1990) 79–116.

[108] S.-J. Park, M.-S. Cho, J.-R. Lee, Studies on the surface free energy of carbon-carbon composites: effect of filler addition on the ILSS of composites, J. Colloid Interface Sci. 226 (2000) 60–64.
[109] V. Fernández, M. Khayet, Evaluation of the surface free energy of plant surfaces: toward standardizing the procedure. Front. Plant Sci. 5 (510) (2015) https://doi.org/10.3389/fpls.2015.00510.
[110] S. Ahadian, M. Mohseni, S. Morawian, Ranking proposed models for attaining surface free energy of powders using contact angle measurements, Int. J. Adhes. Adhes. 29 (2009) 458–469.
[111] G.Z. Xiao, Effects of solvent on the surface properties of oxygen plasma-treatment polyethylene and polypropylene films, J. Adhes. Sci. Technol. 11 (1997) 655–663.
[112] J. Shang, M. Flury, J.B. Harsh, R.L. Zollars, Comparison of different methods to measure contact angles of soil colloids, J. Colloid Interface Sci. 328 (2008) 229–307.
[113] J.M. Schuster, C.E. Schvezov, M.R. Rosenberger, Influence of experimental variables on the measure contact angle in metals using the sessile drop method, Procedia Mater. Sci. 8 (2015) 742–751.
[114] M. Schmitt, F. Heib, High-precision drop shape analysis on inclining flat surfaces: Introduction and comparison of this special method with commercial contact angle analysis, J. Chem. Phys. 139 (2013) 134–201.
[115] R.J. Good, Contact angle, wetting, and adhesion: a critical review, J. Adhes. Sci. Technol. 6 (1992) 1269–1302.
[116] M. Schmitt, R. Hempelmann, S. Ingebrandt, W. Munief, D. Durneata, F. Groß, F. Heib, Statistical approach for contact angle determination on inkling surfaces: slow-moving: analyses of non-axisymmetric drops on a flat silanozed silicon wafer, Int. J. Adhes. Adhes. 55 (2014) 123–131.
[117] F.-C. Wang, J.-T. Feng, Y.-P. Zhao, The head-on colliding process of binary liquid droplets at low velocity: high-speed photography experiments and modeling, J. Colloid Interface Sci. 326 (2008) 196–200.
[118] D.E. Packham, Work of adhesion: contact angles and contact mechanics, Int. J. Adhes. Adhes. 16 (1996) 121–128.
[119] D. Janssen, R. De Palma, S. Verlaak, P. Heremans, W. Dehaen, Static solvent contact angle measurements, surface free energy and wettability determination of various self-assembled monolayers on silicon dioxide, Thin Solid Films 515 (2006) 1433–1438.
[120] M.E. Diaz, J. Fuentes, R.L. Cerro, M.D. Savage, Hysteresis during contact angles measurement, J. Colloid Interface Sci. 343 (2010) 574–583.
[121] M.H.V.C. Adão, B.J. Saramago, A.C. Fernandes, Estimation of the surface properties of styrene-acrylonitrile random copolymers from contact angle measurements, J. Colloid Interface Sci. 217 (1999) 319–328.
[122] C. Della Volpe, S. Siboni, Analysis of dynamic contact angle on discoidal samples measured by the Wilhelmy method, J. Adhes. Sci. Technol. 12 (1998) 197–224.
[123] A. Rudawska, Adhesive properties, in: V. Kazmiruk (Ed.), Scanning Electron Microscope, InTech Publisher, Rijeka, 2012, pp. 101–126 (Chapter 7).
[124] J.M. Schuster, C.E. Schvezov, M.R. Rosenberger, Analysis of the results of surface free energy measurement of Ti6Al4V by different methods, Process. Mater. Sci. 8 (2015) 732–741.
[125] ASTM D 5946-04. Standard Test Method for Corona-Treated Polymer Films Using Water Contact Angle Measurements.
[126] ASTM D 724-99. Standard test method for surface wettability of paper (Angle-of-contact method).

[127] E. Chibowski, L. Hołysz, On the use of Washburn's equation for contact angle determination, J. Adhes. Sci. Technol. 11 (1997) 1289–1301.
[128] L. Galet, S. Patry, J. Dodds, Determination of the wettability of powders by the Washburn capillary rise method with bed preparation by a centrifugal packing technique, J. Colloid Interface Sci. 346 (2010) 470–475.
[129] M.V. Hammes, E.S. Heberle, P.R. da Silva, C.P.Z. Noreña, A.H. Englert, N.S.M. Cardozo, Mathematical modeling of the capillary rise of liquids in partially soluble particle beds, Powder Technol. 325 (2018) 21–30.
[130] F. Awaja, M. Gilbert, G. Kelly, B. Fox, P.J. Pigram, Adhesion of polymers, Prog. Polym. Sci. 34 (2009) 948–968.
[131] C.W. Extrand, A thermodynamic model for contacts angle hysteresis, J. Colloid Interface Sci. 207 (1998) 11–19.
[132] J.D. Eick, R.J. Good, A.W. Neumann, Thermodynamics of contact angles. II. Rough solid surfaces, J. Colloid Interface Sci. 53 (1975) 235–248.
[133] J.F. Joanny, P.G. de Gennes, A model for contact angle hysteresis, J. Chem. Phys. 81 (1984) 552–562.
[134] K. Yamamoto, S. Ogata, 3-D thermodynamic analysis of superhydropbic surface, J. Colloid Interface Sci. 326 (2008) 471–477.
[135] F.M. Fowkes, Determination of interfacial tensions, contact angles, and dispersion forces in surfaces by assuming additivity of intermolecular interactions in surfaces, J. Phys. Chem. 66 (1962) 382.
[136] S. Shael-Levanon, A. Marmur, Validity and accuracy in evaluating surface tension of solid by additive approaches, J. Colloid Interface Sci. 262 (2003) 489–499.
[137] M. Żenkiewicz, J. Gołębiewski, S. Lutomirski, Doświadczalna weryfikacja niektórych elementów metody van Ossa-Gooda (Experimental verification of some elements of the van Oss-Good method), Polimery 44 (1999) 212–217 (in Polish).
[138] T. Tanaka, J. Lee, P.R. Scheller, Interfacial free energy and wettability, Treatise on Process Metallurgy, Volume 2, Elsevier Ltd, 2014 (Chapter 1.5). doi:https://doi.org/10.1016/B978-0-08-096984-8.00025-2, 2018 (Accessed 13 November 2018).
[139] C.J. van Oss, R.J. Good, M.K. Chaudhury, Additive and nonadditive surface tension components and the interpretation of contact angles, Langmuir 4 (1998) 884–891.
[140] F. Hejda, P. Solař, J. Kousal, Surface free energy determination by contact angle measurements—a comparison of various approaches, in: WDS'10 Proceedings of Contributed papers, Part III, 2010, pp. 25–30.
[141] Z.Z. You, J.Y. Dong, Effect of oxygen plasma treatment on the surface properties often-doped indium oxide substrates for polymer LEDs, J. Colloid Interface Sci. 300 (2006) 697–703.
[142] M.-C. Michalski, J. Hardy, B.J.V. Saramago, On the surface free energy of PVC/EVA polymer blends: comparison of different calculation methods, J. Colloid Interface Sci. 208 (1998) 319–328.
[143] L. Hołysz, Surface free energy components of silica gel determined by the thin layer wicking method for different layer thicknesses of gel, J. Mater. Sci. 30 (1998) 445–452.
[144] L. Hołysz, Investigation of the effect of substrata on the surface free energy components of silica gel determined by thin layer wicking method, J. Mater. Sci. 35 (2000) 6081–6091.
[145] Y.-L. Hsieh, S. Xu, M. Hartzell, Effects of acid oxidation on wetting and adhesion properties of ultra-high modulus and molecular weight polyethylene (UHMWPE) fibers, J. Adhes. Sci. Technol. 5 (1991) 1023–1039.

[146] B. Jańczuk, T. Białopiotrowicz, A. Zdziennicka, Some remarks on the components of the liquid surface free energy, J. Colloid Interface Sci. 211 (1999) 96–103.
[147] A. Burdzik, M. Stähler, M. Carmo, D. Stolten, Impact of reference values used for surface free energy determination: an uncertainty analysis, Int. J. Adhes. Adhes. 82 (2018) 1–7.
[148] M. Greiveldinger, M.E.R. Shanahan, A critique of the mathematical coherence of acid/base interfacial free energy theory, J. Colloid Interface Sci. 215 (1999) 170–178.
[149] C. Sperandio, J. Bardon, A. Laachachi, H. Aubiet, D. Ruch, Influence of plasma surface treatment on bond strength behaviour of an adhesively bonded aluminium-epoxy system, Int. J. Adhes. Adhes. 30 (2010) 720–728.
[150] C. Della Volpe, D. Maniglio, M. Brugnara, S. Siboni, M. Morra, The solid surface free energy calculation I. In defense of the multicomponent approach, J. Colloid Interface Sci. 271 (2004) 434–453.
[151] Z. Yao, J.Y.Y. Heng, S. Lanceros-Méndes, A. Pegoretti, M. Xia, J. Tang, W. Wu, Surface free energy and mechanical performance of LDPE/CGBF composites containing toxic-metal free filler, Int. J. Adhes. Adhes. 77 (2017) 58–62.
[152] C.D. Volpe, S. Siboni, Some reflections on acid-base solid surface free energy theories, J. Colloid Interface Sci. 195 (1997) 121–136.
[153] M. González-Martín, L. Labajos-Broncano, B. Jańczuk, J.M. Bruque, Wettability and surface free energy of zirconia ceramics and their constituents, J. Mater. Sci. 341 (1999) 5923–5926.
[154] A. Rudawska, J. Kuczmaszewski, Surface free energy of zinc coating after finishing treatment, Mater. Sci.-Poland 24 (2006) 975–981.
[155] M. Mohseni, M. Mirabedini, M. Hashemi, G.E. Thompson, Adhesion performance of an epoxy clear coat on aluminum alloy in the presence of vinyl and amino-silane primers, Prog. Org. Coat. 57 (2006) 307–313.
[156] W. Polini, L. Sorrentino, Adhesion of a protective coating on a surface of aluminium alloy treated by air cold plasma, Int. J. Adhes. Adhes. 27 (2007) 1–8.
[157] A. Rudawska, I. Danczak, Adhesive properties and adhesive joint strength of aluminium alloys following surface pre-treatment, Recent Pat. Mech. Eng. 6 (2013) 152–165.
[158] P.A. Thompson, W.B. Brinckerhoff, M.O. Robbins, Microscopic studies of static and dynamic contact angles, J. Adhes. Sci. Technol. 7 (1993) 535–554.
[159] N. Zouvelou, X. Mantzouris, P. Nikolopoulos, Interfacial energies in oxide/liquid metal systems with limited solubility, Int. J. Adhes. Adhes. 27 (2007) 380–386.
[160] W.B. Jansen, Overview lecture. The Lewis acid-base concepts: recent results and prospects for the future, J. Adhes. Sci. Technol. 5 (1991) 1–21.
[161] M. Żenkiewicz, New method of analysis of the surface free energy of polymeric materials calculated with Owens-Wendt and Neumann methods, Polimery 51 (2006) 584–587.
[162] Z. Yao, J.Y.Y. Heng, S. Lanceros-Méndez, A. Pegoretti, M. Xia, J. Tang, W. Wu, Surface free energy and mechanical performance of LDPE/CBF composites containing toxic metal free filler, Int. J. Adhes. Adhes. 77 (2017) 58–62.
[163] C. Spadaro, C. Dispenza, C. Sunseri, The influence of the nature of the surface oxide on the adhesive fracture energy of aluminium-bonded joint as measured by T-peel tests, Int. J. Adhes. Adhes. 28 (2008) 211–221.
[164] N.T. Correia, J.J. Ramos, B.J.V. Saramago, J.C.G. Calado, Estimation of the surface-tension of a solid—application to a liquid-crystalline polymer, J. Colloid Interface Sci. 189 (1997) 361–369.
[165] K.L. Mittal, The role of the interface in adhesion phenomena, Polym. Eng. Sci. 17 (1997) 467–473.

[166] D.S. Rimal, L.P. Demejo, R.C. Bowen, Mechanics of particle adhesion, J. Adhes. Sci. Technol. 8 (1994) 1333–1355.

[167] M.K. Chaudhury, G.M. Whitesides, Direct measurement of interfacial interactions between semispherical lenses and flat sheets of poly(dimethylsiloxane) and their chemical derivatives, Langmuir 7 (1991) 1013–1025.

[168] J. Drelich, The significance and magnitude of the line tension in three-phase (solid-liquid-fluid) system, Colloids Surf. A Physicochem. Eng. Asp. 116 (1996) 43–54.

[169] D.H. Kaelble, Dispersion-polar surface tension properties of organic solids, J. Adhes. 2 (1970) 66–81.

[170] M. Żenkiewicz, Analiza głównych uwarunkowań metody van Ossa-Chaunhury'ego-Gooda w badaniach warstwy wierzchniej materiałów polimerowych (Analysis of the main determinants of the van Oss-Chaunhury-Good method in the study of the surface layer of polymeric materials), Polimery 51 (2006) 169–176 (in Polish).

[171] B. Jańczuk, T. Białopiotrowicz, Swobodna energia powierzchniowa niektórych polimerów (Surface free energy of some polymers), Polimery 32 (1987) 269–271 (in Polish).

[172] A.J. Kinloch, G.K.A. Kdokian, On the calculation of dispersion and polar force components of the surface free energy, J. Adhes. 34 (1991) 41–44.

[173] J.R. Dann, Forces involved in the adhesive process: I. Critical surface tensions of polymeric solids as determined with polar liquids, J. Colloid Interface Sci. 32 (1970) 302–320.

[174] R.J. Good, R.R. Stromberg (Eds.), Surface and Colloid Science, vol. 11, Plenum Press, New York, 1979.

[175] C.J. Van Oss, R.J. Good, Surface tension and the solubility of polymers and biopolymers, J. Macromol. Sci. Chem. A26 (1989) 1183–1203.

[176] I.A. Starostina, O.V. Stoyanov, R.Y. Deberdeev, Polymer surfaces and interfaces. Acid-base interactions and adhesion in polymer-metal systems, Apple Academic Press, New York, 2014.

[177] F.M. Fowkes, Role of acid-base interfacial bonding in adhesion, J. Adhes. Sci. Technol. 1 (1987) 7–27.

[178] D.Y. Kwok, The usefulness of the Lifshitz-van der Waals/acid-base approach for surface tension components and interfacial tensions, Colloids Surf. A Physicochem. Eng. Asp. 156 (1999) 191–200.

[179] D.Y. Kwok, Y. Lee, A.W. Neumann, Evaluation of the Lifshitz-van der Waals/acid-base approach to determine interfacial tensions, Langmuir 10 (1994) 1323–1328.

[180] P. Hass, O. Kläusler, S. Schlegel, P. Niemz, Effects of mechanical and chemical surface preparation on adhesively bonded wooden joints, Int. J. Adhes. Adhes. 51 (2014) 95–102.

[181] O. Lunder, F. Lapique, B. Johnsen, K. Nisancioglu, Effect of pre-treatment on the durability of epoxy-bonded AA6060 aluminium joints, Int. J. Adhes. Adhes. 24 (2004) 107–117.

[182] G. Rotella, M. Alfano, T. Schiefer, I. Jansen, Evaluation of mechanical and laser surface pre-treatments on the strength of adhesive bonded steel joints for automotive industry, J. Adhes. Sci. Technol. 30 (2016) 747–758.

[183] A. Rudawska, E. Bociąga, E. Olewnik-Kruszkowska, The effect of primers on adhesive properties and strength of adhesive joints made with polyurethane adhesives, J. Adhes. Sci. Technol. 31 (2017) 327–344.

[184] L. Bouazaoui, A. Li, Analysis of steel/concrete interfacial shear stress by means of pull out test, Int. J. Adhes. Adhes. 28 (2008) 101–108.

[185] M.K. Apalak, A. Engin, Effect of adhesive free-end geometry on the initiation and propagation of damaged zones in adhesively bonded lap joints, J. Adhes. Sci. Technol. 18 (2004) 529–559.
[186] I. Skeist, Handbook of Adhesives, third ed., Chapman & Hall, New York, 1990.
[187] S.R. Hartshorn (Ed.), Structural Adhesives, Chemistry and Technology, Plenum Press, New York/London, 1986.
[188] Y. Gilbert, G. Verchery, Influence of surface roughness on mechanical properties of joints, in: K.L. Mittal (Ed.), Adhesive Joints Formation, Characteristics, and Testing, Plenum Press, New York, 1984, pp. 69–84.
[189] L.D.R. Grant, R.D. Adams, L.F.M. da Silva, Effect of the temperature on the strength of adhesively bonded single lap and T joints for the automotive industry, Int. J. Adhes. Adhes. 29 (2009) 535–542.
[190] J.F. Durodola, Functionally graded adhesive joints, Int. J. Adhes. Adhes. 76 (2017) 83–89.
[191] E.M. Moya-Sanz, I. Ivañez, S.K. Garcia-Castilo, Effect of geometry in the strength of single-lap adhesive joints of composite laminates under uniaxial tensile load, Int. J. Adhes. Adhes. 72 (2017) 23–29.
[192] V. Shenoy, I.A. Ashcroft, G.W. Critchlow, A.D. Crocombe, M.M. Abdel Wahab, An evaluation of strength wearout models for the lifetime prediction of adhesive joints subjected to variable amplitude fatigue, Int. J. Adhes. Adhes. 29 (2009) 639–649.
[193] V. Shenoy, I.A. Ashcroft, G.W. Critchlow, A.D. Crocombe, M.M. Abdel Wahab, Strength wearout of adhesively bonded joints under constant amplitude fatigue, Int. J. Fatigue 31 (2009) 820–830.
[194] M.V. Fernández, M.F.S.F. de Moura, L.F.M. da Silva, A.T. Marques, Characterization of composite bonded joints under pure mode II fatigue loading, Compos. Struct. 95 (2013) 222–226.
[195] S. Azari, M. Papini, J.A. Schroeder, J.K. Spelt, Fatigue threshold behaviour of adhesive joints, Int. J. Adhes. Adhes. 30 (2010) 145–159.
[196] M.V. Fernández, M.F.S.F. de Moura, L.F.M. da Silva, A.T. Marques, Composite bonded joints under mode I fatigue loading, Int. J. Adhes. Adhes. 31 (2011) 280–285.
[197] P.R. Underhill, A.N. Rider, D.L. DuQuesnay, The effect of warm surface treatments on the fatigue life in shear of aluminium joints, Int. J. Adhes. Adhes. 26 (2006) 199–205.
[198] R.D. Adams, J.A. Harris, The influence of local geometry on the strength of adhesive joints, Int. J. Adhes. Adhes. 7 (1987) 69–80.
[199] A.D. Crocomble, Y.X. Hua, W.K. Loh, M.A. Wahab, I.A. Ashcroft, Predicting the residual strength for environmentally degraded adhesive lap joints, Int. J. Adhes. Adhes. 26 (2006) 325–336.
[200] Z. Abdo, H. Aglan, Analysis of aircraft adhesive joints under combined thermal and mechanical cyclic loadings, J. Adhes. Sci. Technol. 11 (1997) 941–956.
[201] R.D. Adams, J. Coppendale, V. Mallick, H. Al.-Hamdan, The effect of temperature on the strength of adhesive joint, Int. J. Adhes. Adhes. 12 (1992) 2185–2190.
[202] I.A. Ashcroft, J. Comyn, S. Tellwright, Adhesives for automotive windscreen replacement: kinetics of cure and bond strength, Int. J. Adhes. Adhes. 29 (2009) 155–166.
[203] N. Brack, A.N. Rider, The influence of mechanical and chemical treatments on the environmental resistance of epoxy adhesive bonds to titanium, Int. J. Adhes. Adhes. 44 (2014) 20–27.
[204] L.F.M. da Silva, R.D. Adams, Adhesive joint at high and low temperatures using similar and dissimilar adherends and dual adhesives, Int. J. Adhes. Adhes. 27 (2007) 216–226.

[205] B. Golaz, V. Michaud, S. Lavanchy, J.-A.E. Månson, Design and durability of titanium adhesive joints for marine applications, Int. J. Adhes. Adhes. 45 (2013) 150–157.
[206] J. Kuczmaszewski, Fundamentals of metal-metal adhesive joint design, Lublin University of Technology, Polish Academy of Sciences, Lublin, 2006.
[207] E.M. Petrie, Handbook of Adhesives and Sealants, McGraw Hill, New York, 1999.
[208] ASTM D3762-03. Standard test method for adhesive-bonded surface durability of aluminum (Wedge Test).
[209] G. Habenicht, Applied Adhesive Bonding, Wiley-VCH Press, Weinheim, 2006.
[210] A.D. Crocombe, R.D. Adams, Peel analysis using the finite element method, J. Adhes. 12 (1981) 127–139.
[211] ASTM E6-15e1. Standard Terminology Relating to Methods of Mechanical Testing.
[212] PN-EN ISO 4624. Farby i lakiery. Próba odrywania do oceny przyczepności (Paints and varnishes. Tear test for adhesion evaluation) (in Polish).
[213] ASTM D 4541-02. Standard test method for pull-off strength of coatings using portable adhesion testers.
[214] PN-EN 28510-1. Kleje. Oznaczanie wytrzymałości na oddzieranie połączeń elementu giętkiego ze sztywnym. Część 1: Oddzieranie pod kątem 90° (Adhesives. Determination of the peel strenght of flexible and rigid joints. Part 1: Peeling at 90°) (in Polish).
[215] EN 1939. Self adhesive tapes. Measurement of peel adhesion from stainless steel or from its own backing.
[216] ASTM D3330/D3330M-04. Standard test method for peel adhesion of pressure-sensitive tape.
[217] Peel strength test standards, https://www.mecmesin.com/peel-strength-test-standards, 2018 (Accessed November 20, 2018).
[218] PN-EN ISO 11339. Kleje. Oznaczanie wytrzymałości na oddzieranie, metodą T, połączeń materiału giętkiego z giętkim (Adhesives. Determination of the peel strength, by the T method, of the combination of flexible and flexible material) (in Polish).
[219] ASTM D1876-08. Standard test method for peel resistance of adhesives (T-Peel Test).
[220] PN-EN ISO 8510-2. Kleje. Oznaczanie wytrzymałości na oddzieranie połączeń elementu giętkiego ze sztywnym. Część 2: Oddzieranie pod kątem 180 stopni (Adhesives. Determination of the peel strenght of flexible and rigid joints. Part 2: Peeling at 180°) (in Polish).
[221] ASTM D903-98. Standard test method for peel or stripping strength of adhesive bond.
[222] PN-EN 1464. Kleje. Oznaczanie wytrzymałości na oddzieranie połączeń klejowych. Metoda rolki pływającej (Adhesives. Determination of the peel strength. Floating roll method) (in Polish).
[223] ASTM D3167-10. Standard test method for floating roller peel resistance of adhesives.

Further reading

[224] C.J. van Oss, Interfacial forces in aqueous media, Dekker, New York, 1994.

Index

Note: Page numbers followed by *f* indicate figures and *t* indicate tables.

S

Kolibri 360 GmbH
Ferdinand-Jühlke-Straße 7
99095 Erfurt, Deutschland
produktsicherheit@kolibri360.de